Theory and Technology of Water-Induced Geological Disaster Prevention and Water Resource Utilization in Mines

Theory and Technology of Water-Induced Geological Disaster Prevention and Water Resource Utilization in Mines

Guest Editors

Helong Gu
Fan Feng
Yuan Zhao

MDPI

Basel • Beijing • Wuhan • Barcelona • Belgrade • Novi Sad • Cluj • Manchester

Guest Editors
Helong Gu
College of Energy and Mining Engineering
Xi'an University of Science and Technology
Xi'an
China

Fan Feng
College of Energy and Mining Engineering
Shandong University of Science and Technology
Qingdao
China

Yuan Zhao
School of Human Settlements and Civil Engineering
Xi'an Jiaotong University
Xi'an
China

Editorial Office
MDPI AG
Grosspeteranlage 5
4052 Basel, Switzerland

This is a reprint of the Special Issue, published open access by the journal *Water* (ISSN 2073-4441), freely accessible at: https://www.mdpi.com/journal/water/special_issues/99Z28567VL.

For citation purposes, cite each article independently as indicated on the article page online and as indicated below:

Lastname, A.A.; Lastname, B.B. Article Title. *Journal Name* **Year**, *Volume Number*, Page Range.

ISBN 978-3-7258-6574-1 (Hbk)
ISBN 978-3-7258-6575-8 (PDF)
https://doi.org/10.3390/books978-3-7258-6575-8

Contents

About the Editors

Helong Gu

Helong Gu is a professor at Xi'an University of Science and Technology's College of Energy and Mining Engineering specializing in coal mine dynamic disaster prevention and control, covering rockbursts, strong strata pressure, and coal gas outbursts. His research focuses on multi-field coupled dynamic fracture mechanics of coal–rock, disaster mechanisms, and efficient prevention technologies. He has won five provincial-level first-class science and technology awards, including the China Gold Association Science and Technology Award. He has authored over 30 academic papers. Additionally, he holds over 30 authorized national patents and helped draft one industry standard.

Fan Feng

Fan Feng is from Shandong University of Science and Technology's College of Energy and Mining Engineering, focusing on rock mechanics, mining engineering, and geotechnical engineering. His research interests include rock joint mechanical behavior under cyclic loading, numerical simulation of rock mass stability, and innovative mining techniques for safer and more efficient underground operations. His key achievements involve developing advanced methodologies for analyzing rock joint shear characteristics under cyclic shear loading, aiding the understanding of rock mass deformation mechanisms in mining.

Yuan Zhao

Yuan Zhao is affiliated with the School of Human Settlements and Civil Engineering at Xi'an Jiaotong University. His research interests include the micro- and macro-scale mechanisms of water-induced geological hazards, as well as the mechanical behavior of granular geomaterials. He specializes in super-resolution characterization of the microstructure of granular geomaterials and discrete element modeling of particulate systems. He has led multiple research projects, including programs funded by the National Natural Science Foundation of China.

Preface

The Reprint focuses on the theory and technology of water-induced geological disaster prevention and water resource utilization in mines. It addresses the complex geological conditions in deep mining areas and surface slopes, which, when combined with extreme precipitation and intensive mining activities, trigger disasters like landslides, water inrush, and rock pillar weakening, threatening safety and causing resource waste. By studying occurrence mechanisms, prediction methods, and prevention measures, this work aims to provide theoretical foundations and practical solutions for disaster mitigation and water resource optimization. The research integrates advanced methodologies such as numerical modeling and physical experiments to enhance early warning systems and response strategies, ultimately minimizing losses and promoting sustainable mining practices. This Reprint is designed for researchers, engineers, policymakers, and industry professionals seeking insights into innovative disaster prevention and water management techniques.

Helong Gu, Fan Feng, and Yuan Zhao
Guest Editors

water

Editorial

Water-Induced Geological Disaster Prevention and Sustainable Water Resource Utilization in Mines

Helong Gu [1,*], Fan Feng [2] and Yuan Zhao [3]

1 College of Energy Engineering, Xi'an University of Science and Technology, Xi'an 710054, China
2 College of Energy and Mining Engineering, Shandong University of Science and Technology, Qingdao 266590, China; fengfan0213@sdust.edu.cn
3 School of Human Settlements and Civil Engineering, Xi'an Jiaotong University, Xi'an 710049, China
* Correspondence: helonggu@xust.edu.cn

1. Introduction

The deep areas and surface slopes of mines exhibit complex geological conditions. When combined with extreme precipitation and high-intensity mining activities, these conditions are highly susceptible to triggering a series of water-induced geological disasters, including landslides, water inrush, and rock pillar weakening [1,2]. These events threaten the safety and stability of mining operations and lead to considerable resource wastage. Studying the occurrence mechanism, disaster prediction methods, and prevention measures of water-induced geological disasters can provide a theoretical basis and effective solutions for disaster prevention and resource utilization [3–5]. Through the study of the fundamental mechanics theory of water-bearing coal rocks, technologies for mine water resource utilization, mechanisms of surrounding rock water disasters, and advanced early warning and prevention technologies for water-induced geological disasters, accurate prediction and rapid response to water-induced mine hazards can be realized [6]. This approach helps minimize losses caused by water-induced mine disasters and facilitates effective utilization of water resources in mines [7,8]. To ensure mine safety production, modern technological methods are becoming increasingly crucial in both mine safety production and water resource management [9–11].

To tackle these issues, researchers have achieved significant advancements by utilizing cutting-edge technological methodologies [12,13]. For instance, they have integrated numerical modeling and physical experiments to analyze the impact resistance of retaining dams in high-altitude cold regions, revealing that prestressed reinforced concrete structures with embedded reinforcement exhibit superior performance in reducing impact depth, resultant force, and acceleration compared to conventional concrete dams [14,15]. This finding offers critical technical insights for enhancing the design of debris flow prevention barriers in such environments. Concurrently, the application of remote sensing and geospatial analysis has enabled comprehensive monitoring of post-mining land movement and groundwater rebound, demonstrating how climate-driven subsidence and mining-induced uplift interact to reshape surface hydrology and vegetation patterns, thereby informing integrated environmental management strategies in reclaimed mining areas [16]. Additionally, studies on the coupled effects of groundwater seepage and non-uniform loading on mine roadway stability have employed scaled physical modeling, acoustic emission monitoring, and numerical simulations to elucidate the deformation stages and failure mechanisms of water-saturated rock masses [17,18]. These investigations provide theoretical frameworks for predicting shear and tensile failure zones, supporting the development of targeted reinforcement measures to mitigate roadway collapse risks. Furthermore, research on the

 https://doi.org/10.3390/w18010086

influence of external factors like impact speed, angle, and projectile shape on structural response has quantified their relationships with mechanical parameters, offering empirical guidelines for optimizing protective designs in dynamic loading scenarios [19,20]. Collectively, these efforts underscore the transformative role of interdisciplinary technological approaches in advancing disaster prevention, resource extraction efficiency, and ecological restoration in geotechnical engineering practice.

This Special Issue research explores the application of modern technology to enhance mine safety and promote sustainable utilization of mine water resources from multiple perspectives. Through advanced numerical modeling, physical experiments, and integrated monitoring techniques, the studies analyze critical issues such as impact resistance of retaining structures, groundwater rebound dynamics, roadway failure mechanisms under seepage loading, and water diffusion patterns in underground tunnels. The findings provide critical theoretical foundations and technical support for ensuring safe mine production, mitigating environmental risks, and optimizing water resource management in mining operations.

2. An Overview of Published Articles

The incubation process, scientific evaluation, and efficient prevention of mine water-induced geological disasters are critical for ensuring the safe and sustainable development of coal mining operations in China. These disasters, including water inrush, slope instability, and groundwater rebound, pose significant threats to mine safety and environmental stability. To address these challenges, advanced technologies such as integrated geophysical monitoring, numerical modeling of seepage dynamics, and risk assessment based on multi-source data fusion are essential for early warning and disaster mitigation. This Special Issue compiles eight innovative studies that delve into the theory and technology of water-induced geological disasters prevention and water resources utilization in mines.

Tie et al. studied the impact resistance of retaining dams in high-altitude cold regions by establishing a three-dimensional numerical model of the retaining dam. Results show that the prestressed reinforced concrete retaining dam with embedded prestressed reinforcement has significantly lower impact depth, resultant impact force, and acceleration than the concrete retaining dam, effectively improving impact resistance; its impact depth and resultant impact force increase with the steel ball's impact speed, angle, and mass, but decrease with the increase of the steel ball's shape coefficient. The findings provide a scientific basis for the design of debris flow prevention barriers in such regions.

Guzy et al. explored the links between groundwater rebound, land movement, and environmental transformation after the flooding closure of the Olkusz-Pomorzany mine (Poland), by integrating EGMS-based land movement data, groundwater levels, meteorological records, and Sentinel-2-derived hydrospectral indices and land cover data. Results indicate climate-driven regional subsidence of less than 1 mm/year, while the mining zone shifts to uplift with a maximum groundwater rebound of 103 m; surface water and wetlands expand, and vegetation activity and moisture modestly increase. The study highlights the need for integrated monitoring in post-mining areas.

Gang et al. investigated the damage and failure mechanisms of mine roadways under the coupled effects of groundwater seepage and non-uniform loading, by fabricating scaled physical models based on geometric similarity principles, conducting uniaxial compression tests on water-saturated sandstone specimens with pore defects (monitored by AE techniques and static strain systems), and performing FLAC^{3D} numerical simulations. Results show that the stress–strain curves of saturated sandstone exhibit four deformation stages; shear failure zones are regionally clustered, and tensile failure zones are evenly

https://doi.org/10.3390/w18010086

distributed. The findings provide guidance for surrounding rock control and reinforcement of water-infiltrated mine roadways.

Fan et al. focused on quantitative prediction of water spreading in coal mine tunnels after emergency water release from underground reservoirs, by constructing a three-dimensional physical model of the complex tunnel network and adopting the VOF Eulerian multiphase flow model for simulation. Results show that water flows to lower-altitude areas, with water level experiencing three stages (rapid rise, slow increase, and stable equilibrium); the water level near the outlet rises sharply within 550 s, while distant tunnels see rapid rise after 13,200 s, and the maximum safe evacuation time within 300 m of the outlet is 1 h. The study enhances underground safety by clarifying water spreading characteristics.

Yuchi et al. evaluated the stability of waterproof coal pillars affected by goaf water in Nanhu Second Mine, through exploration holes to analyze water source and infiltration direction, theoretical analysis of water-affected mechanical parameters of coal and rock, numerical simulation of coal pillar and roadway performance, and in situ monitoring (ground penetrating radar, deformation measurement, and loosing circle detector). Results confirm the goaf water is static from the V aquifer; the coal pillar has a 6–11 m plastic zone and a 6–8 m middle elastic zone, with slight internal cracks, effectively preventing water damage and ensuring safe production.

Wang et al. revealed the quality evolution patterns of mine water in arid mining regions of western China and assessed its ecological irrigation risks, by collecting surface water, groundwater, and mine water samples for hydrochemical analysis and multi-index risk evaluation. Results show mine water TDS exceeds 1000 mg/L, with quality evolution controlled by water-rock interaction, evaporation, and concentration; direct irrigation causes salt accumulation and environmental threats. Three measures (desalination, improved irrigation technologies, and drought-tolerant vegetation selection) are proposed, providing a theoretical basis for scientific utilization of mine water in arid areas.

Wu et al. studied the impact of rainfall on slope instability and stability, by establishing a rainfall infiltration slope model, conducting saturated–unsaturated flow–solid coupling numerical analysis, and combining the strength reduction method to calculate slope safety factors. Results indicate heavy rainfall forms a continuous sliding surface; pre-peak rainfall pattern causes the largest safety factor decrease and earliest slope failure; rainfall intensity is inversely proportional to the safety factor. The findings provide a scientific basis for analyzing rainfall-induced slope instability.

Yang et al. explored the reinforcement mechanism of anti-slide piles on reservoir landslides under water level fluctuations, by establishing a three-dimensional slope model to simulate water level drawdown from 175 m to 145 m, and analyzing displacement, strain fields, and safety factors under stress–seepage coupling. Results show water level drawdown expands strain/displacement zones and reduces safety factors; anti-slide piles significantly reduce maximum deformation, with optimal reinforcement when arranged in the middle of the landslide, spaced four times the pile diameter, and embedded to critical depth. The study supports mitigation of reservoir landslides.

3. Conclusions

The research on water-induced geological disasters and water resource utilization in mining environments integrates multidisciplinary perspectives, including hydrogeological dynamics, rock mechanics, and slope stability analysis. Through systematic examination of phenomena such as inrush propagation patterns, stress–strain behaviors of fractured rock, and rainfall-triggered slope failure mechanisms, this study elucidates the key stages and spatial–temporal characteristics of water movement and structural failure. Concurrently, it evaluates the physicochemical properties of mine water and their environmental implica-

https://doi.org/10.3390/w18010086

tions for irrigation use. By synthesizing these findings, the research not only clarifies the interrelationships between geological conditions, fluid dynamics, and engineering risks but also establishes a comprehensive framework for disaster mitigation strategies and sustainable water management in mining operations. The main conclusions are as follows:

(1) Water flows along the complex roadway network toward lower-altitude areas, with sloping tunnels acting as initial water spread barriers. The water level at each point in the roadway undergoes three stages: rapid rise, slow increase, and stable equilibrium. The water level near the inrush outlet surges within 550 s, while that in distant areas only rises rapidly after 13,200 s. The maximum safe evacuation time for personnel within 300 m is 1 h, and the water storage capacity of the roadway network extends the evacuation window in distant areas.

(2) Under non-uniform loading, the stress–strain curves of saturated sandstone with pore defects show four stages: compaction, elastic, plastic yield, and failure instability. Crack propagation is closely related to stress zones (intense in high-stress zones, gentle in low-stress zones), with shear failure concentrated in high-stress zones and tensile failure controlled by geometric structures. Acoustic emission (AE) signals exhibit four stages (quiet–active–intense–peak) with a bimodal characteristic.

(3) Heavy rainfall causes the transient saturation zone in the slope to move upward, forming a continuous sliding surface and triggering instability of the soil above the slope toe. The pre-peak rainfall pattern is most unfavorable to slope stability (largest safety factor reduction and earliest failure). Rainfall intensity is inversely proportional to the slope safety factor—higher intensity leads to shorter failure time.

(4) Mine water is weakly alkaline, with total dissolved solids (TDS) all exceeding 1000 mg/L. The main ions are Na^+, Ca^{2+}, $SO_4{}^{2-}$, and Cl^-, derived from the dissolution of halite, gypsum, and anorthite. Water quality evolution is mainly controlled by water–rock interaction and affected by evaporation concentration.

(5) Direct irrigation with untreated mine water may cause soil salinization due to excessive soluble sodium percentage (SSP), sodium adsorption ratio (SAR), salinity hazard, and magnesium adsorption ratio (MAR). It inhibits plant growth and poses a slight pollution risk to groundwater.

(6) For mine water inrush, zoned evacuation plans should be formulated based on the distance from the inrush outlet. For ecological irrigation of mine water in arid regions, desalination via nanofiltration combined with reverse osmosis, drip/intermittent irrigation technologies, and planting salt-tolerant and drought-tolerant vegetation are recommended. For slope disasters, focus on rainfall patterns and intensity, and establish targeted early warning mechanisms.

Author Contributions: Conceptualization, H.G.; methodology, H.G.; software, Y.Z.; validation, H.G., Y.Z. and F.F.; formal analysis, H.G.; investigation, H.G.; resources, H.G.; data curation, H.G.; writing—original draft preparation, H.G.; writing—review and editing, H.G.; visualization, F.F.; supervision, Y.Z.; project administration, H.G.; funding acquisition, H.G. All authors have read and agreed to the published version of the manuscript.

Acknowledgments: As Guest Editors of the Special Issue "Theory and Technology of Water-Induced Geological Disasters Prevention and Water Resources Utilization in Mines", we would like to express our deep appreciation to all the authors whose valuable work was published in this issue and who have thus contributed to the success of this edition.

Conflicts of Interest: The authors declare no conflicts of interest.

https://doi.org/10.3390/w18010086

List of Contributions:

1. Tie, Y.B.; Jiang, W.; Wang, M.; Zheng, Y. Numerical Analysis of the Dynamic Response of the PRCRD under the Impact of Debris Flow Block Stones. *Water* **2026**, *18*, 60. https://doi.org/10.3390/w18010060.
2. Guzy, A. Environmental Impacts of Post-Closure Mine Flooding: An Integrated Remote Sensing and Geospatial Analysis of the Olkusz-Pomorzany Mine, Poland. *Water* **2025**, *17*, 3337. https://doi.org/10.3390/w17233337.
3. Wang, H.; Shang, H.B.; Wang, T.T.; Xue, J.K.; Wang, X.D.; Zhou, Z.F.; Wang, Q.M. Hydrogeochemical Evolution and Ecological Irrigation Evaluation of Mine Water in an Arid Coal Region: A Case Study from Northwest China. *Water* **2025**, *17*, 3132. https://doi.org/10.3390/w17213132.
4. Wu, Z.L.; Yang, G.; Li, W.; Chen, X.L.; Liu, F.; Zheng, Y. Numerical Simulation Study of Rainfall-Induced Saturated-Unsaturated Landslide Instability and Failure. *Water* **2025**, *17*, 2229. https://doi.org/10.3390/w17152229.
5. Liu, G.; Zan, Y.L.; Wang, D.W.; Wang, S.X.; Yang, Z.T.; Zeng, Y.; Wei, G.Q.; Shi, X. Numerical simulation of fracture failure propagation in water-saturated sand-stone with pore defects under non-uniform loading effects. *Water* **2025**, *17*, 1725. https://doi.org/10.3390/w17121725.
6. Fan, D.L.; Li, S.B.; He, P.D.; Chen, S.S.; Zou, X.; Wu, Y. Numerical Simulation Study on the Spread of Mine Water In-rush in Complex Roadways. *Water* **2025**, *17*, 1434. https://doi.org/10.3390/w17101434.
7. Yang, G.; Wu, L.Z.; Zhang, L.; Hou, F.J.; Shen, T.; Liu, F.; Zheng, Y. Study on the deformation mechanism of landslide reinforced by anti-slip pile under the effect of reservoir water level change. *Water* **2025**, *17*, 1390. https://doi.org/10.3390/w17091390.
8. Yuchi, X.; Gu, H.; Du, X.; Shu, P. Monitoring and Analysis of Waterproof Coal Pillars Under the Influence of Goaf Water. *Water* **2025**, *17*, 65. https://doi.org/10.3390/w17010065.

References

1. Ao, L.M.; Xu, Q.X.; Zhou, Y.C.; Wu, Y.Q.; Zhang, S.P.; Fu, Z.Y.; Chen, H.S. Rock Surface Flow Accelerates Rill Erosion of Excavated Slopes in Karst Mining Areas. *Catena* **2024**, *236*, 107739. [CrossRef]
2. Yu, X.; Mao, X. A Preliminary Discrimination Model of a Deep Mining Landslide and Its Application in the Guanwen Coal Mine. *Bull. Eng. Geol. Environ.* **2020**, *79*, 485–493. [CrossRef]
3. Siddique, A.; Tan, Z.; Li, J.; Algarni, A.; Zardari, I.; Zeshan, M.; Khurram, S.; Kumar, D. An Inclusive Approach to Disaster Risk Reduction in Water-Related Hazards for the Mining Industry: Prioritizing the Multi-Dimensional Risk Framework for Achieving Sustainable Development. *Mine Water Environ.* **2025**, *44*, 811–830. [CrossRef]
4. Niu, C.; Tian, Q.; Xiao, L.; Xue, X.; Zhang, R.; Xu, D.; Luo, S. Principal Causes of Water Damage in Mining Roofs under Giant Thick Topsoil-Lilou Coal Mine. *Appl. Water Sci.* **2024**, *14*, 129. [CrossRef]
5. Liu, W.; Han, M.; Zhao, J. Prediction of Floor Failure Depth Based on Dividing Deep and Shallow Mining for Risk Assessment of Mine Water Inrush. *Water* **2024**, *16*, 2786. [CrossRef]
6. Chen, G.; Tang, W.; Chen, S.; Wang, E.; Wang, C.; Li, T.; Zhang, G. Damage Effect and Deterioration Mechanism of Mechanical Properties of Fractured Coal-Rock Combined Body under Water-Rock Interaction. *Rock Mech. Rock Eng.* **2025**, *58*, 513–537. [CrossRef]
7. Jia, C.; Lai, X.; Cui, F.; Xu, H.; Zhang, S.; Li, Y.; Zong, C.; Luo, Z. Mining Pressure Distribution Law and Disaster Prevention of Isolated Island Working Face under the Condition of Hard "Umbrella Arch". *Rock Mech. Rock Eng.* **2024**, *57*, 8323–8341. [CrossRef]
8. Zhang, J.; Wang, M.; Zhou, H.; Zhang, D.; Yu, B.; Wang, C.; Wang, Y. Coal and Rock Dynamic Disaster Prevention and Control Technology in the Large Mining Face of a Deep Outburst Mine. *Front. Earth Sci.* **2023**, *17*, 701–712. [CrossRef]
9. Li, Z.; Qiao, D.; Yang, T.; Wang, J.; Chen, H. Event Recognition Technology and Short-Term Rockburst Early Warning Model Based on Microseismic Monitoring and Ensemble Learning. *Sci. Rep.* **2025**, *15*, 12345. [CrossRef] [PubMed]
10. Cao, G.; Ma, L.; Liu, W.; Khan, N.; Osemudiamhen, A.; Gao, Q.; Yu, K.; Guo, Z.; Li, S.; Yang, R.; et al. Characteristics of Infrared Radiation Precursors of Water Inrush from Coal and Rock Fractures under Hydrodynamic Coupling. *Infrared Phys. Technol.* **2025**, *151*, 106143. [CrossRef]
11. Liu, S.; Zhang, P.; Tang, Q.; Wu, S.; Huang, Y. A Novel Safety Early Warning Methodology for Pipelines under Landslide Geological Hazard. *J. Pipeline Syst. Eng. Pract.* **2024**, *15*, 04023074. [CrossRef]
12. Peng, L.; Wang, Z.; Wang, H.; Zhang, Y. Surface Subsidence Prediction Model in Deep Composite Lithologic Gangue Backfill Mining under the Coupling Effect of High Ground Stress and Water Environment. *Adv. Civ. Eng.* **2025**, *2025*, 1013764. [CrossRef]

https://doi.org/10.3390/w18010086

13. Zha, X.; Li, X.; Chen, S.; Li, B.; Cao, X.; Huang, J.; Wu, P. Geochemical Process of Arsenic Source and Fate in Water Environment of Karst Gold Mining Region, Southwestern China. *Environ. Technol. Innov.* **2025**, *38*, 104159. [CrossRef]
14. Wang, B.; Zhang, J.; Wang, X.; Zhang, X. Assessing the Impact of Changing Environment on Coal Mining Water Use in China. *Water-Energy Nexus* **2023**, *6*, 177–186. [CrossRef]
15. Wang, R.; Lin, H.; Cheng, J.; Xu, Z.; Feng, H.; Tang, Y. Optimizing the Water Ecological Environment of Mining Cities in the Yangtze River Economic Belt Using the Cloud Model, CV-TOPSIS, and Goupling Coordination Degree. *Int. J. Environ. Res. Public Health* **2022**, *19*, 2469. [CrossRef] [PubMed]
16. Menshikova, E.; Fetisov, V.; Karavaeva, T.; Blinov, S.; Belkin, P.; Vaganov, S. Reducing the Negative Technogenic Impact of the Mining Enterprise on the Environment through Management of the Water Balance. *Minerals* **2020**, *10*, 1145. [CrossRef]
17. Chen, X.; Zheng, L.; Dong, X.; Jiang, C.; Wei, X. Sources and Mixing of Sulfate Contamination in the Water Environment of a Typical Coal Mining City, China: Evidence from Stable Isotope Characteristics. *Environ. Geochem. Health* **2020**, *42*, 2885–2899. [CrossRef] [PubMed]
18. Doering, C.; Carpenter, J.; Orr, B.; Urban, D. Whole Organism Concentration Ratios in Freshwater Wildlife from an Australian Tropical U Mining Environment and the Derivation of a Water Radiological Quality Guideline Value. *J. Environ. Radioact.* **2019**, *198*, 27–35. [CrossRef] [PubMed]
19. Kuffour, R.A.; Tiimub, B.M.; Agyapong, D. Impacts of Illegal Mining (Galamsey) on the Environment (Water and Soil) at Bontefufuo Area in the Amansie West District. *J. Environ. Earth Sci.* **2018**, *8*, 98–107.
20. Yang, Y.; Guo, T.; Jiao, W. Destruction Processes of Mining on Water Environment in the Mining Area Combining Isotopic and Hydrochemical Tracer. *Environ. Pollut.* **2018**, *237*, 356–365. [CrossRef] [PubMed]

https://doi.org/10.3390/w18010086

water

MDPI

Article

Numerical Analysis of the Dynamic Response of a Prestressed Reinforced Concrete Retaining Dam Under the Impact of Debris Flow Block Stones

Yongbo Tie [1], Wei Jiang [2,*], Min Wang [2] and Yong Zheng [3]

1 Chengdu Center of China Geological Survey (Geosciences Innovation Center of Southwest China), Chengdu 610081, China
2 Chongqing Key Laboratory of Facility Damage Mechanism and Protection in Highland Mountain Environment, PLA Joint Logistics Support Force University of Engineering, Chongqing 401311, China
3 Key Laboratory of Geological Hazards Mitigation for Mountainous Highway and Waterway, Chongqing Municipal Education Commission, Chongqing Jiaotong University, Chongqing 400074, China
* Correspondence: jw202509@hotmail.com

Abstract

Debris flow is a common geological disaster in mountainous areas, characterized by its sudden onset, frequent occurrence, and high destructive power. Retaining dams are one of the most commonly used measures for debris flow prevention and are widely applied in debris flow management projects. This study investigates the impact resistance of retaining dams in high-altitude cold regions by establishing a three-dimensional numerical model of the retaining dam. The results show that the impact depth, resultant impact force, and acceleration of the prestressed reinforced concrete retaining dam with embedded prestressed reinforcement are significantly lower than those of the concrete retaining dam. The prestressed reinforced concrete retaining dam with embedded prestressed reinforcement can improve its impact resistance, effectively mitigating the impact of debris flow block collisions. The impact depth and resultant impact force of the prestressed reinforced concrete retaining dam both increase with the steel ball's impact speed, impact angle, and impact mass, while they decrease with an increase in the shape coefficient of the steel ball. The effects of different parameters of the steel ball on the impact depth and resultant impact force of the barrier vary. The research findings provide a scientific basis for the design of barriers in the prevention and control of debris flows in high-altitude cold regions.

Keywords: debris flow; retaining dam; prestressed reinforcement; impact depth; impact resistance

1. Introduction

Debris flows are a very common natural geological disaster in mountainous areas [1]. They are a special type of torrent characterized by a large number of solid particles and liquid fluid, generated by material sources under the influence of heavy rain or earthquakes [2–4]. Debris flows are known for their sudden onset, violent momentum, short duration, strong impact, difficulty in forecasting and warning, and challenges in defense [5–7]. They are widely distributed in mountainous regions around the world, and their immense impact can cause devastating damage to various buildings and structures. Historical records indicate that between 1950 and 2011, hundreds of major debris flow events occurred worldwide, resulting in tens of thousands of fatalities, reflecting the severe threat posed by such disasters to human safety [8,9]. Notably, the catastrophic debris flow disaster that occurred in Zhouqu, Gansu, China, on 7 August resulted in 1557 fatalities and 284 missing persons, with direct economic losses reaching 400 million yuan, making it one of the most severe geological disasters in the world.

The study of debris flows is relatively complex, as previous research often simplifies them to either pure particulate matter or fluid. The debris flow fluid and the rocks it contains exert significant impact forces, which are often the most direct cause of engineering damage. Consequently, many scholars have focused on studying the impact of debris flows. For example, Scheidl et al. [10] conducted small-scale debris flow impact tests using a flume device to analyze the impact forces of debris flows. Ashwood and others performed physical simulation experiments using quartz sand and rocks to investigate the impact of debris flows. Hu et al. et al. [11] conducted field tests in Jiangjiagou, Yunnan, but real-time recording of impact forces during debris flow disasters is extremely difficult and dangerous. Zanuttigh et al. [12] conducted impact experiments using two types of solid particle flows (well-graded and coarse particles) on models of different shapes of obstacles. They expressed the impact force in the form of static fluid pressure and noted that the magnitude of the resultant impact force is significantly related to the shape of the obstacle. Adam B. Prochaska [13] conducted experiments to study the impact forces from large rocks in debris flows, analyzing that these forces are sensitive to particle size and velocity, and that particle arrangement affects the impact force of the large rocks. Chen et al. [14] proposed a two-dimensional fluid model, simplifying debris flows as a two-phase fluid and establishing a formula for the intensity of impact forces based on Newton's second law. Shan et al. [15] developed a coupling model using Computational Fluid Dynamics–Discrete Element Method (CFD-DEM) to reproduce the transport characteristics of debris flows and their impact effects on downstream structures.

Debris flow barriers are one of the most commonly used measures for debris flow prevention. They can directly reduce the movement speed of debris flows and decrease the effective volume of debris flows, thereby lowering the peak flow rate. Many researchers have extensively studied the performance of debris flow barriers under impact loads. Han and Ou [16] investigated the blocking efficiency of comb dams for non-cohesive mudflows, discussing the impact of the gap between dam teeth and the maximum particle size ratio of debris flow on clogging patterns. Huang et al. [17] conducted a series of impact tests on debris flow barriers and comb dams, finding that their values followed an elastic collision model. Shieh et al. [18] studied the impact of mudflows on barriers of different shapes, discovering that under the same impact conditions, the impact force on curved barriers is smaller both theoretically and experimentally compared to diagonal and vertical barriers. Remaître and Malet [19] utilized one-dimensional numerical simulation methods to analyze the effects of the number and spacing of debris flow barriers on debris flow prevention. Kim et al. [20] studied the role of impact-absorbing structures in reducing the impact force of debris flows on barriers, finding that such structures can significantly dissipate the impact energy of debris flows, with triangular structures performing better than rectangular ones. Collins et al. [21] found that multiple comb dams can both weaken the recovery of debris flow mobility after passage and allow larger particles to form supportive arches at the openings, effectively intercepting the debris flow.

As mentioned above, extensive research has been conducted on the impact loads generated by debris flows and the impact resistance of retaining dams. However, studies on the impact of rock blocks from debris flows on retaining dams in high-altitude cold regions are relatively scarce. This paper takes the Tianmo Valley debris flow in southeastern Tibet, China, as a case study and employs finite element numerical analysis to simulate and analyze the process of rock block impacts on the retaining dam.

https://doi.org/10.3390/w18010060

It should be noted that since mortar and rubble are readily available locally, most debris flow barrier dams are constructed using concrete. However, concrete barrier dams (CRD) struggle to withstand the impact force of debris flows, as evidenced by the past failure of a barrier dam during the Zhouqu debris flow disaster. Therefore, this paper will focus on discussing the dynamic response of prestressed reinforced concrete retaining dam (PRCRD) under the impact load of boulders in debris flows. Firstly, two types of debris flow barriers were designed: a concrete retaining dam (CRD) and a prestressed reinforced concrete retaining dam (PRCRD). The numerical calculation model for the impact of debris flow block stones on a barrier dam was established using the finite element software Abaqus/Explicit 6.14 [22]. The numerical results of stress, displacement, and acceleration at the impact points were analyzed, comparing the impact-resistance performance of the CRD and PRCRD. Subsequently, the impact resistance of the PRCRD under different shape coefficients, impact masses, impact velocities, and impact angles was studied. At the end of this paper, the sensitivity indicator was introduced to quantitatively evaluate the parameter sensitivity of different influencing factors. The findings of this study have certain reference value for the application of a retaining dam in debris flow prevention engineering.

2. Methodology

2.1. Description of Tianmo Valley Area

The Tianmo Valley debris flow is located in Songrao Village, Guxiang County, Bomi, in the Nyingchi region of Tibet, within the eastern part of the Parlung Zangbo River basin, northeast of the bend of the Yarlung Tsangpo River. It is situated on the deeply cut fault zone of the Yigong Zangbo–Parlung Zangbo. This area is also one of the regions in China where modern glaciers and marine glaciers are well developed. The main valley has a V shape, with a catchment area of 17.4 km^2. Modern glaciers, with an area of 8.63 km^2, develop at elevations above 3800 m in the upper reaches of the valley, as shown in Figure 1a. The main valley is approximately 4 km long, with a longitudinal slope of about 256‰, and the highest elevation in the basin reaches 5560 m, resulting in a relative height difference of 3100 m. Under the influence of glacial development, Tianmo Valley has formed a large amount of glacial deposits, rockfall debris, and residual slope deposits, with a solid loose material storage of over 5×10^7 m^3. The steep terrain, underlying geological conditions, and glacier distribution are the main controlling factors of geological disaster processes in Tianmo Valley, as illustrated in Figure 1b. Previously, small landslides and rockfalls occurred on both slopes and were deposited in the river channel. Figure 1a shows the mass distribution in Tianmo Valley and two visible landslides. When landslides recur and transform into long-distance flowing debris flows, the sediments may reach downstream areas of human activity (Figure 1c,d). Since 2007, Tianmogo has experienced seven glacial debris flows, damaging National Highway G318 and severely impacting residents, with agriculture and road infrastructure also suffering losses.

2.2. Mathematical Formulation

The entire motion process of debris flow block stones from initiation to completion follows the conservation of mass and momentum. The mass conservation is expressed as follows:

$$\frac{\partial p}{\partial t} + \frac{\partial (pu)}{\partial x} + \frac{\partial (pv)}{\partial y} + \frac{p(pw)}{\partial z} = 0 \tag{1}$$

https://doi.org/10.3390/w18010060

where u, v, and w represent the components of fluid velocity along the x, y, and z axes, respectively, and p is the density of debris flow. The momentum conservation equations in the x, y, and z axes are expressed as follows:

$$\frac{\partial(pu)}{\partial t}+\frac{\partial(pu^2)}{\partial x}+\frac{\partial(puv)}{\partial y}+\frac{\partial(puw)}{\partial z}=pg_x+(\frac{\partial\tau_{xx}}{\partial x}+\frac{\partial\tau_{yx}}{\partial y}+\frac{\partial\tau_{zx}}{\partial z}) \quad (2)$$

$$\frac{\partial(pv)}{\partial t}+\frac{\partial(pv^2)}{\partial y}+\frac{\partial(puv)}{\partial x}+\frac{\partial(puw)}{\partial z}=pg_y+(\frac{\partial\tau_{xx}}{\partial x}+\frac{\partial\tau_{yx}}{\partial y}+\frac{\partial\tau_{zx}}{\partial z}) \quad (3)$$

$$\frac{\partial(pw)}{\partial t}+\frac{\partial(pw^2)}{\partial z}+\frac{\partial(puw)}{\partial x}+\frac{\partial(puw)}{\partial y}=pg_z+(\frac{\partial\tau_{xx}}{\partial x}+\frac{\partial\tau_{yx}}{\partial y}+\frac{\partial\tau_{zx}}{\partial z}) \quad (4)$$

where t is time; τ_{ij} represents the stress components in different directions; g_x, g_y, and g_z are, respectively, the components of the gravitational acceleration g along the x, y, and z axes.

Figure 1. (**a**) Map of the topography and landforms of Tianmo Valley; (**b**) Initiation zone of debris flow; (**c**) The middle reaches of Tianmo Valley; (**d**) Deposition area at the outlet of Tianmo Valley; (**e**) Clastic sediment sample.

2.3. Model Geometry

As mentioned earlier, retaining dams are one of the most commonly used measures for debris flow prevention. To avoid or reduce the hazards posed by debris flows in Tianmo Valley to the lives and properties of residents, especially along transportation routes, it is proposed to design barriers in the debris flow movement area to either accumulate or slow down the flowing debris. This paper conducted a numerical simulation analysis of the impact resistance of retaining dams in the management engineering of debris flow disasters in the Tianmo Valley.

The retaining dam has a top width of 1 m and a bottom width of 2.8 m, with a length and height of 10 m and 6 m, respectively, as shown in Figure 2. The prestressed reinforced concrete retaining dam (PRCRD) is primarily composed of prestressed steel

https://doi.org/10.3390/w18010060

bars and concrete. To facilitate the placement of the prestressed steel bars, the flow surface impacted by the rock blocks is vertical, with a slope of 1:0.067. Since the rocks in the debris flow mostly have irregular shapes, in the finite element model, the rocks are equivalent to circular steel balls. Based on on-site geological survey data, the maximum length of the rock blocks in the debris flow ranges from 0.3 m to 2 m, and in this paper, the diameter of the steel balls is set to 0.5 m. Because the movement of the rocks is powered by the debris flow slurry, this paper assumes that the maximum speed of the larger rock block will not exceed the maximum speed of the debris flow slurry, and the effect of gravity on the impact speed of the rock blocks is neglected.

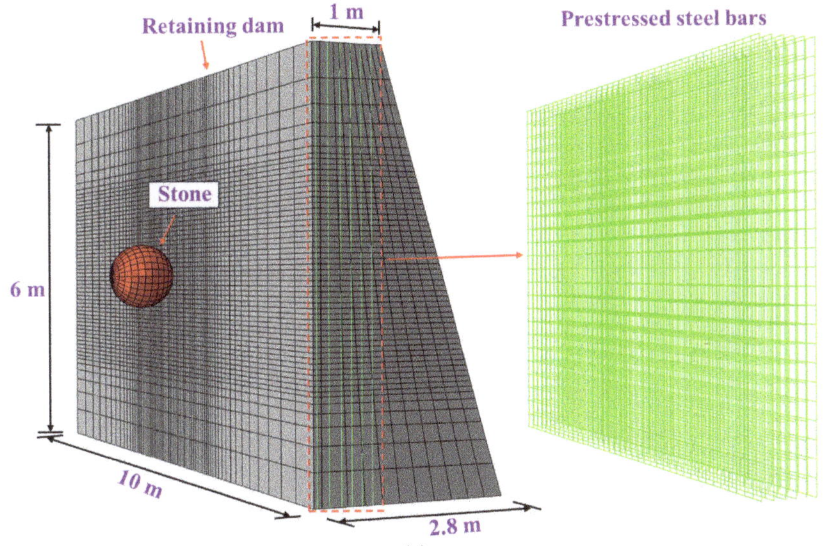

(a)

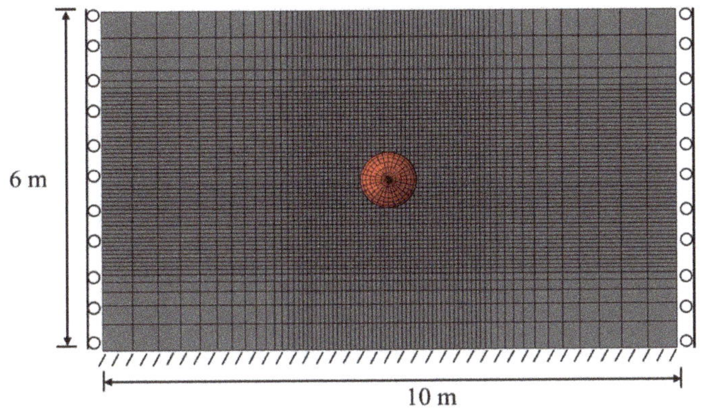

(b)

Figure 2. Finite element model: (**a**) Retaining dams and prestressed steel bars; (**b**) Model boundary condition.

2.4. Mesh and Boundary Conditions

The meshes for the steel balls (i.e., rock block) and the retaining dam are created using tetrahedral free meshing, with the element type being eight-node three-dimensional linear reduced integration solid elements (C3D8R). The element type for the steel bars is two-node

https://doi.org/10.3390/w18010060

linear three-dimensional truss elements (C3D2). To reduce the number of model elements and save computation time, a graded mesh is employed for the debris flow barrier. The mesh near the impact zone is refined, while the mesh further away from the impact zone is coarser, as shown in Figure 2. The bottom of the retaining dam is set as a fixed constraint, with the prestressed steel bars rigidly connected to the bottom of the retaining dam. The top of the steel bars and the retaining dam are connected using tie connections, and prestress is applied to the prestressed steel bars using a cooling method.

2.5. Model Parameters

The focus of this study is not on the destruction or fragmentation of the impacting rock blocks, but rather on the impact resistance of the barrier. Therefore, the steel balls are modeled as rigid bodies. The concrete strength used in the retaining dam is C30. Concrete constitutive behavior in the numerical analysis was approximated by the Concrete Damaged Plasticity (CDP) constitutive model using the material properties listed in Table 1. The CDP model is a continuous, plastic-based damage model for concrete that combines the concept of isotropic damage elasticity with isotropic tensile and compressive plasticity to express the inelastic behavior of concrete. On one hand, it introduces a damage coefficient to reduce the unloading stiffness of concrete, reflecting the characteristic that the unloading stiffness decreases with increased damage during the unloading process. On the other hand, it incorporates a non-associated hardening criterion to characterize the plastic deformation of concrete under compression. Therefore, the CDP model is suitable for simulating and analyzing the impact resistance of the retaining dam discussed in this paper.

Table 1. Properties of the Concrete.

Parameter	Symbol	Unit	Value
Density	p	kN/m^3	25.0
Elasticity modulus	G	MPa	3×10^4
Poisson's ratio	v		0.20
Dilatancy angle	ψ	°	38
Eccentricity ratio	e		0.1
Stress ratio	f_{b0}/f_{c0}		1.16
Constant stress	K		0.67
Viscosity parameter	v		1×10^{-5}

The bond slip between the prestressed steel reinforcement and the concrete is neglected, embedding the former within the concrete. In this paper, the prestressed steel reinforcement embedded in the barrier is a rate-dependent material (rate-dependent refers to the change in the material's mechanical properties with strain rate during dynamic loading), and its parameters are shown in Table 2.

Table 2. Parameters of prestressed steel reinforcement.

Parameter	Symbol	Unit	Value
Density	p	kN/m^3	7800
Elasticity modulus	G	MPa	2×10^5
Poisson's ratio	v		0.20
Yield strength	σ_y	MPa	460
Ultimate strength	σ_u	MPa	690
Ultimate tensile strain	ε_{ut}		0.13

https://doi.org/10.3390/w18010060

2.6. Interface Condition

In simulating the impact of a steel ball on the retaining dam, the interface condition between the steel ball and the retaining dam is a key focus of the analysis. In this paper, the impact load is applied through the interface between the steel ball and the retaining dam. The contact between the steel ball and the concrete is modeled using a 'surface-to-surface contact' approach, with the inner surface of the ball as the master surface and the surface of the barrier as the slave surface, assuming that the interface is frictionless.

2.7. Simulation Scheme

The specific numerical analysis scheme is presented in Table 3. The results of Scheme 1 and Scheme 2 are mainly intended to compare the impact resistance of the PRCRD with that of the concrete retaining dam (CRD). Scheme 3, Scheme 4 and Scheme 5, and 6 discuss the effects of steel ball impact speed V, impact angle θ, impact mass m, and shape coefficient Ψ on the impact resistance of the PRCRD.

Table 3. Numerical analysis scheme.

Scheme	Impact Velocity V (m/s)	Impact Angle θ (°)	Impact Mass m (kg)	Shape Coefficient Ψ	Type of Retaining Dam
Scheme 1	15	90	4000	1	CRD
Scheme 2	15	90	4000	1	PRCRD
Scheme 3	5, 10, 15, 20	90	4000	1	PRCRD
Scheme 4	15	30, 50, 70, 90	4000	1	PRCRD
Scheme 5	15	90	1500, 2250, 3250, 4000	1	PRCRD
Scheme 6	15	90	4000	0.4, 0.6, 0.8, 1	PRCRD

3. Results and Discussions

3.1. Analyze the Impact Resistance Performance of the Retaining Dam

Figure 3 shows the stress distribution of a CRD during a steel ball impact. As can be seen from this figure, at t = 0.001 s, the steel ball has just come into contact with the CRD, and significant deformation occurs at the impact point. At this moment, the stress at the impact point immediately reaches the compressive strength of the concrete, indicating that the concrete at the impact point has already failed. As the steel ball impact continues, the stress wave generated by the steel ball rapidly propagates outward from the impact point, causing an increase in stress in the surrounding area, as shown in Figure 3b,c. Subsequently, the velocity of the steel ball gradually decreases, and the stress also diminishes, as shown in Figure 3d. The duration of the steel ball's impact on the CRD is relatively short, with the entire impact process lasting 0.011 s.

Figure 4 shows the stress distribution nephogram of the PRCRD during steel ball impact. As can be seen from Figure 4, the stress at the impact point of the PRCRD is not as concentrated as that of the CRD. The stress propagation range at the impact point is also significantly smaller than that of the CRD, with relatively lower stress magnitude. This is because the prestressed steel bars embedded inside the PRCRD provide a restoring force, dissipating most of the impact energy generated by the steel ball. Consequently, both the stress propagation range and the stress values in the PRCRD remain smaller than those in the CRD. Furthermore, by comparing Figures 3 and 4, it can be observed that the failure range of the PRCRD under steel ball impact is also smaller than that of the CRD. This demonstrates that the PRCRD exhibits superior resistance to steel ball impact compared to the CRD.

https://doi.org/10.3390/w18010060

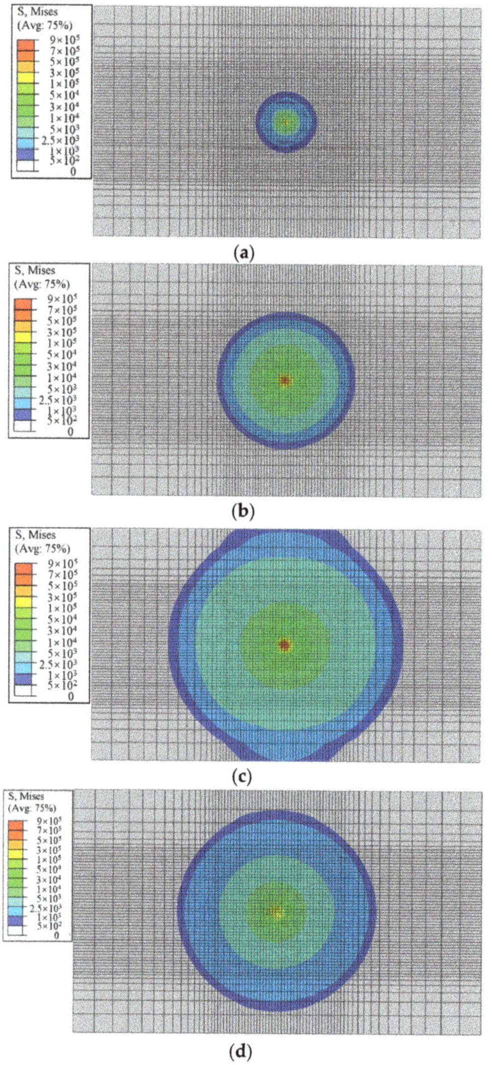

Figure 3. Mises stress distributions of the CRD: (**a**) t = 0.001 s; (**b**) t = 0.0023; (**c**) t = 0.005 s; (**d**) t = 0.0062 s.

Figure 5 illustrates the time history curve of normal displacement u_x at the impact point. From Figure 5, it is evident that, whether for the CRD or the PRCRD, the u_x curve initially rises rapidly to a peak and then declines sharply. This behavior occurs because, when the steel ball contacts the barrier, energy propagates as stress waves from the impact point to the surrounding area, causing the barrier's acceleration to reach its maximum in a very short time. As the deformation of the barrier increases and resistance builds, the acceleration gradually decreases. At this point, the elastic potential energy stored near the impact point causes a reverse motion during the release process, leading to a reduction in u_x. Whether for the CRD or the PRCRD, after the steel ball impact concludes, the u_x at the impact point oscillates around zero. This is likely due to the free vibrations of the barrier following the impact. Additionally, from Figure 5, it can be seen that the maximum normal displacement $u_{x,\ max}$ at the impact point of the CRD is 63.5 mm, while the $u_{x,\ max}$ at the

https://doi.org/10.3390/w18010060

impact point of the PRCRD is 34 mm, indicating that the $u_{x,\ max}$ at the impact point of the CRD is nearly twice that of the PRCRD. This reduction in displacement for the PRCRD is due to the embedded prestressed steel reinforcement, which acts as a support under the impact load of the steel ball, effectively countering a portion of the impact force without causing plastic failure. As a result, the displacement of the PRCRD is smaller compared to the CRD.

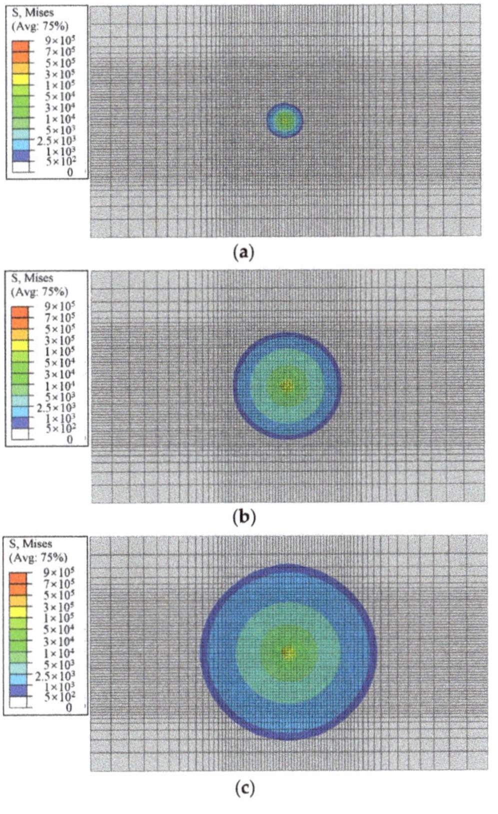

Figure 4. Mises stress distributions of a PRCRD: (**a**) t = 0.00091 s; (**b**) t = 0.0013 s; (**c**) t = 0.00185 s.

Figure 6 shows the normal displacement distribution of CRD and PRCRD. It should be noted that the u_x reaches its maximum value at this time. From the figure, it can be observed that there is a significant u_x displacement near the impact point of the CRD, forming an impact crater with a diameter of approximately 0.5 m. When using a PRCRD, the diameter of the impact crater decreases to 0.2 m. Additionally, as shown in Figure 6, despite the fact that the impact crater diameter is smaller for PRCRD than for CRD, its displacement field range is slightly larger than that of CRD. This may be because, although the embedded prestressed reinforcement in PRCRD effectively counters a portion of the impact force under the impact load, it also transmits some of the uncountered impact force into the barrier, resulting in a slightly larger displacement field range for PRCRD compared to CRD.

According to D'Alembert's principle, under the action of impact loads, the inertial forces generated by acceleration and external loads will put the structure in a state of dynamic equilibrium. Therefore, the acceleration response of the structure, particularly the maximum acceleration, is a very important parameter. From Figure 7, it can be observed that after being impacted, the PRCRD experiences an instantaneous maximum acceleration of 2300 km/s^2 at the impact point, which then rapidly decreases to 0. In contrast, the CRD reaches a maximum acceleration of 1130 km/s^2 after impact, followed by fluctuations around zero acceleration. The maximum acceleration at the impact point of the PRCRD is 2.05 times that of the CRD.

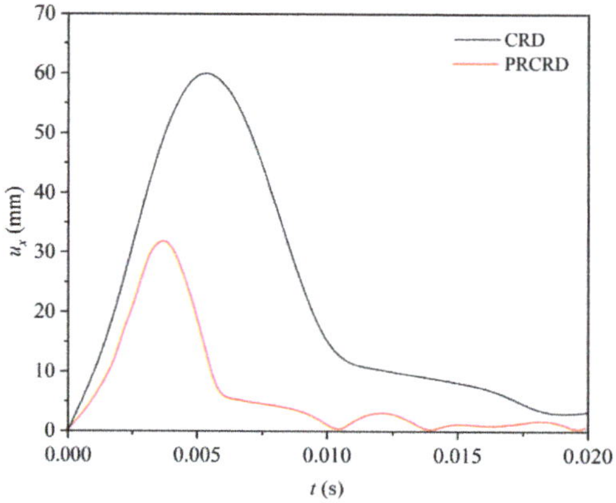

Figure 5. The time–history curve of normal displacement at the impact point.

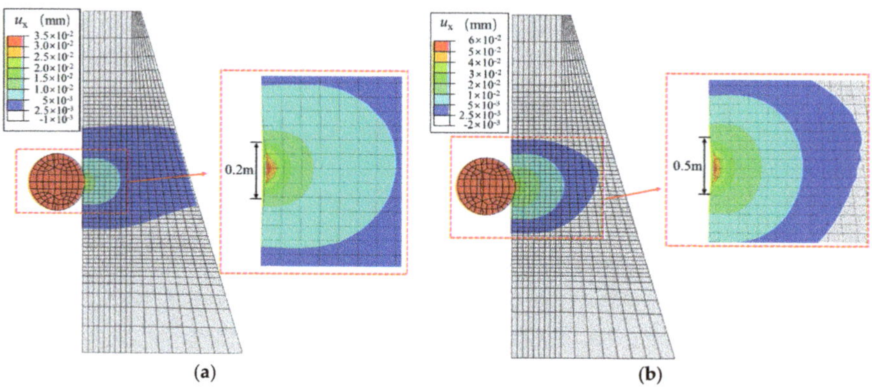

Figure 6. The normal displacement distribution for (**a**) PRCRD; (**b**) CRD.

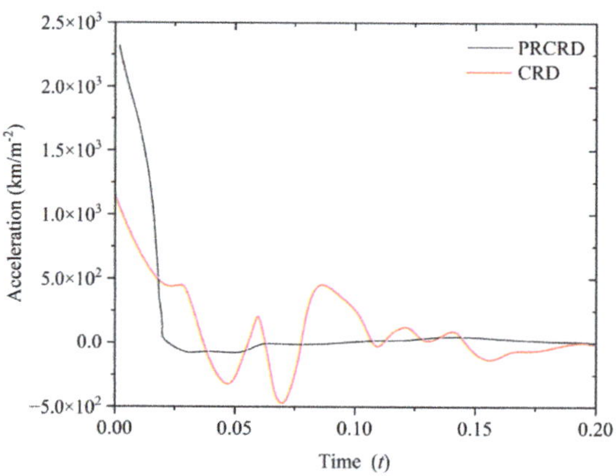

Figure 7. The acceleration time–history curve at the impact point.

https://doi.org/10.3390/w18010060

3.2. Parameter Analysis

From the numerical analysis results in Section 3.1, it is clear that the impact resistance of the PRCRD is superior to that of the CRD. This section will provide a detailed analysis of the impact resistance of the PRCRD under different parameters of the steel ball, which mainly include impact speed, impact mass, impact angle, and shape coefficient.

(a) Impact angle

Figure 8 shows the stress distribution contour of the PRCRD with different θ; it is clear that the expansion range of the stress wave exhibits a distinct directionality. When the θ are 30° and 50°, the tangential impact force F_t acting on the PRCRD is relatively high, causing the stress wave to preferentially extend toward the base of the PRCRD. In contrast, when the θ are 70° and 90°, the normal impact force F_n acting on the PRCRD is higher, leading the stress wave to preferentially extend toward the backside of the PRCRD.

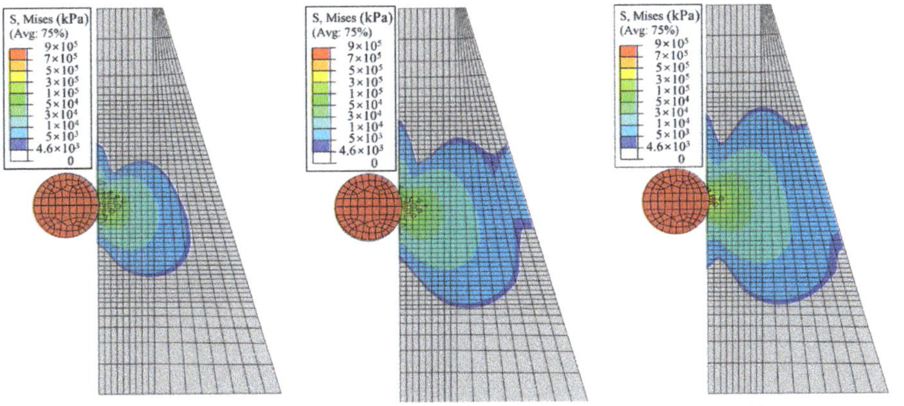

Figure 8. The stress distribution of PRCRD under different impact angles: (**a**) $\theta = 30°$; (**b**) $\theta = 50°$; (**c**) $\theta = 70°$.

Figure 9 shows the calculated results of the normal impact force F_n, tangential impact force F_t, and resultant impact force F_r at the impact point for different θ. From Figure 9a, it can be seen that the F_n increases with the increase in θ, with the maximum resultant impact force $F_{r,\,max}$ corresponding to $\theta = 30°, 50°, 70°$, and $90°$ being 0.52 MN, 0.74 MN, 0.84 MN, and 0.92 MN, respectively. As shown in Figure 9b, the F_t at the impact point exhibits an initial increase followed by a decrease as the θ increases, with the F_t occurring at an $\theta = 50°$. At $\theta = 90°$, the F_t approaches zero. Comparing Figures 9a and 10b, it can be observed that the F_n and F_t begin simultaneously, but they do not reach their peak values at the same time; the former lags slightly behind the latter. For example, at $\theta = 50°$, the peak F_n occurs at 0.003 s, while the F_t reaches its maximum at 0.0035 s.

Figure 10 shows the normal displacement distribution of the barrier for different θ. Clearly, the θ also significantly influences the normal displacement distribution, exhibiting a clear directionality. Figure 11 illustrates the relationship curve between θ and impact depth $u_{x,\,\mathrm{max}}$ (i.e., the maximum normal displacement at the impact point). There is a positive correlation between the θ and the $u_{x,\,\mathrm{max}}$; as the θ increases from 30° to 90°, $u_{x,\,\mathrm{max}}$ increases from 17 mm to 34 mm.

(b) Impact velocity

According to the findings in reference [23], the V of debris flows generally does not exceed 20 m/s. Therefore, this study selects V = 5 m/s, 10 m/s, 15 m/s, and 20 m/s, while

https://doi.org/10.3390/w18010060

keeping other conditions constant, to investigate the effect of V on the impact resistance of PRCRD.

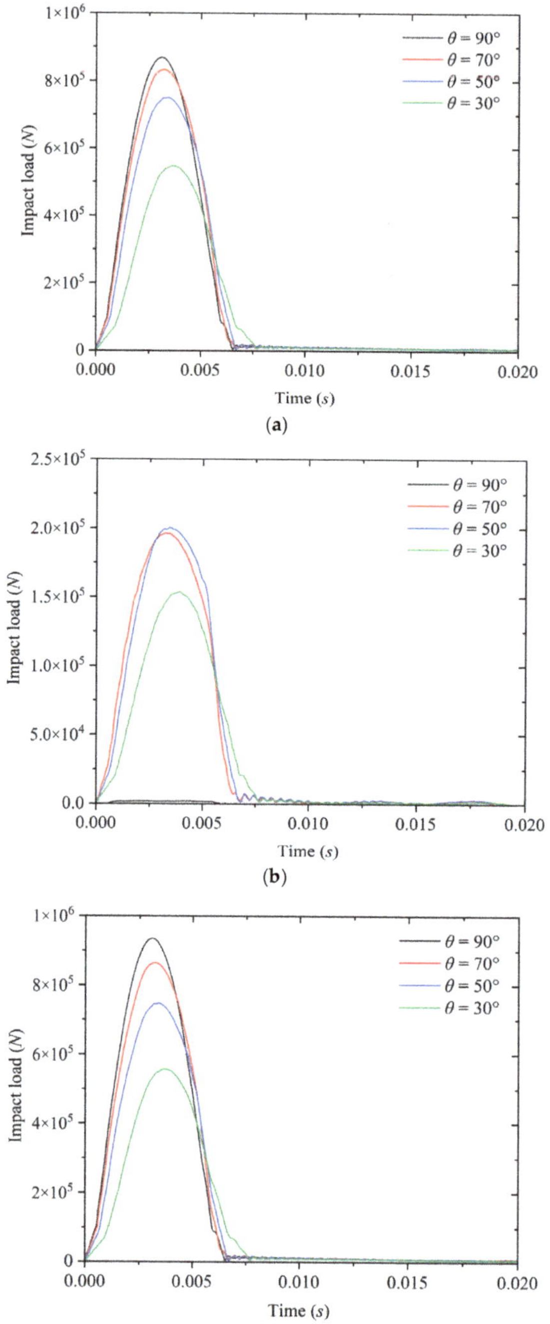

Figure 9. Relationship between impact force and θ: (**a**) normal impact force; (**b**) tangential impact force; (**c**) resultant impact force.

https://doi.org/10.3390/w18010060

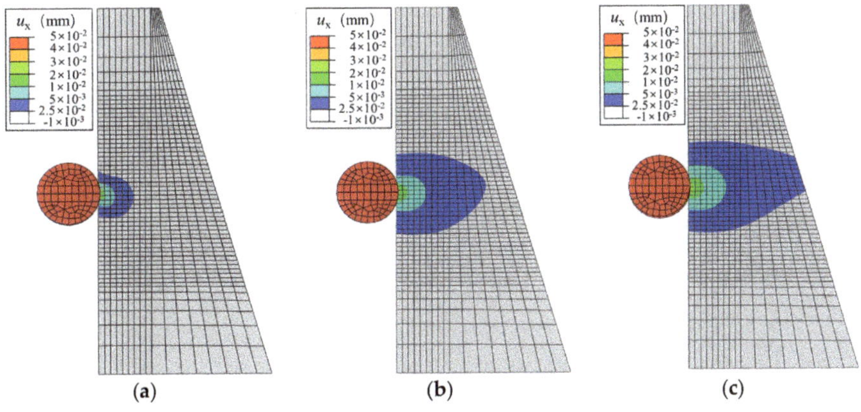

Figure 10. The normal displacement distribution for (a) $\theta = 30°$; (b) $\theta = 50°$; (c) $\theta = 70°$.

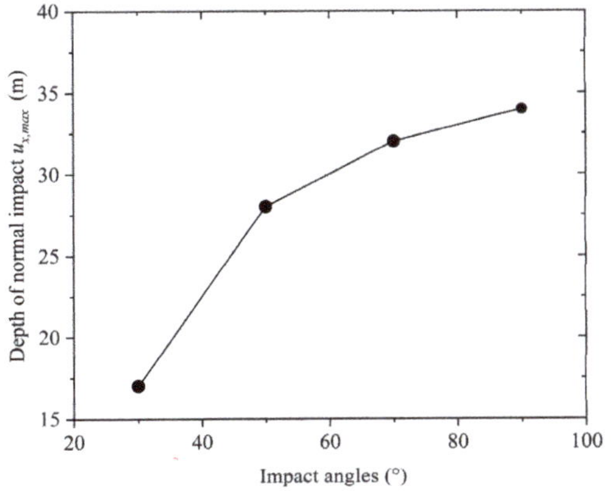

Figure 11. The relationship between impact angle and impact depth.

Figure 12 shows the stress distribution of the PRCRD for different V. From the figure, V has a significant effect on the stress distribution of the PRCRD. As the V increases, the range of the stress wave gradually expands. When V is 5 m/s, the range of the stress wave is relatively small. However, when V increases to 20 m/s, the stress wave extends to the backside, top, and base of the PRCRD, with a significant increase in the stress values at the contact surface between the steel ball and the PRCRD. This is because the kinetic energy of the steel ball increases with its velocity, which in turn increases the energy transmitted to the PRCRD upon impact.

Figure 13 shows the time–history curves of impact force for different V at $\theta = 90°$. From the figure, the impact force increases with the rise in V. The $F_{r,\ max}$ corresponding to $V = 5$ m/s, 10 m/s, 15 m/s, and 20 m/s are 0.37 MN, 0.61 MN, 0.86 MN, and 1.07 MN. When the V increases from 5 m/s to 20 m/s, the $F_{r,\ max}$ increases by a factor of 2.9.

Figure 14 shows the contour of normal displacement of the PRCRD for different impact speeds. The V has a significant effect on the displacement field of the PRCRD. When the V is 5 m/s, the displacement field is limited to the vicinity of the impact point. As the V increases to 10 m/s, the displacement field begins to expand toward the backside, top, and shoulder of the PRCRD. The curve showing the relationship between $u_{x,\ max}$ and V in Figure 15 also illustrates this point. When the V increases from 5 m/s to 20 m/s, the $u_{x,\ max}$ increases from 17 mm to 46 mm, representing a growth of 2.7 times.

https://doi.org/10.3390/w18010060

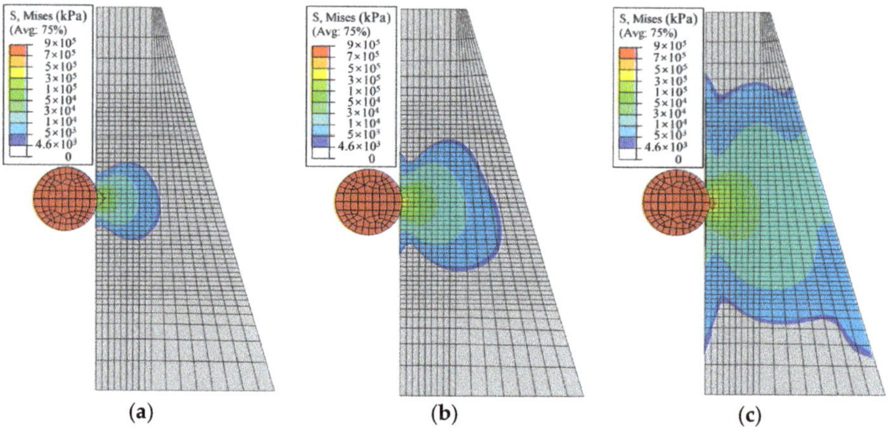

Figure 12. The stress distribution for (**a**) V = 5 m/s; (**b**) V = 10 m/s; (**c**) V = 20 m/s.

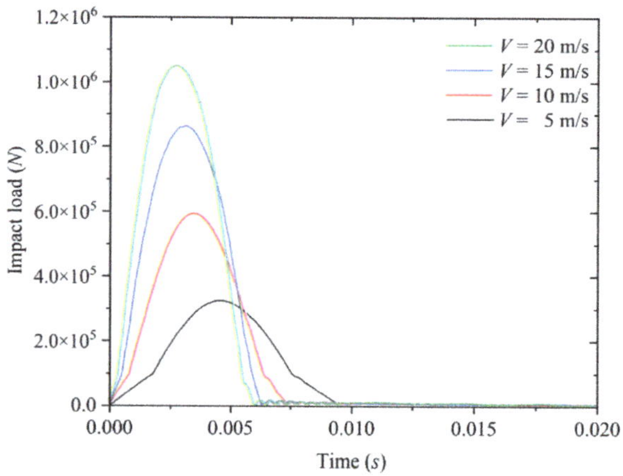

Figure 13. Time–history curve of impact load with different impact velocities.

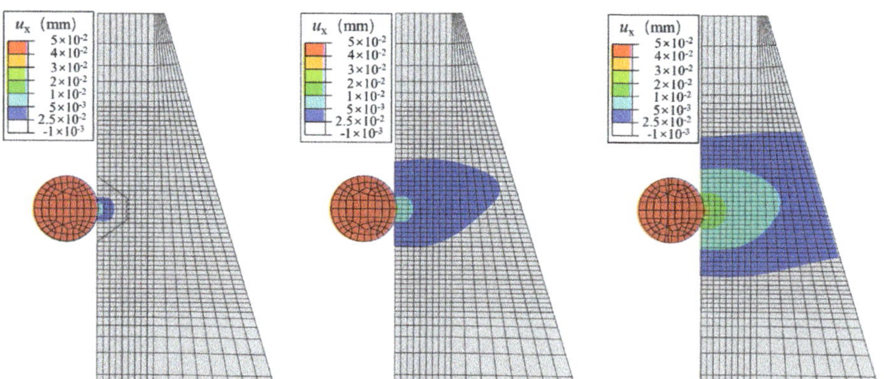

Figure 14. The normal displacement distribution for (**a**) V = 5 m/s; (**b**) V = 10 m/s; (**c**) V = 20 m/s.

https://doi.org/10.3390/w18010060

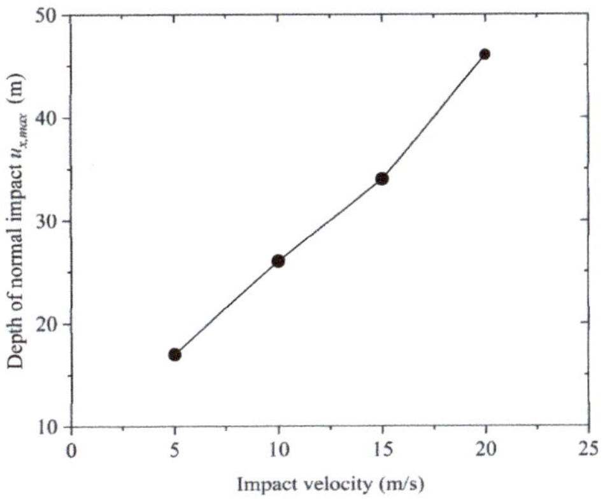

Figure 15. The relationship between impact velocity and impact depth.

(c) Impact mass

Figure 16 illustrates the stress contour map of the PRCRD at the peak resultant impact force, which helps to explain this phenomenon. From Figure 16, it is evident that as the m increases, on one hand, the peak stress at the contact surface between the steel ball and the PRCRD significantly increases. On the other hand, the range of stress influence also gradually expands. When m = 1500 kg, the stress range does not penetrate the entire PRCRD. However, when the m increases to 2250 kg, 3150 kg, and 4000 kg, the stress range penetrates the entire PRCRD.

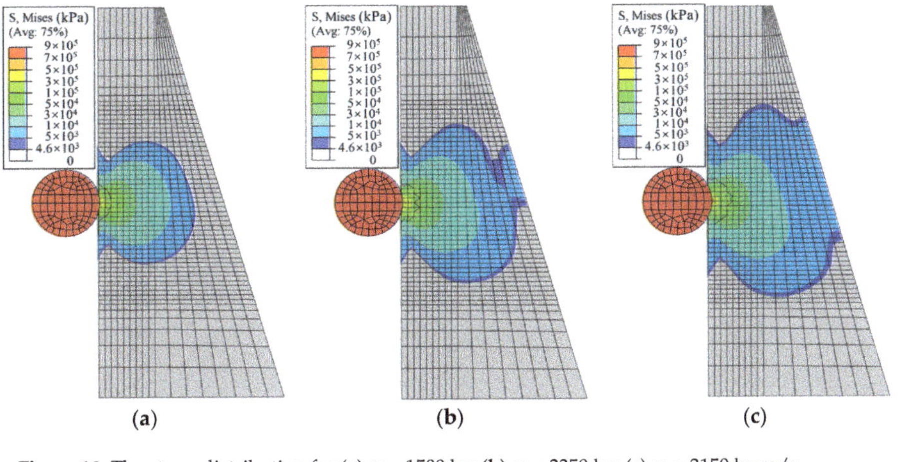

Figure 16. The stress distribution for (**a**) m = 1500 kg; (**b**) m = 2250 kg; (c) m = 3150 kg m/s.

Figure 17 shows the time–history curves of the resultant impact force for different m. From the figure, it can be observed that the $F_{r,\,max}$ increases with the increase in m. For m = 1500 kg, 2250 kg, 3150 kg, and 4100 kg, the corresponding $F_{r,\,max}$ = 0.57 MN, 0.69 MN, 0.79 MN, and 0.87 MN, respectively. This increase in m leads to an increase in impact energy, which in turn results in a $F_{r,\,max}$.

Figure 18 shows the displacement field contour map of the PRCRD for different m. Similarly to the stress distribution, the range of the displacement field gradually expands with an increase in m. Figure 19 illustrates the $u_{x,\,max}$ for different m. When the m increases from 1500 kg to 4000 kg, the $u_{x,\,max}$ increases from 25.5 mm to 34 mm.

https://doi.org/10.3390/w18010060

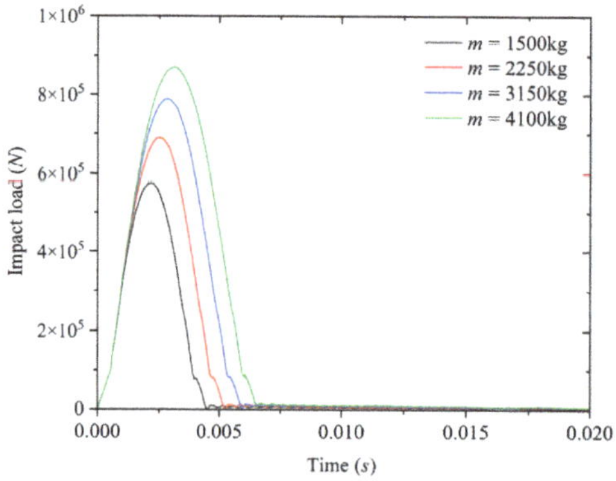

Figure 17. Time–history curves of impact resultant force at different impact masses.

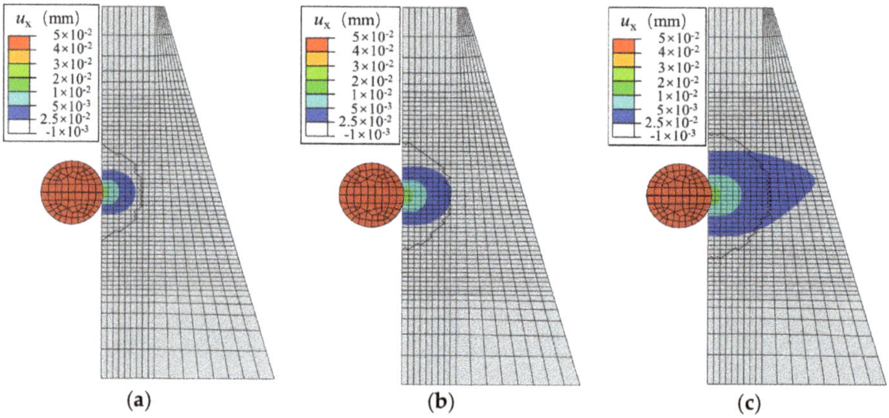

Figure 18. The normal displacement distribution for (**a**) m = 1500 kg; (**b**) m = 2250 kg; (**c**) m = 3150 kg m/s.

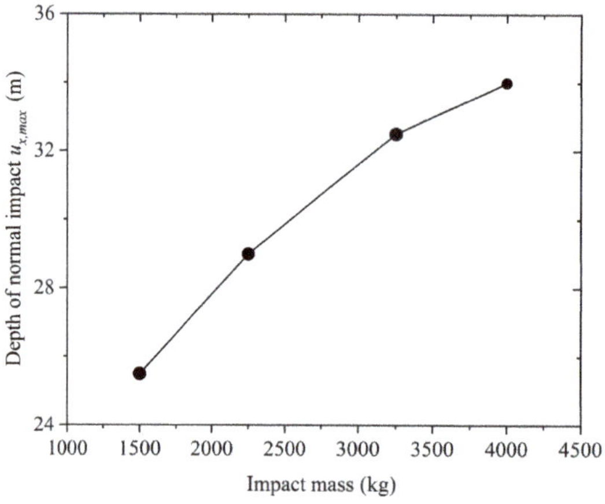

Figure 19. The relationship between impact mass and impact depth.

(d) Shape coefficient

Numerous field investigations of debris flows indicate that there is significant variability in the shape of rock blocks [24]. Simplifying all shapes of rock blocks to spherical ones

https://doi.org/10.3390/w18010060

leads to considerable discrepancies with actual conditions. Therefore, this study considers equating some rock block to ellipsoidal shapes, using the dimensions in three directions of the ellipsoid to describe the irregularity of the rock block, as shown in Figure 20. To facilitate comparative analysis, a shape coefficient, ψ, is given by Equation (5):

$$\psi = \frac{\sqrt{bc}}{a} \tag{5}$$

where a, b, and c represent the semi-lengths of the long axis, middle axis, and short axis of the rock block, respectively. When the lengths of the three axes are equal, $\psi = 1$, which means the rock block is spherical.

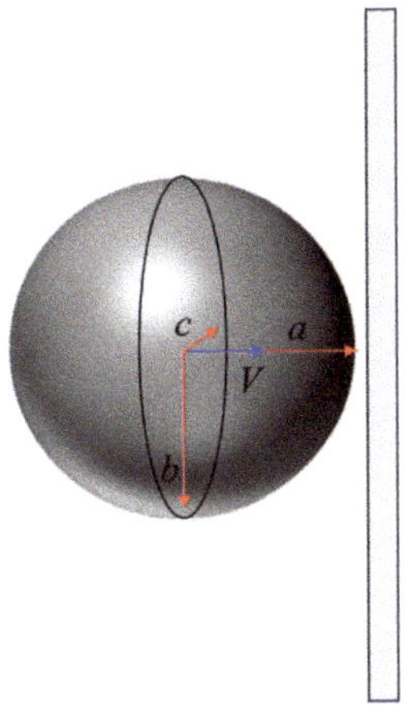

Figure 20. Schematic diagrams of stone shape.

Figure 21 shows the stress contour of the PRCRD for different ψ. From the figure, it is evident that the ψ has a significant impact on the stress values near the impact point; the smaller the ψ, the greater the impact force. To further clarify this phenomenon, Figure 22 presents the time–history curves of the resultant impact force for the impact point with different ψ. Clearly, when $\Psi = 0.4$, the $F_{r,\ max}$ induced by the ellipse ball is 1.32 times that of a spherical ball under the same conditions. This is because the impact energy and impact speed are the same, so the impact energy of steel balls with different Ψ is also identical. The smaller the Ψ, the smaller the contact area between the steel ball and the PRCRD, which results in a greater amount of energy being transmitted to the impact point. Consequently, this leads to a higher resultant impact force. Additionally, from Figure 22, it can also be seen that the ψ has almost no effect on the range of stress wave expansion. This is because the change in Ψ does not alter the amount of energy transmitted to the PRCRD by the steel ball.

Figure 23 shows the normal displacement distribution of the PRCRD for different Ψ. Similarly to the stress field, the ranges of the normal displacement for different Ψ are comparable, with the main difference being the $u_{x,\ max}$ at the contact point between the steel ball and the PRCRD. From Figure 24, it can be observed that the $u_{x,\ max}$ increases as the Ψ decreases. When the Ψ decreases from 1 to 0.4, the $u_{x,\ max}$ increases from 34 mm to 43 mm, representing a 26% increase.

https://doi.org/10.3390/w18010060

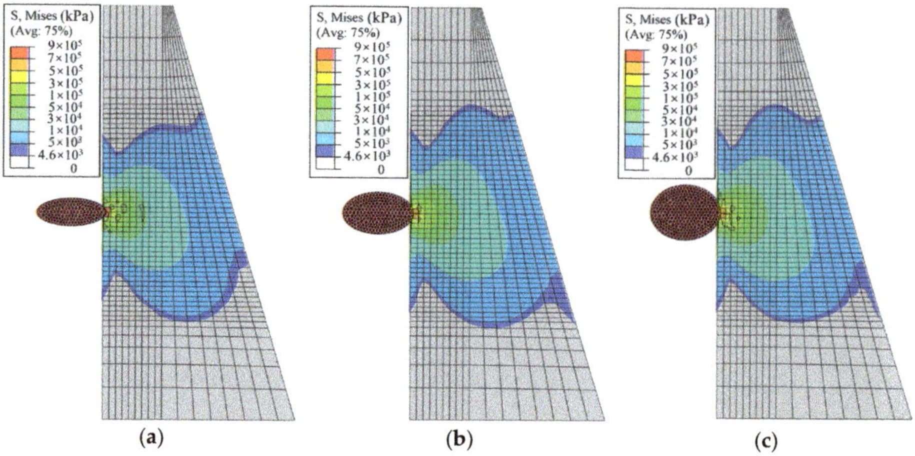

Figure 21. The stress distribution of the PRCRD for (**a**) Ψ = 0.4; (**b**) Ψ = 0.6; (**c**) Ψ = 0.8.

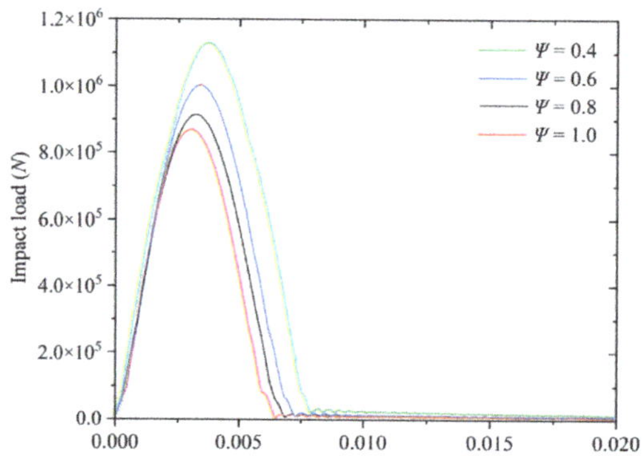

Figure 22. Time–history curves of impact resultant force at different shape coefficient.

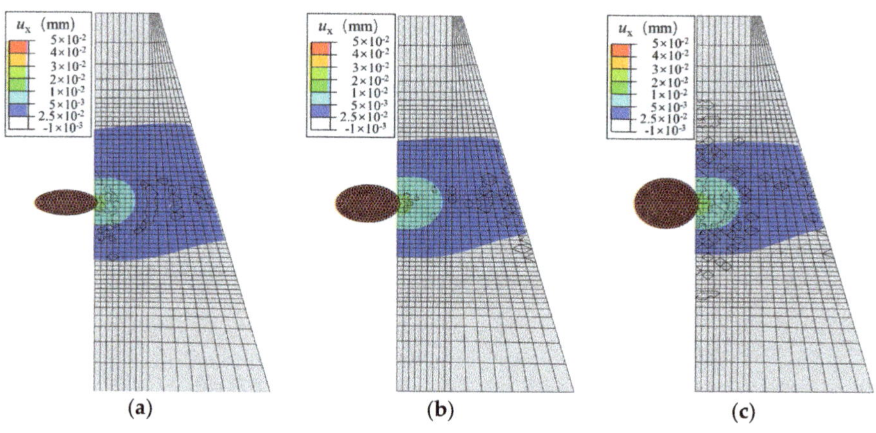

Figure 23. The normal displacement distribution for (**a**) Ψ = 0.4; (**b**) Ψ = 0.6; (**c**) Ψ = 0.8.

https://doi.org/10.3390/w18010060

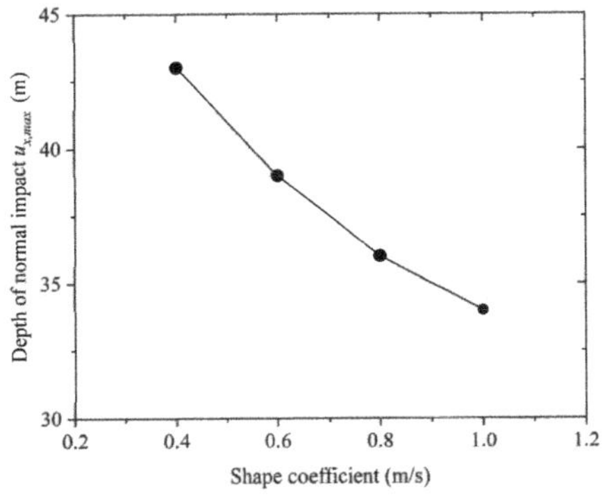

Figure 24. The relationship between shape coefficient and impact depth.

3.3. Sensitivity Analysis of Various Influencing Factors

From the results of the parameter analysis in Section 3.2, it can be seen that the influence of various factors on the impact resistance of the new-type barrier varies. To quantify the degree of influence of each parameter, a sensitivity indicator is introduced to quantitatively evaluate parameters such as impact speed, impact angle, and ball mass. The sensitivity of each influencing parameter is calculated using the following Equation (6):

$$S_{\omega}^{x} = |\Delta\omega / \omega| / |\Delta\delta / \delta| \quad (6)$$

where ω represents the dependent variable in the sensitivity analysis, which in this context refers to the performance parameter characterizing the impact resistance of the PRCRD; x represents the independent variable in the sensitivity analysis, referring to the parameters considered to affect impact resistance. According to the definition of sensitivity, a larger sensitivity value indicates a greater influence of the examined parameter on the impact resistance of the PRCRD.

In this section, the $u_{x,max}$ and the $F_{r,\,max}$ at the impact point of the PRCRD were selected as characterization parameters. Thus, the $u_{x,\,max}$ and the $F_{r,\,max}$ serve as control indicators for the impact resistance of the PRCRD from both deformation and stress perspectives. Based on these two indicators, the sensitivity of the PRCRD impact resistance to various influencing parameters is analyzed. According to the numerical calculation results in Section 3.2, Table 4 lists the sensitivity calculation results for each influencing parameter. From Table 4, it can be seen that for the three influencing factors, V, m and Ψ, the sensitivity of $F_{r,\,max}$ is greater than that of $u_{x,\,max}$, while the effect of the θ is the opposite, with the sensitivity of $u_{x,max}$ being greater than that of $F_{r,max}$.

Table 4. Sensitivity calculation results of each parameter.

Influencing Factors	**Parameter Sensitivity: $u_{x,max}/F_{r,max}$**			
θ	30° 0.75/0.55	50° 0.50/0.31	70° 0.26/0.16	Mean value 0.50/0.34
V	5 m/s 0.77/0.95	10 m/s 0.77/0.93	15 m/s 0.86/0.62	Mean value 0.80/0.83
m	1500 kg 0.45/0.54	2250 kg 0.38/0.46	3150 kg 0.25/0.40	Mean value 0.36/0.47
Ψ	0.40 0.28/0.52	0.60 0.29/0.41	0.80 0.14/0.29	Mean value 0.27/0.41

https://doi.org/10.3390/w18010060

4. Conclusions

This study uses the finite element software Abaqus/Explicit 6.14. to numerically simulate the impact resistance of prestressed reinforced concrete barriers under debris flow block impact loads and the effect of the shape coefficient, impact mass, impact velocity, and impact angle on the impact resistance of the PRCRD. The main findings are summarized as follows:

(1) Under impact from a steel ball, both the stress propagation range and the stress magnitude in a CRD are greater than those in a PRCRD. The maximum normal displacement and maximum acceleration at the impact point of CRD are 1.87 times and 2.05 times those of PRCRD, respectively. Therefore, the impact resistance of RCRD is superior to that of conventional CRD.

(2) The impact resistance of the PRCRD is closely related to the impact angle, impact speed, impact mass, and shape coefficient of the steel ball. The $u_{x,\ max}$ and $F_{r,\ max}$ at the PRCRD impact point increase with a higher impact angle, velocity, and mass. Conversely, a higher steel ball coefficient decreases both.

(3) The sensitivity indicator was introduced to quantitatively evaluate the parameter sensitivity of the influencing factors, such as shape coefficient, impact mass, impact velocity, and impact angle. Among the three influencing factors (V, m and Ψ), the $F_{r,\ max}$ demonstrates higher sensitivity than the $u_{x,\ max}$. In contrast, the influence of the impact angle θ shows an opposite trend, with $u_{x,\ max}$ being more sensitive than $F_{r,\ max}$. In the practical engineering phase, the relevant design of the PRCRD can be based on parameter sensitivity to enhance the impact resistance of the PRCRD.

It is noteworthy that this study considers only the key factor of debris flow block stones, thereby excluding other important factors such as the hydrodynamic effects and the shape of the barrier dam. Undoubtedly, a comprehensive consideration of these factors could significantly enrich and enhance the practical significance of the findings presented in this paper. Nevertheless, the main purpose of this paper is to reveal the stability analysis of the barrier dam under the impact of debris flow block stones. More specific issues, including how the coupling effects of hydrodynamic forces and the dynamic effects of debris flow block stones impact the stability of the barrier dam, remain to be addressed in future research.

Author Contributions: Methodology, Software, Investigation, Validation, Data curation, Funding acquisition, Y.T.; Methodology, Supervision, Resources, Writing—review & editing, W.J.; Writing—review & editing, Software, Funding acquisition, M.W.; Writing—review & editing, Y.Z. All authors have read and agreed to the published version of the manuscript.

Funding: This study is financially supported by the Key Research and Development Program of the Xizang Autonomous Region (No. XZ202401ZY0029) and the Key Program of the Natural Science Foundation of the Xizang Autonomous Region (No. XZ202401ZR0073).

Data Availability Statement: The original contributions presented in this study are included in the article. Further inquiries can be directed to the corresponding author.

Conflicts of Interest: The authors declare no conflicts of interest.

References

1. Badoux, A.; Graf, C.; Rhyner, J.; Kuntner, R.; McArdell, B.W. A debris-flow alarm system for the Alpine Illgraben catchment: Design and performance. *Nat. Hazards* **2009**, *49*, 517–539. [CrossRef]
2. Brighenti, R.; Segalini, A.; Ferrero, A.M. Debris flow hazard mitigation: A simplified analytical model for the design of flexible barriers. *Comput. Geotech.* **2013**, *54*, 1–15. [CrossRef]

3. Cheng, P.; Liu, Y.; Hu, J.; Fang, K.; Hu, Y. Experimental study on dynamic response and failure mode of bedding rock slope with cracks under earthquake. *Bull. Eng. Geol. Environ.* **2025**, *84*, 81. [CrossRef]
4. Liu, Y.; Cheng, P.; Liu, E.; Hu, J.; Cheng, Y.; Su, L. Seismic response and dynamic failure mode of a class of bedding rock slope subjected to freeze–thaw cycles. *Rock Mech. Rock Eng.* **2025**, *58*, 3289–3307. [CrossRef]
5. Bugnion, L.; McArdell, B.W.; Bartelt, P.; Wendeler, C. Measurements of hillslope debris flow impact pressure on obstacles. *Landslides* **2012**, *9*, 179–187. [CrossRef]
6. Chen, X.; Zhang, J.; Hu, Y.; Wang, W.; Liu, Y. Random large-deformation modelling on face stability considering dynamic excavation process during tunnelling through spatially variable soils. *Can. Geotech. J.* **2025**, *62*, 1–21. [CrossRef]
7. Chen, X.; Ren, S.; Guo, X.; Wang, Y.; Liu, F.; Nguyen, H.; Sousa, R.L. Comparative modelling of retrogressive landslide runout: 2D and 3D random large-deformation analyses using coupled Eulerian-Lagrangian method. *Int. J. Min. Sci. Technol.* **2025**, *35*, 2011–2030. [CrossRef]
8. Chai, B.; Tao, Y.Y.; Du, J.; Huang, P.; Wang, W. Hazard assessment of debris flow triggered by outburst of Jialong Glacial Lake in Nyalam County, Tibet. *Earth Sci.* **2020**, *45*, 4630–4639.
9. Chen, J.-G.; Chen, X.-Q.; Wang, T.; Zou, Y.-H.; Zhong, W. Types and causes of debris flow damage to drainage channels in the Wenchuan earthquake area. *J. Mt. Sci.* **2014**, *11*, 1406–1419. [CrossRef]
10. Scheidl, C.; Chiari, M.; Kaitna, R.; Müllegger, M.; Krawtschuk, A.; Zimmermann, T.; Proske, D. Analysing debris-flow impact models, based on a small scale modelling approach. *Surv. Geophys.* **2013**, *34*, 121–140. [CrossRef]
11. Hu, K.; Wei, F.; Li, Y. Real-time measurement and preliminary analysis of debris-flow impact force at Jiangjia Ravine, China. *Earth Surf. Process. Landf.* **2011**, *36*, 1268–1278. [CrossRef]
12. Zanuttigh, B.; Lamberti, A. Experimental analysis of the impact of dry avalanches on structures and implication for debris flows. *J. Hydraul. Res.* **2006**, *44*, 522–534. [CrossRef]
13. Prochaska, A.B.; Santi, P.M.; Higgins, J.D. Relationships between size and velocity for particles within debris flows. *Can. Geotech. J.* **2008**, *45*, 1778–1783. [CrossRef]
14. Chen, H.; Tang, H.; Chen, Y. Research on method to calculate velocities of solid phase and liquid phase in debris flow. *Appl. Math. Mech.* **2006**, *27*, 399–408. [CrossRef]
15. Shan, T.; Zhao, J. A coupled CFD-DEM analysis of granular flow impacting on a water reservoir. *Acta Mech.* **2014**, *225*, 2449–2470. [CrossRef]
16. Wenbing, H.; Guoqiang, O. Efficiency of slit dam prevention against non-viscous debris flow. *Wuhan Univ. J. Nat. Sci.* **2006**, *11*, 865–869. [CrossRef]
17. Huang, H.P.; Yang, K.C.; Lai, S.W. Impact force of debris flow on filter dam. *Momentum* **2007**, *9*, 03218.
18. Shieh, C.L.; Chia-Hsien, T.; Hung-Wen, P.A.N. Impulsive force of debris flow on a curved dam. *Int. J. Sediment Res.* **2008**, *23*, 149–158. [CrossRef]
19. Remaître, A.; Malet, J.P. The effectiveness of torrent check dams to control channel instability: Example of debris-flow events in clay shales. In *Check Dams, Morphological Adjustments and Erosion Control in Torrential Streams*; Nova Science Publishers, Inc.: Hauppauge, NY, USA, 2010; pp. 211–237.
20. Kim, Y.; Nakagawa, H.; Kawaike, K.; Zhang, H. Study on hydraulic characteristics of sabo dam with a flap structure for debris flow. *Int. J. Sediment Res.* **2017**, *32*, 452–464. [CrossRef]
21. Collins, T.K. Debris flows caused by failure of fill slopes: Early detection, warning, and loss prevention. *Landslides* **2008**, *5*, 107–120. [CrossRef]
22. *Abaqus*, Abaqus 6.14 standard; Dassault Systemes Simulia Corp: Providence, RI, USA, 2014.
23. Guo, Y.; Feng, Z.; Wang, L.; Tian, Y.; Chen, L. Hazard assessment of debris flow: A case study of the huiyazi debris flow. *Water* **2024**, *16*, 1349. [CrossRef]
24. Gong, X.-L.; Chen, K.-T.; Chen, X.-Q.; You, Y.; Chen, J.-G.; Zhao, W.-Y.; Lang, J. Characteristics of a debris flow disaster and its mitigation countermeasures in Zechawa Gully, Jiuzhaigou Valley, China. *Water* **2020**, *12*, 1256. [CrossRef]

https://doi.org/10.3390/w18010060

water

MDPI

Article

Environmental Impacts of Post-Closure Mine Flooding: An Integrated Remote Sensing and Geospatial Analysis of the Olkusz-Pomorzany Mine, Poland

Artur Guzy

Faculty of Geo-Data Science, Geodesy and Environmental Engineering, AGH University of Krakow, 30-059 Krakow, Poland; aguzy@agh.edu.pl

Abstract

Mine closure by flooding initiates hydrogeological changes that affect land stability, soil moisture, and surface ecosystems, further shaped by regional climatic trends that increase pressure on water resources. This study examines the Olkusz–Pomorzany mine (Poland), flooded between 2021 and 2022, focusing on the links between groundwater rebound, land movement, and environmental transformation after closure. This analysis combines EGMS-based land movement (2018–2023), groundwater levels (2022–2024), meteorological records (1981–2024), and Sentinel-2-derived Normalized Difference Vegetation Index, Normalized Difference Water Index, and Moisture Index time series (2016–2024). Land cover changes were assessed using Sentinel-2 data for 2019–2024. Results show climate-driven subsidence of less than 1 mm/year across the area and a shift to uplift within the mining zone, with maximum groundwater rebound of 103 m in the central depression cone and uplift of up to 3.6 mm/year. Climatic water balance remained negative, with Vertical Water Exchange averaging −11.6 mm/month in 2022–2024. Hydrospectral indices indicate seasonal variability and modest increases in vegetation activity and moisture after flooding. Land cover analysis shows an expansion of surface water and wetlands where historical drainage and rebound overlap. These findings confirm that groundwater recovery is already reshaping surface conditions and highlight the need for integrated monitoring in post-mining areas.

Keywords: post-mining areas; groundwater rebound; climate change; land movement; hydrospectral indices; land cover change

1. Introduction

Across Europe, many underground mines are being decommissioned for economic or climate policy reasons [1,2]. Over the past decade, this trend has been particularly evident in the hard coal mining sector, reflecting the implementation of European Union climate and environmental policies aimed at reducing greenhouse gas emissions under the European Green Deal [3,4]. However, closures also increasingly affect other long-operating mines extracting different raw materials, where deposits have been depleted or further extraction is no longer economically justified [5,6]. Mine closure is frequently carried out by controlled flooding, which restores groundwater levels within aquifer systems previously drained during mining. As hydraulic heads rise within the former depression cone, increasing hydrostatic pressure can induce land uplift [7,8]. These subsurface geomechanical processes often manifest at the surface through the formation of open water bodies and wetlands, changes in soil moisture and vegetation [9,10]. Furthermore, local geological hazards such as sinkholes or landslides may occur, particularly in areas undergoing rapid groundwater rebound in post-mining settings characterized by shallow workings or complex geological and hydrogeological conditions [11,12].

At the same time, Europe has been experiencing progressive climate change, including warming and steppe-formation processes, which intensify moisture deficits [13,14]. Recent decades have brought higher mean air temperatures, more frequent but short-lived intense rainfall events, longer dry periods, and shorter snow seasons [15,16]. In this context, post-mining areas are subject to two overlapping pressures. On the one hand, climatic trends drive regional drying and reduced water availability [17]. On the other hand, mine closure induces local hydrogeological changes that may raise groundwater levels and alter surface hydrology. These processes can, at least partially, counteract regional water deficits. Consequently, water-based reclamation, such as the formation of pit lakes and wetlands, has become an increasingly common management strategy, offering both ecological and recreational benefits [18,19]. However, its effective implementation depends on a thorough understanding of groundwater-surface water interactions to support sustainable planning and risk mitigation.

Such an understanding is well established for the operational phase of mining [20]. Mining-induced subsidence and aquifer-system dewatering during operation have been extensively documented and are routinely monitored using geodetic and remote sensing techniques, including terrestrial laser scanning [21], unmanned aerial vehicle [22,23] and satellite radar interferometry (InSAR) [24,25], complemented by dense networks of piezometers [26]. Moreover, numerous reliable modeling approaches have been developed, enabling credible forecasts of subsidence and mining-related dewatering [27–29].

In contrast, post-closure groundwater rebound and associated land uplift are less frequently studied and more difficult to quantify [30–32]. These processes are particularly challenging to analyze due to the long-term geomechanical alterations of the aquifer system caused by mining, which have modified the original geological and hydrogeological conditions [26,33,34]. Previous studies indicate that uplift rates associated with groundwater rebound are typically lower than subsidence during active drainage, and that the uplift zone often extends beyond the formal mining boundary into areas previously affected by groundwater drawdown [32,35,36]. Although land uplift related to groundwater rebound can be tracked using InSAR, and groundwater rise can be monitored with piezometers, in practice dynamic environmental changes and surface transformations make effective monitoring difficult, especially in areas with dense vegetation where signal quality is low [34,37]. At the same time, after a mine closure, piezometer networks are often poorly maintained, providing limited temporal and spatial information on the rebound process. Additionally, the superposition of hydrogeological processes and regional climate-driven storage changes makes it difficult to separate the contribution of rebound-induced uplift from background subsidence or climate-related effects [35,38]. In many post-mining settings, land deformation represents a superposition of direct mechanical subsidence caused by ore extraction and indirect compaction of aquifer systems due to prolonged dewatering, followed by rebound related to mine flooding and re-saturation of the rock mass [26,39–42].

To monitor environmental changes related to variations in surface water and soil moisture, remote sensing, particularly multispectral satellite imagery, remains a valuable research tool [43,44]. Such data make it possible to derive hydrospectral indices that serve as proxies for vegetation condition, near-surface soil moisture, and the presence of surface water [45]. Commonly applied indices include the Normalized Difference Vegetation Index (NDVI), which reflects vegetation vigor and canopy density; the Normalized Difference Water Index (NDWI), sensitive to surface water and vegetation water content; and the

Normalized Difference Moisture Index, used to assess soil and vegetation moisture status [46]. Time series of these indices can capture both gradual and seasonal variations in hydrological and ecological conditions, and their spatial distribution often corresponds closely to hydrogeological gradients, which are typically well expressed in post-mining landscapes [43,47,48]. The main advantages of this approach include the ability to analyze large areas, reconstruct temporal dynamics over long periods, and detect subtle environmental shifts that may be difficult to observe directly.

Nevertheless, relying solely on hydrospectral indices does not allow for a comprehensive assessment of environmental transformations in post-mining settings. These indicators provide indirect information and cannot fully capture subsurface processes or their timing. Many previous assessments have analyzed land deformation, groundwater levels, and hydrospectral signals separately [47,49,50], without integrating them into a broader climatic and environmental framework. In reality, changes induced by mine closure are strongly interconnected, namely groundwater rebound drives land uplift, alters hydrological conditions, and affects land cover. Only an integrated analysis of these parameters enables a more complete understanding of post-mining environmental transformations following mine flooding [51].

This study addresses this gap for the Olkusz–Pomorzany zinc–lead underground mine in southern Poland, which was closed and flooded between 2021 and 2022 after several decades of operation [52]. Since the closure, the area has experienced rapid groundwater rebound, land uplift, the emergence of surface water bodies, and growing public concern over environmental impacts and safety [53]. At the same time, the region has been subject to climatic changes typical of Central Europe, characterized by growing pressure on water resources [54].

This study integrates in situ meteorological records, groundwater head observations, InSAR-based vertical land motion, and Sentinel-2-derived hydrospectral indices and land cover data to analyze the temporal sequence and spatial coherence of environmental transformations before and after mine closure in the context of climate change. Specifically, the objectives of the study are as follows:

- Characterize the long- and short-term climatic conditions in the area of the former mine and their relevance for regional hydroclimatic context;
- Quantify the vertical land movement within the mine and in a control area outside the mining impact zone, distinguishing background subsidence from post-closure uplift and regional climate-related land elevation changes;
- Analyze the groundwater rebound using piezometric monitoring inside and outside the former depression cone and relate its timing and magnitude to land movement and climatic drivers;
- Assess the environmental response to these hydrogeological changes using hydrospectral indices and land cover classification, with particular emphasis on the emergence of open water and wetlands.

This integrated approach provides a more complete basis for understanding the relationships between groundwater rebound, land surface deformation, and environmental change under changing climate conditions.

2. Study Area

The Area of Interest (AOI) is located in southern Poland in the historical zinc and lead ore mining district around Olkusz. The AOI covers approximately 400 km^2 and encompasses the entire zone of past and recent mining activities, which extends over roughly 130 km^2 (Figure 1A,B).

The landscape of the AOI is dominated by forested and semi-natural areas, while agricultural and urban land use is relatively limited, and surface water bodies cover only a minor fraction of the territory [55]. One of its most distinctive natural features is the

Błędów Desert, the largest inland dune field in Central Europe. Soils are generally of low fertility, developed mostly on postglacial sands and skeletal rendzinas on limestone, with localized patches of brown soils on loess [56,57].

Elevation ranges from approximately 300 to 430 masl, gradually increasing toward the east and south-east [57]. Erosional valleys cross the southern and south-eastern parts of the AOI. Due to the high permeability of the carbonate bedrock, surface drainage is poorly developed; many valleys are ephemeral, with flow occurring only after heavy precipitation events. Springs and karst outlets are relatively common [55,56].

The AOI lies in one of the most economically important regions in Poland, positioned between Cracow Metropolitan Area and the Upper Silesian–Zagłębie Metropolis (GZM), together inhabited by over 3.5 million people (Figure 1A,B) [57]. Within the AOI, the towns of Olkusz, Bukowno, and Sławków, along with several smaller villages, are home to a combined population of more than 150,000 [57]. Owing to its central location, the AOI functions as a major transportation corridor, traversed by a national road, a national railway line, and the Broad Gauge Metallurgy Line [57], which connects the Polish–Ukrainian border with Sławków and the Katowice Steelworks, one of the largest steel producers in Poland [58].

2.1. Geological Conditions

The AOI is situated on the northern and north-eastern margin of the Upper Silesian Coal Basin (USCB; Figure 1A), a geologically complex region in southern Poland that hosts one of the world's largest Mississippi Valley-Type zinc and lead ore provinces [59,60].

At the base, the geological structure is dominated by Permian–Mesozoic carbonate formations, forming a monoclinal sequence overlying Paleozoic sedimentary rocks. This sequence is intersected by a major fault zone extending from the north-west to the south-east, along which Carboniferous–Permian acidic and alkaline magmatic and volcanic rocks occur [61]. The stratigraphic succession ranges from Devonian to Jurassic units, with mineralization hosted primarily within Middle Triassic dolomites [62] (Figure 1C).

The zinc and lead ore bodies occur as stratiform lenses, pseudo-beds, and irregular cavity fillings associated with the ore-bearing dolomites of the Muschelkalk [59,60]. Mineralization developed through a combination of dolomitization, brecciation, and dissolution processes, strongly influenced by karst development and tectonic evolution, including the formation of numerous fault zones. The principal ore minerals are polymorphic varieties of zinc, lead, and iron sulfides. Economically significant mineralization is concentrated in carbonate rocks of the Triassic sequence, particularly between the Röt and Muschelkalk units. The ore-bearing series is overlain by diplopore dolomites, Jurassic limestones, and Keuper clay formations [59].

Above the Jurassic formations, Quaternary sediments of glacial and fluvial origin occur. These deposits are generally poorly consolidated and mechanically weak, which makes them particularly susceptible to deformation processes associated with underground mining exploitation [62].

2.2. Hydrogeological Conditions

The hydrogeological system within the AOI comprises five aquifer systems: Quaternary, Jurassic, Triassic, Carboniferous, and Devonian [63]. The Quaternary aquifer occurs in heterogeneous sands and gravels of variable thickness that infill erosional forms at the top of older geological formations. These sediments form a porous aquifer with high permeability. The Jurassic aquifer is developed in fractured and karstified Upper Jurassic limestones and forms a fissure and karst groundwater reservoir. The Triassic aquifer, which is the most important in the context of mining, consists of carbonate rocks of the Röt and Muschelkalk formations. It forms a complex system that is porous, fissured, and karstified, with high transmissivity and hydraulic connectivity; this system historically controlled mine inflows and now governs groundwater rebound after mine closure. The Carboniferous and Devonian aquifers are less well studied and occur at greater depths [64].

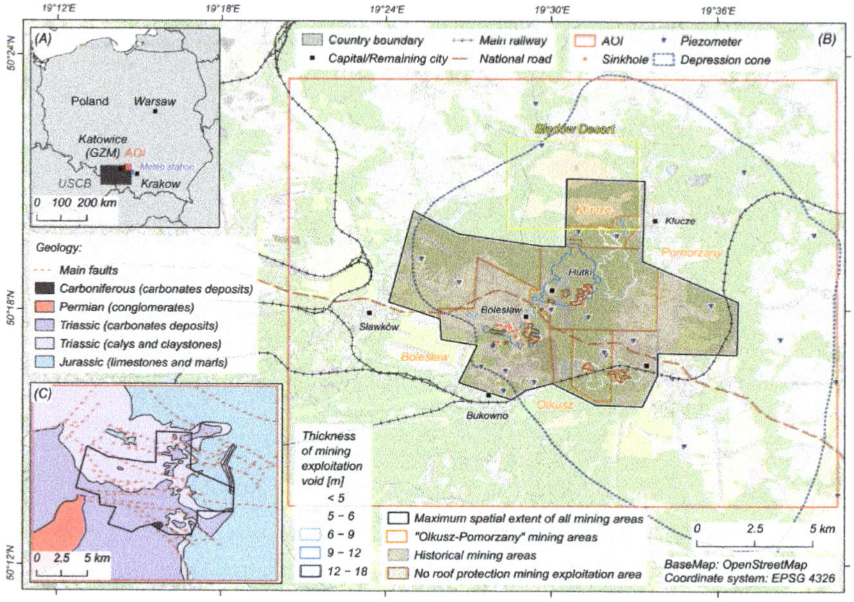

Figure 1. (**A**) Location of the AOI within Poland and position of the meteorological station. (**B**) Map of the AOI showing landscape features, urban areas, major roads, railway lines, rivers, and mining-related factors, including the spatial extent of the depression cone prior to mine closure in 2019, the network of piezometers monitoring groundwater head in the Jurassic aquifer, and the location of sinkholes. (**C**) Main geological formations and spatial distribution of faults within the AOI. Data sources: National Geoportal of Poland [57]; Polish Geological Institute—National Research Institute [65]. Basemap: OpenStreetMap contributors [66].

Between the Triassic and Paleozoic aquifers lies a Permian aquitard composed mainly of conglomerates, which locally grade into clay deposits. This unit generally acts as a hydraulic barrier separating the younger Triassic aquifer from the deeper Paleozoic system. However, in some zones, tectonic disturbances and mining-induced fracturing created local permeable pathways, allowing for limited hydraulic connections. During the development of underground workings at the "Olkusz–Pomorzany" mine, several water inflows from these fractured Permian strata were recorded, reaching several tens of liters per minute [67].

Groundwater flow within the AOI is influenced by both natural and anthropogenic hydraulic connections between aquifers [68]. Natural pathways are formed mainly by erosional features within karstified strata and by fault zones [69], while anthropogenic pathways developed through mining-induced fracturing and deformation of the rock mass [67]. During the operation of the "Olkusz-Pomorzany" mine, these connections controlled the volume and spatial variability of inflows. Following mine closure, they control the rate and spatial pattern of groundwater rebound within the former depression cone [52].

2.3. Mining Exploitation in the Olkusz Region

Mining of zinc and lead ores in the Olkusz region dates back to at least the 12th century, with the first records from 1257 [70]. Early mining focused on shallow, near-surface deposits of lead-silver ores. Large-scale industrial operations began after World War II. The "Olkusz-Pomorzany" mine was established in 1974, exploiting several ore fields, including "Bolesław", "Olkusz", "Klucze", and "Pomorzany". Ore extraction at "Bolesław" ended in 1996, at "Olkusz" in 2001, and finally at "Pomorzany" in 2020. Over the entire operational period, approximately 71.8 million tons of ore were mined, yielding about 3.0 million tons of zinc and 0.9 million tons of lead [71].

The deposits are irregular and structurally complex. Mining was carried out using room-and-pillar and cut-and-fill methods with hydraulic backfill to reduce surface deformation [70,71]. Ore extraction required intensive drainage of the aquifer system due to substantial water inflow into the underground workings. Dewatering depths reached about 65 m at "Bolesław", 95 m at "Olkusz", and 140 m at "Pomorzany". This drainage significantly altered both groundwater and surface water flow [72]. By 2019, the depression cone in the Triassic aquifer covered an area of approximately 400 km^2 [52]

Between December 2021 and January 2022, the staged flooding of the mine began, and groundwater pumping was gradually stopped at successive shafts [52]. This initiated groundwater rebound and progressive infilling of the depression. The rebound rate varies spatially and is mainly controlled by a large recharge area of the depression cone, hydraulic connections through erosional windows in Upper Triassic clays linking Quaternary sands with Triassic carbonates, and karst conduit systems flushed during mine drainage [69]. Additional recharge from precipitation and surface watercourses further accelerates the process [52]. After pumping ceased, several rivers previously sustained by mine drainage dried up [52,73]. At the same time, rising groundwater levels have led to the formation of new inundation zones and wetlands, particularly in closed depressions such as former sand pits located within the former mining area (Figure 2A,B, Supplementary Video) [74,75]. In some places, water levels in these ponds have been rising at rates of several centimeters per day [75]. In early 2024, rising water inundated a newly built bypass road in Bolesław, forcing its closure just weeks after opening (Figure 2C,D) [75]. These processes have been accompanied by changes in water chemistry [76] and reduced geotechnical stability of the ground, leading to local slope failures and small landslides [53].

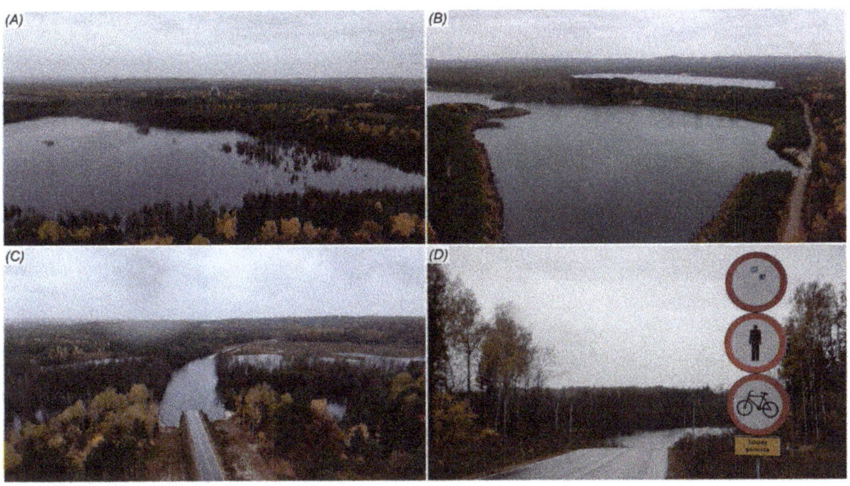

Figure 2. Flooded areas in the AOI following the closure and inundation of the Olkusz–Pomorzany mine, based on the situation as of 27 October 2025. Flooding near the village of Hutki fills former sand open-pit excavations, causing the dieback of surrounding vegetation as the water level continues to rise. To the south-west, the Dąbrówka shaft is visible (**A**), while to the south-east, a local landslide has developed along the road (**B**). The flooded area near the bypass of the village of Bolesław resulted in road submergence and closure to traffic (**C**,**D**). Also see the Supplementary Video. Photo credit: Artur Guzy.

Long-term mining has also resulted in extensive geotechnical impacts. The relatively shallow depth of workings, in many areas less than 100 m, combined with unfavorable geological conditions and caving mining methods, has led to the formation of numerous sinkholes, some of which were already observed during active stage of mining (Figure 1B) [77,78]. The rapid groundwater rebound has led to land uplift of up to several tens of millimeters per year and has intensified suffusion processes, further weakening the rock mass already disturbed by decades of mining activity [35]. Since 2022, many new sinkholes have appeared, with diameters of several dozen meters and depths reaching up to 20 m [53]. These features pose direct hazards to land use, infrastructure, and residents. A recent inventory identified 1260 sinkholes, including more than 200 located within 20 m of buildings or roads [53]. Many of them are relict structures that were previously backfilled during mining. Recent collapses occur mainly above historical shafts and shallow workings, where insufficient infill fails, potentially due to increased hydrostatic pressure and groundwater flow [12]. Similar mechanisms of overburden failure, cumulative damage, and fracture propagation have also been documented in other mining regions affected by repeated exploitation and stress redistribution within multi-layered rock systems [79–82]. Overall, deformation processes in the "Olkusz–Pomorzany" mine result from the combined effects of shallow mining-induced subsidence, gradual compaction of dewatered aquifers, and rebound caused by mine flooding and re-saturation of the depleted aquifer system. These mechanisms together explain the complex spatial pattern of vertical ground movement observed in the AOI. At present, no active subsidence is recorded, given the time elapsed since the cessation of ore extraction in most parts of the mine [53,77,83]. Only minor residual displacements may still occur locally above former workings, while ongoing deformation is mainly associated with groundwater rebound and the increase in pore pressure after mine closure [52,53].

3. Materials and Methods

The methodology of this study consisted of three main components.

In the first stage, meteorological time series were analyzed to characterize climatic conditions in the AOI. The analysis focused on air temperature, total and maximum daily precipitation, the number of precipitation days, snow cover duration, cloud cover, relative humidity, and wind speed. Based on these data, reference evapotranspiration (ET_0) and Vertical Water Exchange (VWE) were calculated to quantify the regional climatic water balance. As the data came from a single measurement station, they were considered representative of the entire AOI. This assumption is justified because mining activities do not influence climatic parameters, which are controlled by regional atmospheric processes. These variables and derived indices have a direct impact on the water balance and are important for interpreting hydrological and hydrogeological conditions.

The second stage was based on InSAR data to identify and analyze vertical land movement within the AOI. This made it possible to distinguish between displacements caused by mining (e.g., residual subsidence due to the convergence of mine voids and mining-induced dewatering) and those linked to broader regional or climatic factors (e.g., hydrological drought). Additionally, time series of vertical land movement were cross-correlated with groundwater head measurements from piezometers to assess how groundwater level fluctuations influence ground deformation.

In the third stage, the analysis focused on hydrospectral indices sensitive to surface and near-surface water content, namely NDVI, NDWI, and Moisture Index. These indices are useful for tracking changes in vegetation water content and soil moisture, which are relevant for assessing drought processes related to both mining and regional climatic variability. Land cover maps combined with Sentinel-2 imagery were also used to detect changes in inundation and wetland areas associated with groundwater rebound after mine closure.

The InSAR and hydrospectral indices–land cover change analyses were conducted separately for the "Olkusz-Pomorzany" mine area and a control area located outside the zone of mining influence. This approach allowed for environmental changes related to mining or post-mining processes to be distinguished from those driven by broader regional trends.

The datasets used in this study covered different temporal ranges. Meteorological data were available for the longest period (January 1981–December 2024), providing a basis for long-term climate assessments. InSAR data were available for March 2018–December 2022 and March 2019–December 2023, while groundwater head measurements from piezometers covered January 2022–December 2024. NDVI, NDWI, and Moisture Index time series were available from July 2015 to December 2024, and annual land cover maps for 2019–2024. Considering these data availability ranges and the timing of mine closure (December 2021–01.2022), three analytical periods were defined.

- Meteorological data

To examine climatic variability at different temporal scales, three time intervals were used: January 1981–December 2024 for long-term trends; January 2016–December 2024 for conditions before and after mine closure; and January 2022–December 2024 for post-closure changes, enabling direct comparison with remote sensing datasets.

- InSAR and groundwater head data

Two InSAR intervals (March 2018–December 2022 and March 2019–December 2023) were used to capture vertical land movement patterns before and after mine closure. The latter period was also combined with groundwater head data (January 2022–December 2024) to examine the relationship between ground deformation and groundwater rebound.

- Hydrospectral indices and land cover

The periods January 2016–December 2024 and January 2022–December 2024 were selected to assess pre-closure, transitional, and post-closure conditions, enabling evaluation of how groundwater rebound and related hydrological changes are reflected in vegetation, surface water dynamics, and land cover.

3.1. Meteorological Data

To assess the influence of both long- and short-term meteorological trends on environmental change, time series of meteorological data were analyzed from the Polish Institute of Meteorology and Water Management—National Research Institute synoptic station in Kraków-Balice [84]. This is the closest first-order synoptic station to the AOI, located approximately 23 km away (Figure 1A). The station is part of the national meteorological observation network, which includes 981 stations in total: 63 first-order synoptic stations, 220 second-order climatological stations, 690 precipitation stations, and 8 special stations.

Although the analysis relied on data from a single meteorological station, this approach is justified because Kraków-Balice represents the same regional climatic zone as the AOI. Mining operations do not influence atmospheric conditions, and thus the Balice station provides reliable regional climatic input for hydrological interpretation. Consequently, these data were used to characterize the regional net hydraulic balance and to assess how broader climatic variability may have influenced water conditions within the AOI.

Historical meteorological data from this network are homogenized, so that the recorded variability reflects actual climatic changes rather than artifacts related to measurement or site modifications. Observations are performed hourly at eight standard times (00,

03, 06, 09, 12, 15, 18, and 21 UTC). The following parameters were used in the analysis: mean, maximum and minimum monthly air temperature, monthly total and maximum daily precipitation, mean monthly cloud cover, number of days with rainfall, snowfall and snow cover, mean monthly relative humidity, wind speed at 2 m, and water vapor pressure [84]. Air temperature was measured using automated sensors placed 2 m above ground level. Daily maximum and minimum temperatures were reported for the standard 24 h period from 18:01 UTC of the previous day to 18:00 UTC of the current day. Precipitation was recorded at 00, 06, 12, and 18 UTC using rain gauges with a 200 cm^2 catchment area placed 1 m above ground. Daily totals were calculated for the period from 06:01 UTC to 06:00 UTC the following day. Cloud cover was expressed in oktas (eighths of the sky dome), ranging from 0 (clear sky) to 8 (overcast). Wind speed was measured at 2 m using an automated cup anemometer integrated with the synoptic station system. Relative humidity and water vapor pressure were recorded by hygrometric sensors operating at the same level. The vapor pressure data were derived from continuous humidity and temperature readings and expressed as monthly means in hPa [65].

In addition to the descriptive analysis, the dataset was used to quantify climatic drivers of hydrological balance by ET_0 and VWE. The ET_0 was computed using the FAO-56 Penman–Monteith method [85], according to Equation (1):

$$ET_0 = \frac{0.408\Delta(R_n - G) + \gamma\frac{900}{T+273}u_2(e_s - e_a)}{\Delta + \gamma(1 + 0.34u_2)} \tag{1}$$

where ET_0 is the reference evapotranspiration [mm/day]; R_n is net radiation at the surface [MJ/(m^2day)]; G is soil flux heat density [MJ/(m^2day)]; T is mean air temperature [°C]; u_2 is wind speed at 2 m height [m/s]; e_s and e_a are the saturation and actual vapor pressure [kPa]; Δ is the slope of vapor pressure-temperature curve [kPa/°C]; and γ is the psychrometric constant [kPa/°C].

The actual vapor pressure e_a was derived directly from the observed monthly mean water vapor pressure, which improved the accuracy of ET_0 estimates. Monthly values of ET_0 were obtained by multiplying daily means by the number of days in each month.

To assess the climatic water balance, the VWE was calculated as Equation (2):

$$VWE = P - ET_0 \tag{2}$$

where P is the monthly total precipitation [mm]. Positive values of VWE indicate potential groundwater recharge (water surplus), while negative values denote a climatic deficit and potential drought stress.

The calculated VWE represents the regional net hydraulic balance and provides a basis for assessing the impact of climate on the hydrological system within the AOI. Long-term VWE trends were used to establish the regional background of water deficit and potential climate-driven subsidence, serving as a reference for comparing climatic and mine-related influences in subsequent analyses.

The analysis covered the time intervals defined in Section 3, with emphasis on both long-term climate variability and the period surrounding mine closure. Linear trends were fitted to the time series of selected variables to identify changes over multi-decadal and shorter periods. Linear functions were applied because they offer a consistent means of describing gradual climatic trends. All periods were defined from January to January to minimize the influence of seasonal variability on trend estimation [86].

Meteorological information from these analyses was then used to interpret hydrological and hydrogeological changes, including vertical land movement, groundwater rebound, and variations in hydrospectral indices and land cover patterns.

3.2. EGMS-Based Vertical Land Movement and Groundwater Head Monitoring Data

Land movement was analyzed using data from the European Ground Motion Service (EGMS), which provides harmonized, high-resolution information on land movement across Europe with millimetric accuracy [87,88]. The dataset is derived from interferometric analysis of Sentinel-1 radar imagery acquired at six-day intervals, offering dense spatial coverage and consistent temporal resolution. Data are updated annually and generated using Persistent Scatterer Interferometry and Distributed Scatterer Interferometry techniques. EGMS is the first publicly available continental-scale land movement product and has been validated in multiple studies across Europe, confirming its reliability [89–92].

EGMS data were downloaded in CSV format and included time series of land movement, mean velocity, standard deviation, mean quadratic error, and spatial coordinates of observation points. The analysis used the Ortho product, which provides vertical and horizontal land movement components referenced to the European Terrestrial Reference Frame and resampled to a 100 m grid [87,88]. Given the low magnitude of horizontal land movement in the AOI, the analysis focused on the vertical component, which typically exhibits greater amplitude and clearer temporal patterns [93]. Spatial and temporal patterns of land movement were analyzed using Geographic Information System (GIS) environment.

To complement this analysis, groundwater level data were obtained from the piezometric monitoring network established in the AOI (Figure 1B) [52]. This network consists of observation wells installed in former shafts and technological boreholes, designed based on long-term hydrogeological monitoring experience. A total of 36 piezometers were monitored, 26 located within the range of the former depression cone and 10 outside it. Most of them are screened in the Triassic aquifer, while four are screened in Upper Jurassic limestones. Groundwater head measurements were performed quarterly by accredited personnel from the "Olkusz-Pomorzany" mine using an electronic well whistle. Measurement errors were minimized through repeated readings [52]. Groundwater level measurements were carried out according to certified industrial procedures commonly applied in the Polish mining sector. Although the exact instrumental accuracy is not publicly documented, the magnitude of the observed groundwater head variations substantially exceeds potential measurement uncertainty, confirming the reliability of the recorded data.

For this study, groundwater head analysis included 15 piezometers for which verified and temporally consistent records were available [52]. The selected wells are distributed across the central part of the former "Olkusz–Pomorzany" mine, where the most intensive dewatering occurred, as well as in peripheral parts of the former depression cone and outside its range. This spatial configuration enables the assessment of both mine-related groundwater rebound and climate-driven groundwater fluctuations in zones weakly or not affected by mining drainage.

To investigate interrelations between groundwater head and land movement, each of the 15 piezometers was paired with the nearest EGMS observation point. Linear trends in vertical land movement were then estimated separately for the pre-closure (March 2019–December 2021) and post-closure (January 2022–December 2023) periods. Subsequently, average groundwater level dynamics were calculated for the same post-closure period to facilitate joint interpretation. This allowed for assessing the effects of groundwater rebound on vertical land movement, including potential time lags depending on the distance from the drainage center [41]. Moreover, analyzing land movement patterns before mine flooding provided a baseline for identifying subsequent uplift related to groundwater rebound [35].

3.3. Hydrospectral Indices and Land Cover Data

Changes in near-surface water conditions in the AOI are driven by two main factors: mining-related processes, including long-term dewatering and subsequent groundwater rebound, and regional climatic variability. These processes affect surface and subsurface hydrology, resulting in shifts in soil moisture, vegetation condition, and the extent of surface water bodies. Such changes can be effectively monitored using hydrospectral indices derived from satellite remote sensing [46,94].

In this study, three hydrospectral indices were analyzed: NDVI, NDWI, and Moisture Index.

NDVI reflects the condition and density of vegetation by contrasting reflectance in the near-infrared (NIR) and red bands. Its values range from −1 to +1, with negative values corresponding to water surfaces, values near zero indicating bare ground, and high positive values indicating dense, healthy vegetation. It is widely used to assess vegetation dynamics and their relationship with soil moisture and climatic variability [95].

NDWI, using green and NIR bands, enhances the spectral response of open water bodies due to their strong absorption of NIR radiation. It is particularly useful for tracking changes in the extent of lakes, ponds, wetlands, and inundation zones, especially in areas affected by hydrological shifts such as groundwater rebound [96].

Moisture Index, based on NIR and shortwave infrared (SWIR) bands, is sensitive to vegetation water content and canopy moisture. Because SWIR reflectance decreases with increasing water content in leaves, this index is commonly used to detect vegetation drought stress and surface moisture variations [97].

Together, these indices provide complementary information on surface water distribution, soil moisture, and vegetation condition, supporting the interpretation of hydrological drought and groundwater rebound effects in post-mining environments [43,98].

Time series of NDVI, NDWI, and Moisture Index were derived from Copernicus Browser using Sentinel-2 Level 1C imagery [99]. The Sentinel-2 mission provides global optical data at 10–20 m spatial resolution and 5–10 day temporal resolution, acquired by the Multi-Spectral Instrument with 13 spectral bands [100].

To minimize atmospheric and radiometric noise, only cloud-free scenes acquired over the AOI were used. This approach is commonly applied in long-term vegetation and moisture trend studies to ensure stable and comparable reflectance values over time [101,102]. After filtering, 115 scenes were used for the "Olkusz–Pomorzany" mine area (Table 1) and 65 for the control area outside the mine (Table 2).

The analysis covered the time intervals defined in Section 3, with emphasis on both long-term environmental variability and the period surrounding mine closure. Linear trends were fitted to the time series of NDVI, NDWI, and Moisture Index to identify gradual changes over multi-year periods and shorter intervals. All periods were defined from January to January to minimize the influence of seasonal variability on trend estimation.

To better understand how these hydrological changes are expressed at the landscape scale, annual land cover data were incorporated into the analysis. Land cover maps for 2019–2024 were obtained from Polish Space Agency [103]. These maps are produced using the Sentinel-2 Global Land Cover methodology developed by Space Research Centre of the Polish Academy of Sciences. This approach applies automated classification of multi-temporal Sentinel-2 imagery to generate high-resolution (10 m) maps with ten land cover classes, including built-up areas, agricultural land, deciduous and coniferous forests, grasslands, wetlands, peatlands, bare ground, and surface water [104].

Raster datasets were processed in GIS to calculate the surface area of each land cover class for each year. This allowed for the quantification of wetland expansion, formation of new water bodies, and other land cover transformations associated with hydrological changes induced by groundwater rebound in the AOI.

Table 1. Cloud cover statistics for Sentinel-2 imagery acquired over the "Olkusz–Pomorzany" mine area (2015–2025). A filtered subset was used in the NDVI, NDWI, and Moisture Index analyses.

Cloud Coverage [%]	2015	2016	2017	2018	2019	2020	2021	2022	2023	2024	2025	Total
0	6	5	13	15	15	16	10	14	11	16	14	135
(0; 10>	5	15	19	23	31	34	23	28	27	32	38	275
(10; 20>	0	4	9	14	10	7	9	11	12	8	9	93
(20; 30>	1	4	6	7	10	11	11	9	7	10	10	86
(30; 40>	2	5	7	10	6	8	18	14	9	7	6	92
(40; 50>	1	1	7	13	8	8	4	5	3	8	5	63
(50; 60>	1	3	2	6	10	4	4	8	8	11	6	63
(60; 70>	1	2	8	9	6	6	7	9	4	2	3	57
(70; 80>	2	5	5	3	3	10	16	8	9	5	8	74
(80; 90>	1	2	7	9	8	8	9	10	5	4	5	68
(90; 100>	3	22	46	35	38	32	35	30	50	41	33	365
Total	23	68	129	144	145	144	146	146	145	144	137	1371

Table 2. Cloud cover statistics for Sentinel-2 imagery acquired over the control area outside the "Olkusz–Pomorzany" mine area (2015–2025). A filtered subset was used in the NDVI, NDWI, and Moisture Index analyses.

Cloud Coverage [%]	2015	2016	2017	2018	2019	2020	2021	2022	2023	2024	2025	Total
0	1	2	4	6	13	15	6	5	9	5	5	71
(0; 10>	10	15	28	31	31	28	26	34	29	41	43	316
(10; 20>	0	7	6	11	11	13	4	11	4	9	11	87
(20; 30>	2	3	8	11	11	12	14	13	11	9	7	101
(30; 40>	2	6	7	8	6	6	15	5	10	10	11	86
(40; 50>	1	2	6	13	6	11	10	11	6	7	4	77
(50; 60>	0	3	7	9	10	9	5	9	7	6	8	73
(60; 70>	2	3	4	8	8	4	16	10	6	6	6	73
(70; 80>	2	2	10	6	9	10	9	12	12	8	1	81
(80; 90>	0	4	5	8	5	8	11	8	9	6	10	74
(90; 100>	3	21	44	33	35	28	30	28	42	37	31	332
Total	23	68	129	144	145	144	146	146	145	144	137	1371

Since surface water changes were most significant in the central part of the former "Olkusz–Pomorzany" mine area, this zone was selected for additional visual inspection using Sentinel-2 true-color imagery. These images were obtained from Copernicus Browser and originate from the same Sentinel-2 Level 1C dataset used to derive NDVI, NDWI, and Moisture Index time series [99]. One cloud-free image was selected for each year in mid-June to provide a consistent reference point within the growing season and to minimize seasonal variability in visual interpretation. The following scenes were analyzed: 12 June 2019, 13 June 2020, 21 June 2021, 26 June 2022, 1 June 2023, and 15 June 2024. This step enabled detailed mapping of new inundation zones, wetland expansion, and other surface changes that may not be fully captured through automated classification alone.

4. Results and Discussion

4.1. Observed Climatic Trends and Variability

The AOI's climate reflects typical Central European conditions, with distinct seasonality, cold winters, warm summers, and most precipitation occurring during the warmer part of the year. Snow cover usually persists for several months, with most snowfall between December and March, contributing to spring groundwater recharge.

The AOI is situated in a moderately cool temperate climate zone on the Silesian–Kraków Upland. Long-term meteorological records from the synoptic station in Kraków–Balice (1981–2024) indicate that the mean annual air temperature is 8.83 °C, with July being the warmest month (19.05 °C) and January the coldest (−1.67 °C). Mean monthly temperatures typically range from about −2 °C in winter to +19 °C in summer (Figure 3). The mean annual precipitation during this period is 656.6 mm, with the highest monthly values recorded in July (90.6 mm) and June (81.1 mm), and the lowest in February (31.9 mm) and January (37.7 mm). Precipitation is strongly seasonal, with the wettest months typically occurring between May and August (Figure 4).

Over the full 1981–2024 period, air temperature shows a clear seasonal cycle with a gradual warming trend of 0.0168 °C per year. Although the increase may appear small, it represents a steady rise in temperature over several decades. The mean monthly temperature for this period is 8.83 °C, with mean minimum and maximum values of 4.42 °C and 13.73 °C, respectively. In the short-term intervals corresponding to the final years of mine operation (2016–2024) and the post-closure rebound phase (2022–2024), mean temperatures are higher. For 2022–2024, the mean monthly temperature reached 10.13 °C, with minimum and maximum values of 5.77 °C and 14.95 °C, exceeding the long-term averages by approximately 1.3 °C for the mean and showing similar increases for the extremes (Figure 3).

These warm anomalies are consistent with broader climatic trends across Central Europe. The rise in minimum temperatures indicates milder winters and fewer cold extremes, while higher maximum temperatures increase the frequency of warm days in spring and summer. Together, these shifts can shorten the snow season, accelerate spring melt, and intensify summer evapotranspiration, all of which affect the local water balance [38,105].

Precipitation has remained relatively stable over the 1981–2024 period, with no significant long-term trend in total amounts. In the shorter intervals, mean annual precipitation was approximately 682 mm for 2016–2024 and 678 mm for 2022–2024, slightly above the long-term average (Figure 4).

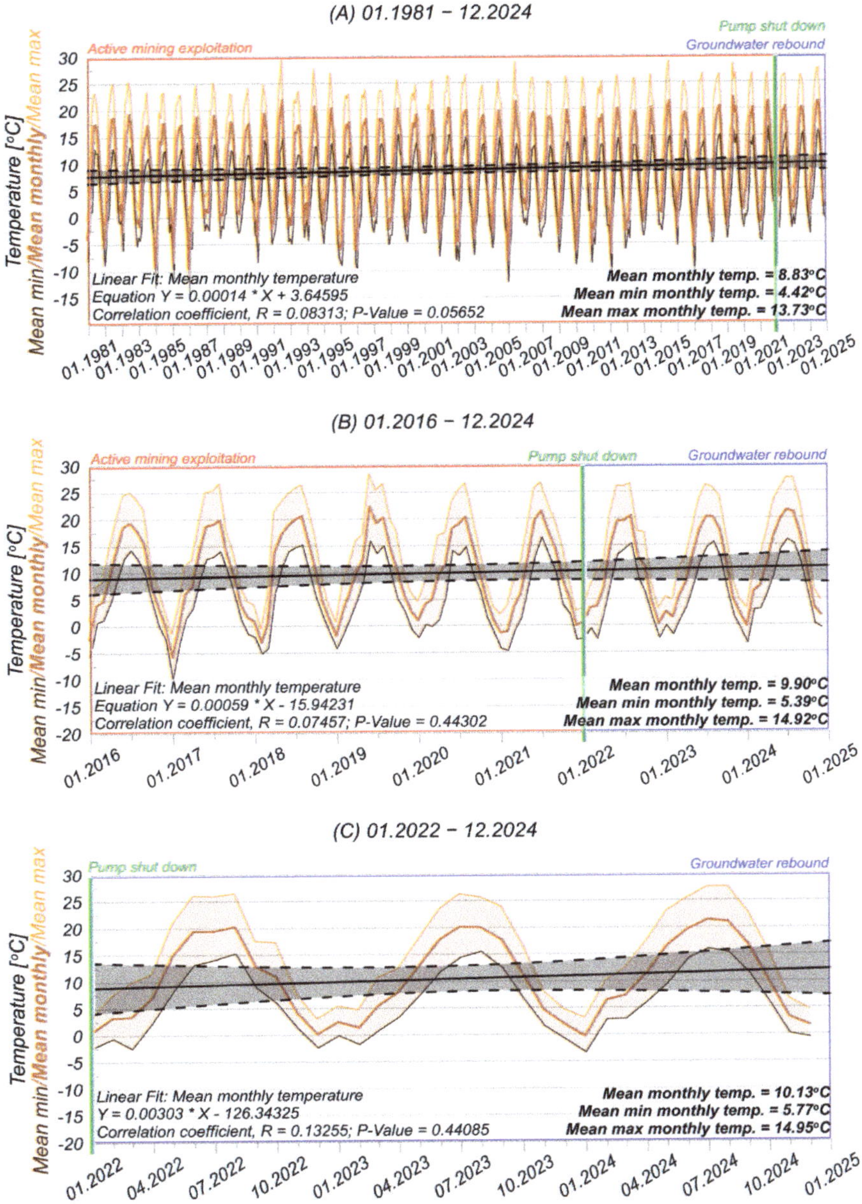

Figure 3. Time series of mean monthly, minimum monthly, and maximum monthly air temperature for (**A**) January 1981–December 2024, representing the long-term climatic context encompassing both active mining and the onset of groundwater rebound; (**B**) January 2016–December 2024, covering the late mining and closure phases; and (**C**) January 2022–December 2024, corresponding to the post-closure flooding and groundwater rebound period. Linear trends (black solid lines) with 95% confidence intervals (black dashed lines; grey areas) are shown for the mean monthly temperature.

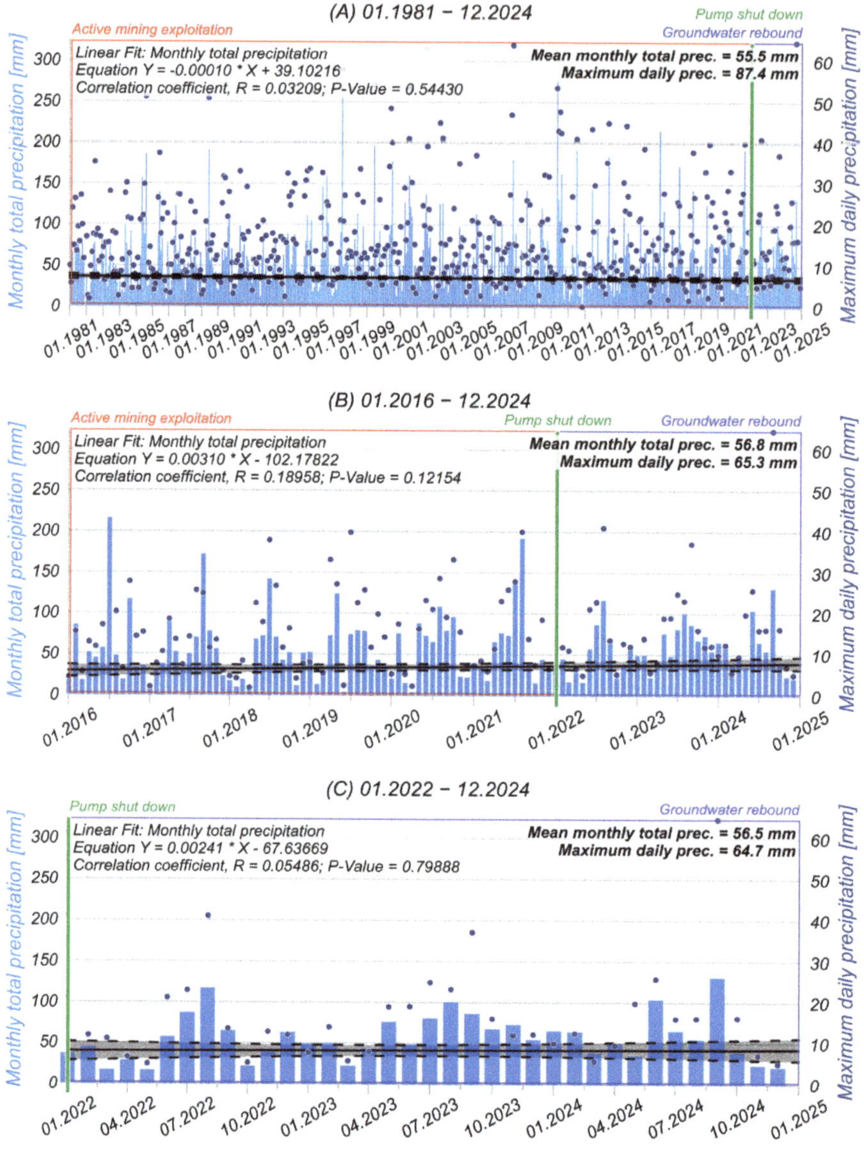

Figure 4. Time series of mean monthly total precipitation and maximum daily precipitation for (**A**) January 1981–December 2024, representing the long-term climatic context encompassing both active mining and the onset of groundwater rebound; (**B**) January 2016–December 2024, covering the late mining and closure phases; and (**C**) January 2022–December 2024, corresponding to the post-closure flooding and groundwater rebound period. Linear trends (black solid lines) with 95% confidence intervals (black dashed lines; grey areas) are shown for the monthly total precipitation.

This indicates that, despite year-to-year variability, the overall precipitation regime has remained largely unchanged. Rainfall remains strongly seasonal, with most precipitation between May and August. Short but intense rainfall events are also hydrologically important. Maximum daily precipitation reached 87.4 mm during the full record, and 65.3 mm and 64.7 mm in 2016–2024 and 2022–2024, respectively. Although the frequency of such events shows no clear trend, their hydrological impact is considerable. Notably, extreme rainfall episodes in 1997 and 2010 coincided with major floods across Central Europe, including Poland, and resulted in unusually high water inflows to the "Olkusz–Pomorzany" mine [52]. Even if infrequent, such events can temporarily dominate groundwater recharge and surface water dynamics.

Supplementary meteorological indicators provide further context (Figure 5). Average cloud cover remained stable at around 5 oktas throughout the 1981–2024 period. The average number of rainy days per month increased slightly, from 10.3 in the long-term record to 11.0 in 2022–2024, suggesting slightly wetter conditions in recent years despite stable total precipitation. In contrast, the number of days with snowfall and snow cover has gradually declined. On average, there were 3.9 days with snowfall and 4.5 days with snow cover per month in the long-term period, compared to 3.1 and 2.8 days, respectively, in the post-closure phase. These changes indicate milder winters and a shorter snow season, consistent with broader regional climatic shifts in Central Europe [106,107].

Additional parameters describing atmospheric moisture and air circulation further illustrate temporal variability relevant to hydrological processes (Figure 6). Mean monthly moisture in the AOI shows strong annual seasonality, with the highest values observed in summer months and the lowest in winter. Overall, mean monthly moisture decreased slightly across all analyzed periods, averaging 78.8% for 1981–2024, 77.7% for 2016–2024, and about 77.1% for 2022–2024. In the two earlier intervals, a weak downward trend was observed, whereas in the most recent period (2022–2024) a slight upward tendency is evident. However, this increase is minor and observed over a short time interval, which limits the possibility of drawing robust conclusions about long-term tendencies.

Wind speed also exhibits clear seasonality, with the highest values in winter and the lowest in late summer (Figure 6). Despite this cyclic pattern, a slight but statistically significant upward trend was identified, with mean values increasing from 2.8 m/s in the long-term record (1981–2024) to 3.1 m/s in both 2016–2024 and 2022–2024. These changes affect evaporation and surface–atmosphere water exchange, with stronger winds in recent years enhancing potential evapotranspiration, particularly during warmer months [108]. Due to ongoing climate change, extreme wind events are becoming more frequent, and higher wind speeds tend to increase evaporation rates by transporting water vapor away from the surface, lowering local humidity and facilitating further vapor loss [108,109]. However, no consistent long-term trend in wind speed is evident across Europe, as regional observations indicate both increasing extreme events and a general decrease in mean wind speeds in parts of Central Europe, suggesting a complex and spatially variable pattern [110].

To evaluate the climatic water balance, ET_0 and VWE were calculated (Figure 7). The long-term mean ET_0 shows strong annual seasonality, with the highest values observed in summer and the lowest in winter, following a temporal pattern similar to precipitation and other analyzed meteorological parameters (Figures 3–6). The calculated mean monthly VWE, defined as the difference between precipitation and ET_0, averaged -5.7 mm/month for 1981–2024, indicating a persistent regional water deficit. In 2016–2022, this deficit deepened to -10.2 mm/month, and in 2022–2024 it reached -11.6 mm/month, suggesting that despite comparable precipitation totals, enhanced evapotranspiration under changing climatic conditions reduced the potential for effective groundwater recharge within the AOI.

These climatic shifts have direct implications for the local water balance. Shorter snow seasons and warmer winters limit spring recharge of aquifers, which historically depended substantially on snowmelt infiltration. Higher temperatures increase evapotranspiration, leading to soil moisture deficits and more intense summer droughts. Extreme rainfall events, although sometimes abundant in total volume, often produce rapid surface runoff rather than infiltration, further constraining groundwater replenishment. Similar patterns have been documented elsewhere in Poland, where rising temperatures and altered precipitation regimes have caused shallow aquifer depletion and measurable land subsidence [35]. Long-term, climate-driven groundwater decline, as observed in the AOI, can lead to aquifer compaction rates of several millimeters per year even in the absence of active dewatering [38,54,111].

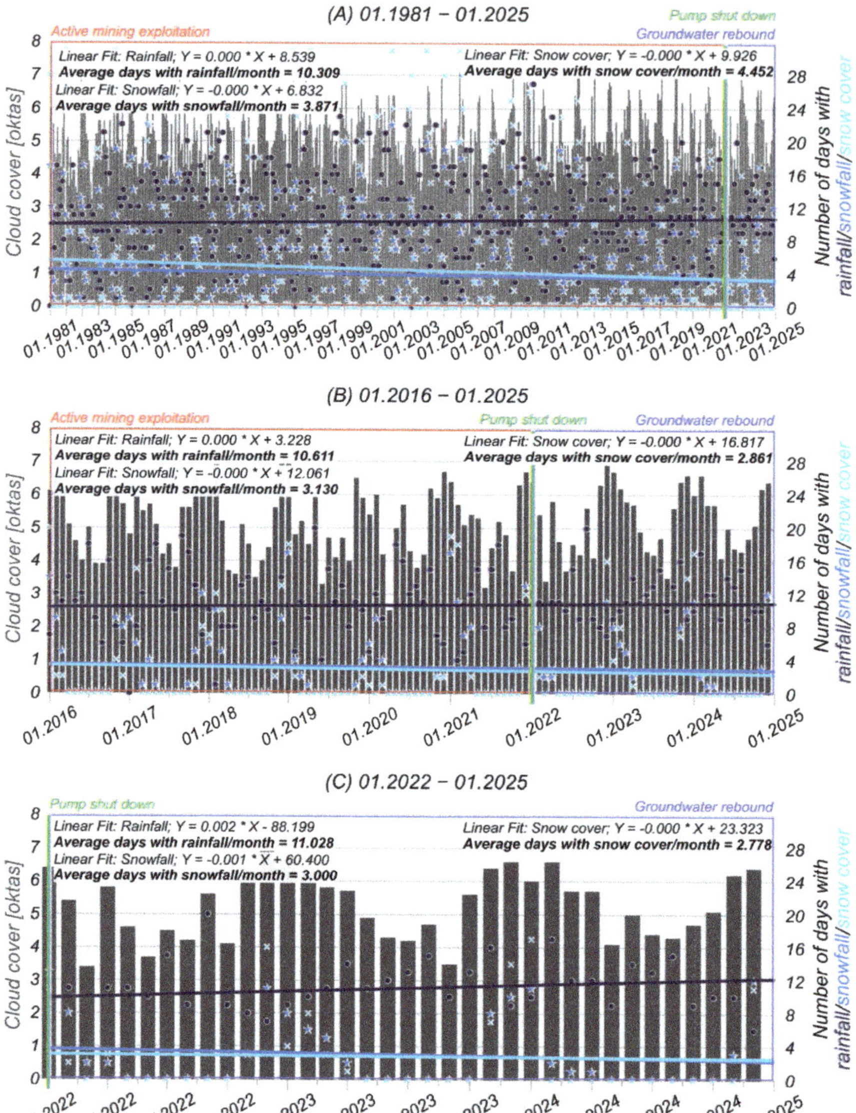

Figure 5. Time series of cloud cover and number of days with rainfall, snowfall, and snow cover for (**A**) January 1981–December 2024, representing the long-term climatic context encompassing both active mining and the onset of groundwater rebound; (**B**) January 2016–December 2024, covering the late mining and closure phases; and (**C**) January 2022–December 2024, corresponding to the post-closure flooding and groundwater rebound period. Linear trends (blue solid lines) are shown for the number of days with rainfall, snowfall, and snow cover.

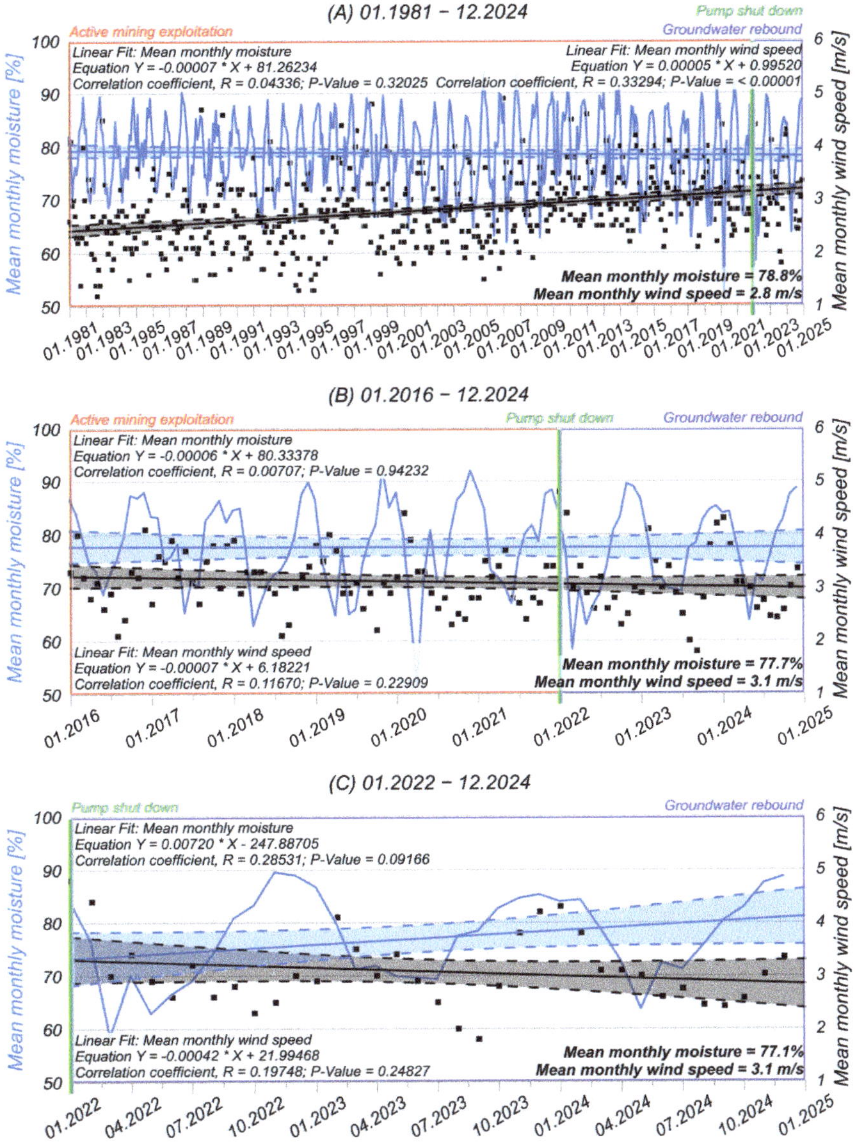

Figure 6. Time series of mean monthly moisture and mean monthly wind speed for (**A**) January 1981–December 2024, representing the long-term climatic context encompassing both active mining and the onset of groundwater rebound; (**B**) January 2016–December 2024, covering the late mining and closure phases; and (**C**) January 2022–December 2024, corresponding to the post-closure flooding and groundwater rebound period. Linear trends (blue and black solid lines) with 95% confidence intervals (blue and black dashed lines; blue and grey areas) are shown for both parameters.

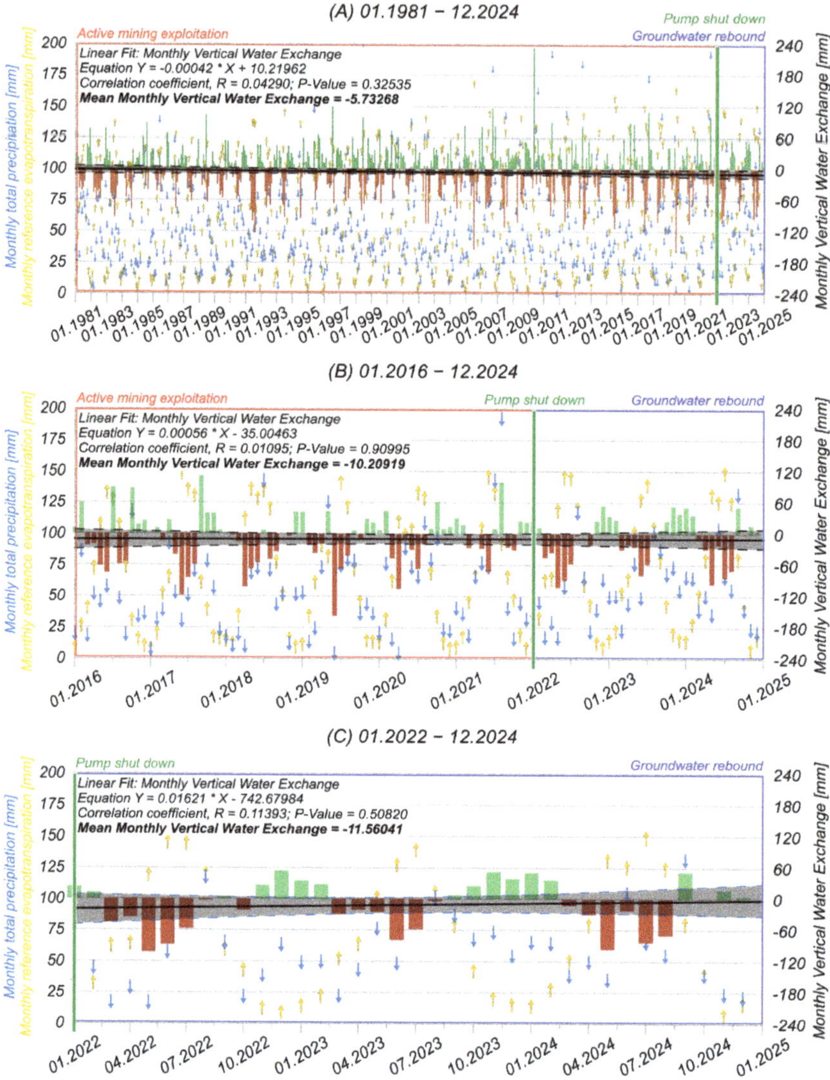

Figure 7. Time series of monthly total precipitation (blue downwards markers), monthly reference evapotranspiration (ET_0) (yellow upward markers) and Vertical Water Exchange (VWE) for (**A**) January 1981–December 2024, representing the long-term climatic context encompassing both active mining and the onset of groundwater rebound; (**B**) January 2016–December 2024, covering the late mining and closure phases; and (**C**) January 2022–December 2024, corresponding to the post-closure flooding and groundwater rebound period. Positive VWE values indicate potential groundwater recharge, while negative values denote climatic water deficit. Linear trends (black solid lines) with 95% confidence intervals (black dashed lines; grey areas) are shown for the VWE.

In the AOI, these climatic influences interact with post-mining processes. The warming trend, changing precipitation regime, and reduced snow cover act together with groundwater rebound, shaping both hydrogeological conditions and surface deformation. This creates a complex environmental response that requires distinguishing regional climatic effects from mining-related impacts. These interactions are examined in the following sections through the analysis of vertical land movement and hydrospectral indices derived from satellite data for both the mining-affected zone and the external control area.

4.2. Vertical Land Movement Patterns and Groundwater Rebound Dynamics

4.2.1. General Characteristics of Vertical Land Movement

Vertical land movement in the AOI is generally low, ranging from −17.8 to +17.5 mm/year during 2018–2022 and from −18.5 to +13.5 mm/year during 2019–2023. The mean rate of vertical movement for the entire AOI was −0.8 mm/year in 2018–2022 and −0.6 mm/year in 2019–2023, indicating minor but continuous subsidence. Similar values were recorded in the control area (−0.8 mm/year and −0.7 mm/year, respectively) (Tables 3 and 4, Figure 8). This suggests that land surface beyond the mining zone experienced slow and ongoing subsidence, likely related to regional climatic effects described in Section 4.1.

Table 3. Summary statistics of vertical land movement derived from EGMS data for the AOI, the "Olkusz–Pomorzany" mine area, and the control area outside the mine for the period March 2018–December 2022.

Research Area	Area	EGMS-Based Points		Vertical Land Movement [mm/year]				
		Number	Density [1/km^2]	Minimum	Maximum	Median	Mean	Standard Deviation
AOI	440.12	9158	20.8	−17.8	17.5	−0.7	−0.8	1.03
"Olkusz-Pomorzany" mine	85.98	2538	29.5	−17.8	14.0	−0.8	−0.9	1.22
Control area outside the "Olkusz-Pomorzany" mine	354.14	6620	18.7	−9.7	17.5	−0.7	−0.8	0.95

Table 4. Summary statistics of vertical land movement derived from EGMS data for the AOI, the "Olkusz–Pomorzany" mine area, and the control area outside the mine for the period March 2019–December 2023.

Research Area	Area	EGMS-Based Points		Vertical Land Movement [mm/year]				
		Number	Density [1/km^2]	Minimum	Maximum	Median	Mean	Standard Deviation
AOI	440.12	9231	21.0	−18.5	13.5	−0.6	−0.6	1.26
"Olkusz-Pomorzany" mine	85.98	2582	30.0	−18.5	11.2	−0.5	−0.4	1.65
Control area outside the "Olkusz-Pomorzany" mine	354.14	6649	18.8	−15.2	13.5	−0.6	−0.7	1.05

In contrast, land deformation patterns within the "Olkusz–Pomorzany" mine area show clear spatial variability between the two analyzed periods (Tables 3 and 4). During 2018–2022, vertical movement rates in the mine area were comparable to those outside the mine, showing a generally uniform and slow subsidence (mean −0.9 mm/year). In the later period (2019–2023), the spatial pattern changed significantly. The mean rate of vertical movement in the mine area was −0.4 mm/year, lower than the mean for the AOI (−0.6 mm/year) and the control area (−0.7 mm/year). A broad uplift zone developed over the central part of the "Olkusz–Pomorzany" mine, extending northward toward the Błędów Desert. The maximum uplift reached almost +2 mm/year in the former drainage center, where the largest historical groundwater drawdown had occurred (Tables 3 and 4, Figure 8).

The geometry of this uplift zone closely corresponds to the extent of the former depression cone within the Triassic aquifer, elongated north–north-east in the direction of regional groundwater flow following mine flooding in late 2021–early 2022 (Figures 1B and 6). Outside this zone, vertical movement rates remained similar to those from the earlier observation period, indicating stable but slight ongoing subsidence across areas unaffected by mining.

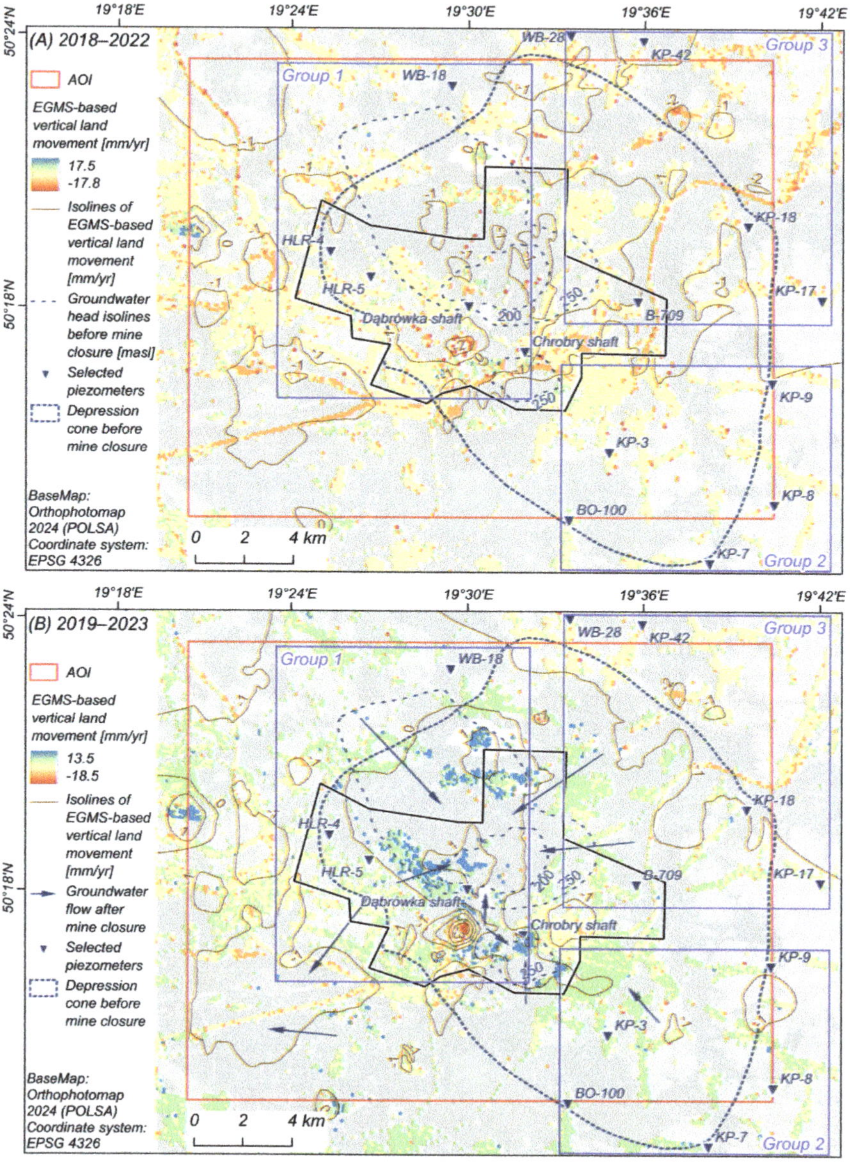

Figure 8. EGMS-based vertical land movement in the AOI for the periods (**A**) March 2018–December 2021 and (**B**) March 2019–December 2023. (**A**) Areas of the greatest groundwater head drawdown before the flooding of the "Olkusz–Pomorzany" mine at the turn of 2021 and 2022 (groundwater head below 250 m a.s.l.); (**B**) direction and intensity of groundwater inflow toward the former drainage center after the onset of groundwater rebound. Basemap: Polish Space Agency [103].

Apart from these regional subsidence and uplift trends, in both periods, a small zone of localized subsidence (around −2 to −3 mm/year) was observed south of the Dąbrówka shaft. This area corresponds to former tailings ponds, where loose post-processing materials continue to compact. A small uplift zone (up to +3 mm/year) was also recorded at the western edge of the AOI, near a wastewater treatment facility, likely reflecting continuous material deposition over time. Their magnitudes and extents remained nearly unchanged between the two analyzed intervals (Figure 8).

The density of EGMS measurement points was comparable in both periods, averaging around 19 points/km^2 in the control area and 30 points/km^2 in the mine area (Tables 3 and 4) [103]. Spatial coverage was uneven, however, particularly in forested regions, which may obscure the full extent of observed uplifts in 2019–2023. Incomplete EGMS coverage could lead to under- or overestimation of deformation amplitudes after spatial interpolation, especially within the uplift zone (Figure 6). Although most of the observed deformation rates fall within the nominal precision range of EGMS (±1 mm/year), these values refer to the velocity precision under optimal conditions rather than absolute accuracy [92]. Nevertheless, the spatial coherence of the detected uplift pattern, its correspondence with the extent of the former depression cone, and the clear temporal shift from pre-closure subsidence to post-closure uplift strongly suggest that the observed deformation reflects a genuine physical process rather than random measurement noise. Similar results were independently validated in the nearby Trzebinia mining area (approximately 15 km south-east of Olkusz), where EGMS-derived velocities showed very good agreement with precise leveling data [11]. Therefore, the EGMS data used in this study are considered reliable within the expected uncertainty range. Despite the relatively small uplift values visible in the interpolated EGMS maps (Figure 8), localized deformation within the former mining zone is higher, as shown in the piezometer-based time series discussed in Section 4.2.2.

4.2.2. Groundwater Rebound and Related Surface Uplift

Groundwater monitoring data from fifteen piezometers show spatial differences in groundwater rebound and the link with vertical land movement after the closure and flooding of the "Olkusz–Pomorzany" mine (Figures 9–11). On the basis of their spatial distribution, three groups of piezometers were distinguished, corresponding to the main hydrogeological zones of the AOI (Figure 8) [52]. Group 1 covers the central, western, and north-western parts of the former mine, where historical groundwater drawdown was greatest and where rebound is now strongest. Group 2 includes piezometers in the south-eastern part of the AOI, in the peripheral zone of the former depression cone. Group 3 comprises sites in the eastern and north-eastern parts of the area, largely outside the cone of depression, where the impact of mining on groundwater conditions has been small.

The strongest rebound and highest uplift rates occur in the central part of the AOI, in the former shafts "Dąbrówka" and "Chrobry", where dewatering was most intensive (Figures 8A and 9A,B). Between January 2022 and December 2024, groundwater heads rose by about 96.2 m (32.1 m/year) in the Dąbrówka shaft and by about 103.7 m (34.6 m/year) in the Chrobry shaft. The groundwater rise follows an exponential recovery curve that gradually flattens towards the end of 2024. The related land uplift reaches about 1.3 mm/year at Dąbrówka and about 3.6 mm/year at Chrobry. Small uplift (0.5–1.2 mm/year) was already visible before mine flooding, most likely due to early recovery of the groundwater system after ore extraction ceased in 2020, when drainage was gradually reduced but pumping was still in operation [52].

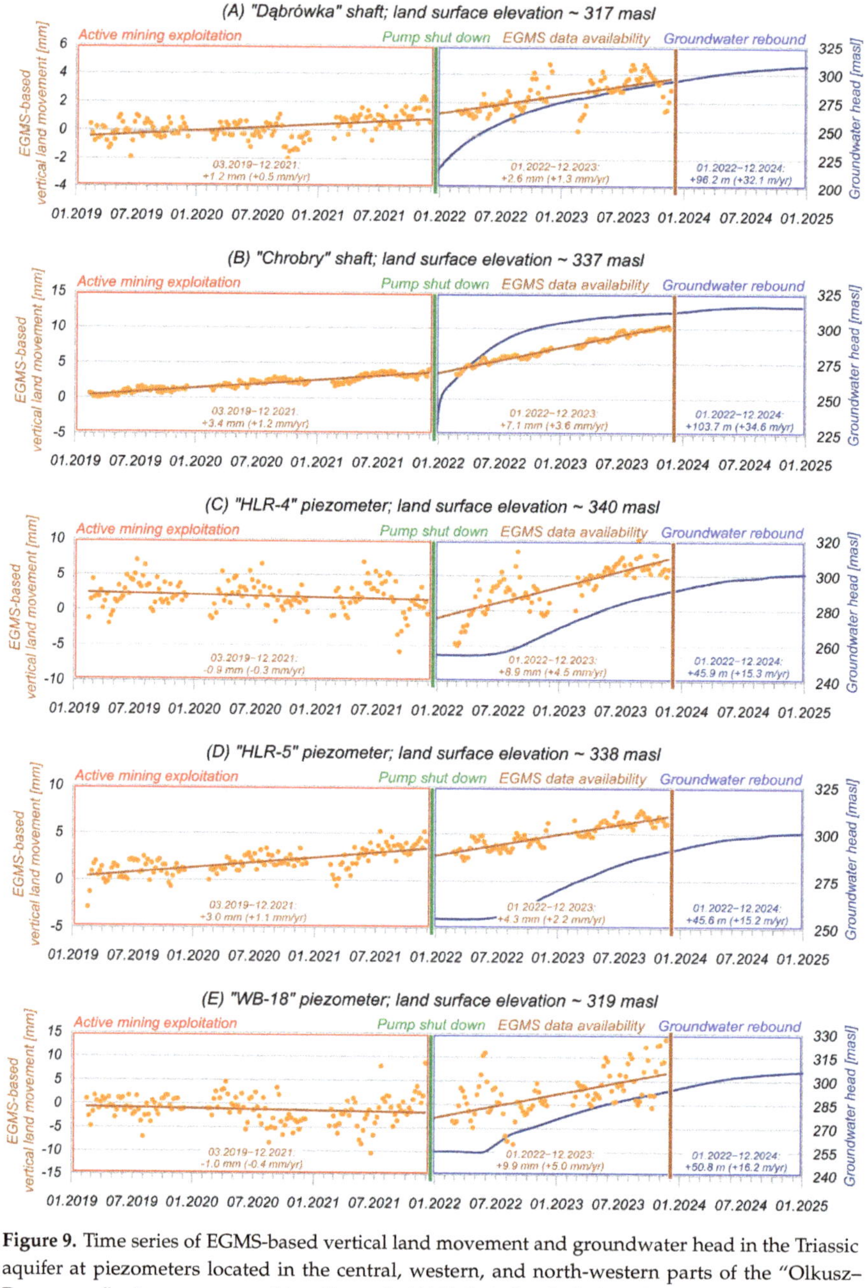

Figure 9. Time series of EGMS-based vertical land movement and groundwater head in the Triassic aquifer at piezometers located in the central, western, and north-western parts of the "Olkusz–Pomorzany" mine area, comprising Group 1: (**A**) Dąbrówka shaft, (**B**) Chrobry shaft, (**C**) HLR-4, (**D**) HLR-5, and (**E**) WB-18. Linear trends in vertical land movement are shown for the periods before mine closure (March 2019–December 2021) and after mine closure (January 2022–December 2023), with 95% confidence intervals. See Figure 8 for the location of the piezometers.

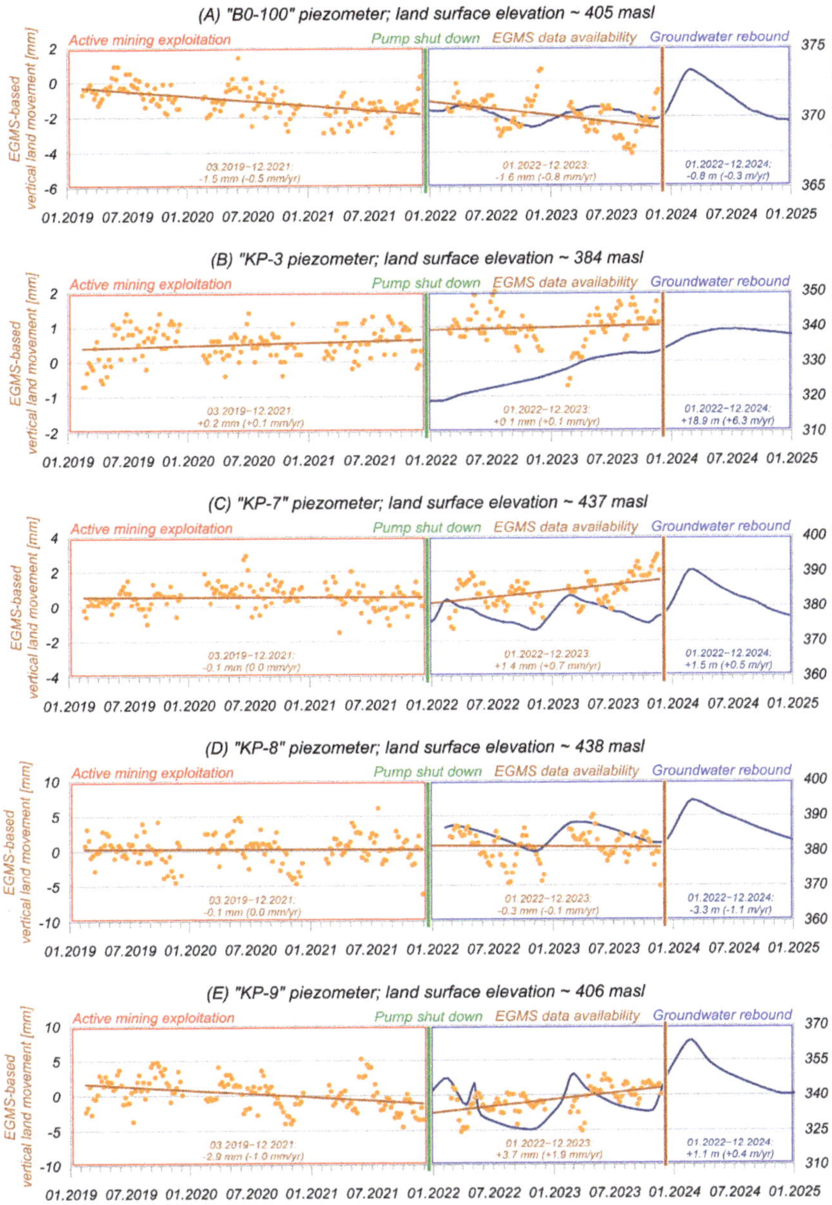

Figure 10. Time series of EGMS-based vertical land movement and groundwater head in the Triassic aquifer at piezometers located in the south-eastern part of the "Olkusz–Pomorzany" mine area, comprising Group 2: (**A**) B0-100, (**B**) KP-3, (**C**) KP-7, (**D**) KP-8, and (**E**) KP-9. Linear trends in vertical land movement are shown for the periods before mine closure (March 2019–December 2021) and after mine closure (January 2022–December 2023), with 95% confidence intervals. See Figure 8 for the location of the piezometers.

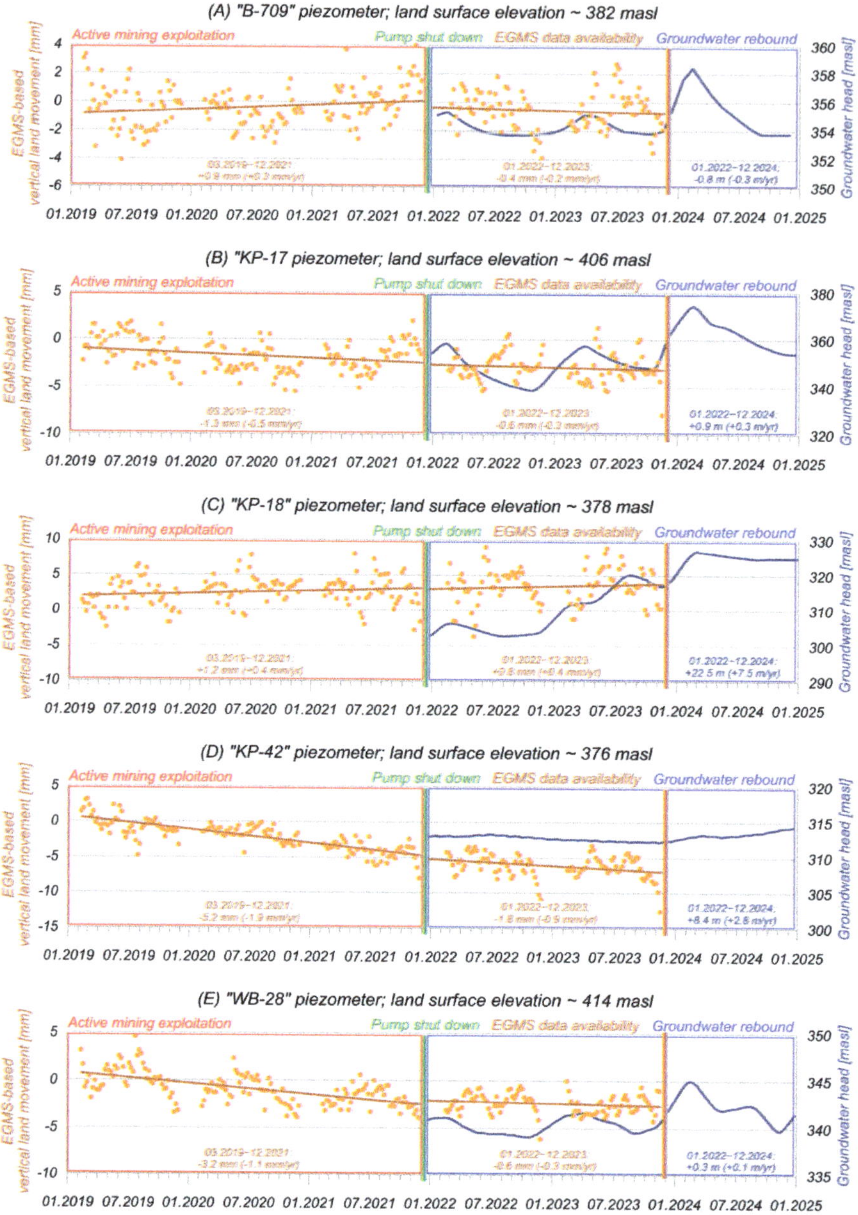

Figure 11. Time series of EGMS-based vertical land movement and groundwater head in the Triassic aquifer at piezometers located in the north-eastern part of the "Olkusz–Pomorzany" mine area, comprising Group 3: (**A**) B-709, (**B**) KP-17, (**C**) KP-18, (**D**) KP-42, and (**E**) WB-28. Linear trends in vertical land movement are shown for the periods before mine closure (March 2019–December 2021) and after mine closure (January 2022–December 2023), with 95% confidence intervals. See Figure 8 for the location of the piezometers.

In the western part of the mine area, at piezometers HLR-4 and HLR-5, groundwater rebound is around 45–46 m (roughly 15 m/year) (Figures 8 and 9C,D). At HLR-4, located farther from the former drainage center, post-closure uplift is about 4.5 mm/year, whereas at HLR-5, closer to the mine center, uplift is about 2.2 mm/year. This difference probably reflects local geological variability that controls aquifer system compressibility and vertical stress transfer. Before mine flooding, HLR-4 showed a slight subsidence trend (about −0.3 mm/year) and HLR-5 a small uplift (about +1 mm/year), which points to generally stable conditions during the final phase of drainage.

Farther north, at WB-18, located beyond the formal depression cone but within a zone of strong regional groundwater inflow, groundwater rebound reached about 50.8 m (16.2 m/year) (Figures 8 and 9E). This rise was accompanied by land uplift of about 5.0 mm/year, more than five times higher than before mine closure (−0.4 mm/year). The parallel development of groundwater rebound and uplift shows a strong link between aquifer system recovery and surface deformation in this part of the AOI. Overall, piezometers in Group 1 show the clearest and most direct response to mine flooding, with uplift magnitude closely following the local rate of groundwater rebound (Figure 9). Here, mine-induced pore pressure recovery in the Triassic aquifer, which had been most heavily drained, is the dominant control.

Group 2 includes piezometers in the south-eastern and southern parts of the AOI, where the depression cone gradually passes into unaffected conditions (Figure 8). The largest rebound in this group is observed at KP-3, the piezometer closest to the former drainage center (Figures 8 and 10B). Between 2022 and 2024, groundwater heads rose by about 19 m (6 m/year), while land uplift remained very small, about 0.1 mm/year. The time series shows a slight increase in uplift towards the end of 2023, which may reflect a delayed surface response to increasing pore pressure. Such time lags are expected and depend on local geological, hydrogeological, and geomechanical factors, including aquifer thickness and compressibility [7,8].

In the remaining Group 2 piezometers (BO-100, KP-7, KP-8, KP-9), groundwater levels show moderate but clearly seasonal changes, from about −1 m/year to +0.5 m/year (Figure 10). This pattern points to a strong hydraulic connection with the surface and a major role of precipitation and vertical recharge (see Section 4.1). Before mine closure, EGMS data indicate small subsidence of up to about −1 mm/year; after closure, the trends change slightly. KP-3, KP-7, and KP-9 now show small uplifts of up to about +1.9 mm/year, while BO-100 and KP-8 still show minor subsidence (down to about −0.8 mm/year). Deformation in this zone therefore seems to depend mainly on local hydrogeology and seasonal moisture balance, with only a weak signal from mine-related rebound (Figure 10).

Piezometers in Group 3 are located in the eastern and north-eastern parts of the AOI, where the influence of mining dewatering was limited or absent (Figure 8). The largest rebound in this group occurs at KP-18, near the edge of the former depression cone, where groundwater heads rose by about 22.5 m (around 7.5 m/year) after mine closure (Figure 11C). The associated uplift is about 0.4 mm/year, similar to the pre-closure trend, which suggests that rebound is damped here by lower local transmissivity.

At B-709, the piezometer closest to the mining-induced drainage center within Group 3, groundwater rebound is very small (about +0.9 m between 2022 and 2024, or 0.3 m/year), and the time series is dominated by seasonal changes driven by meteorological conditions (Figure 11A). Vertical land movement is also small, with rates of about +0.3 mm/year before flooding and −0.2 mm/year after, therefore this location is not affected by regional rebound.

The northernmost piezometers, KP-42 and WB-28, located beyond the depression cone, show substantial seasonal variability in groundwater head together with a consistent downward trend of the land surface (Figures 8 and 11D,E). Mean vertical movement here is between −1 and −2 mm/year both before and after mine flooding, which points to slow consolidation of near-surface deposits under climatic forcing. In KP-42, the deeper groundwater level (around 310–315 m a.s.l.) and higher surface elevation (about 376 m a.s.l.) likely weaken short-term coupling between head changes and surface motion, leading to gradual but steady subsidence. The smaller uplift values compared to other mines also reflect the geological conditions of the "Olkusz–Pomorzany" mine. The rebound occurs primarily within shallow Triassic carbonate formations of low compressibility, which limits vertical deformation even under substantial pore-pressure recovery [30,35,112–114]. In addition, EGMS data cover the period only until December 2023, while groundwater rebound began in early 2022; thus the deformation signal likely represents an early stage of the post-closure response and may increase as rebound progresses.

Nevertheless, taken together, the three groups show that the strength of land-surface uplift decreases with distance from the former drainage center and with diminishing groundwater rebound. In the central and north-western parts of the AOI, uplift is controlled mainly by mine flooding and pore pressure recovery, whereas towards the margins and beyond the former depression cone, deformation is increasingly driven by climatic and lithological factors only [52].

4.3. Hydrospectral Indices and Land Cover Change

4.3.1. Surface Moisture and Vegetation Response Derived from Hydrospectral Indices Before and After the Mine Closure

The temporal variability of NDVI, NDWI, and Moisture Index reflects seasonal vegetation dynamics and changes in surface moisture conditions in both the Olkusz–Pomorzany mine area and the control area outside the mine (Figures 12–14). All three indicators show a seasonal cycle, with peak values during late spring and summer and minima in the winter months, which is consistent with the regional climatic regime. This seasonal pattern corresponds well with the groundwater-level fluctuations observed in the peripheral piezometers (Groups 2 and 3 in Figures 10–12), where head variations follow the same climatic cycle described in Section 4.1.

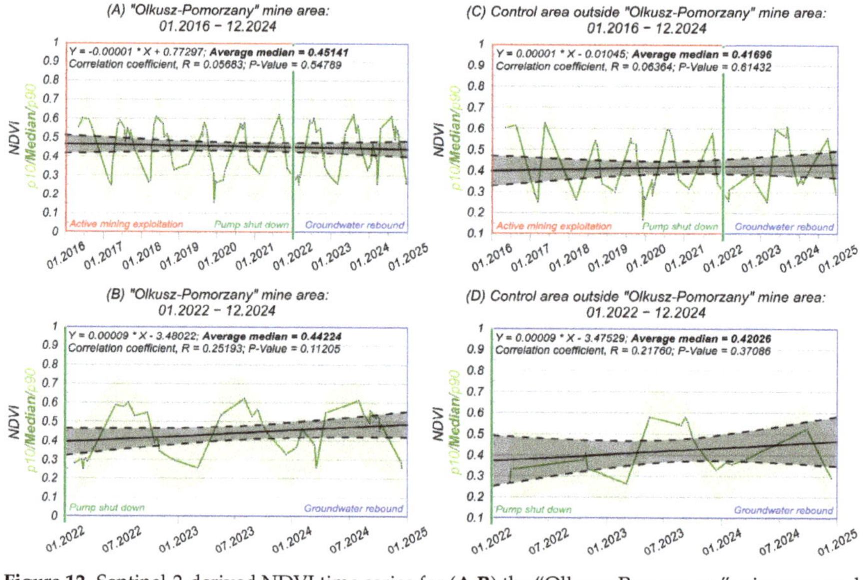

Figure 12. Sentinel-2-derived NDVI time series for (**A**,**B**) the "Olkusz–Pomorzany" mine area and (**C**,**D**) the control area outside the mine. (**A**,**C**) Entire study period (January 2016–December 2024); (**B**,**D**) period following groundwater rebound after mine closure and subsequent flooding (January 2022–December 2024). Linear trends (black solid lines) are shown with 95% confidence intervals (black dashed lines; grey areas).

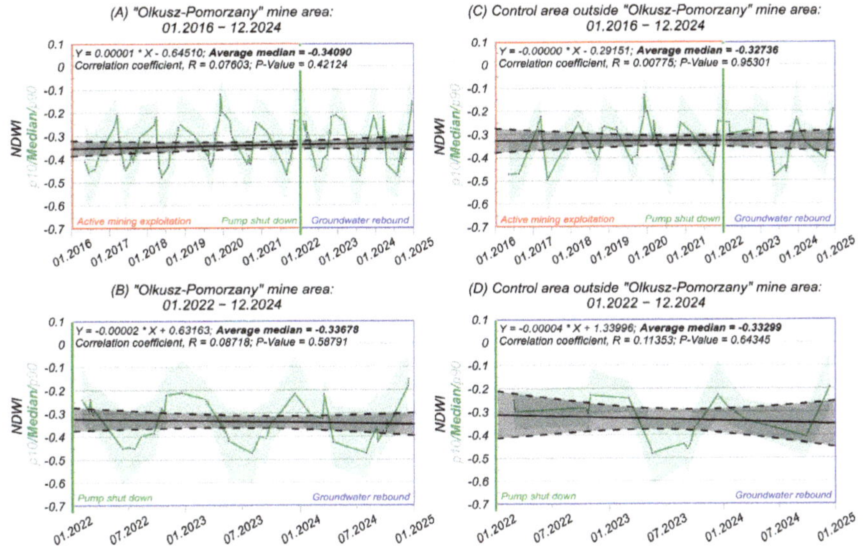

Figure 13. Sentinel-2-derived NDWI time series for (**A**,**B**) the "Olkusz–Pomorzany" mine area and (**C**,**D**) the control area outside the mine. (**A**,**C**) Entire study period (January 2016–December 2024); (**B**,**D**) period following groundwater rebound after mine closure and subsequent flooding (January 2022–December 2024). Linear trends (black solid lines) are shown with 95% confidence intervals (black dashed lines; grey areas).

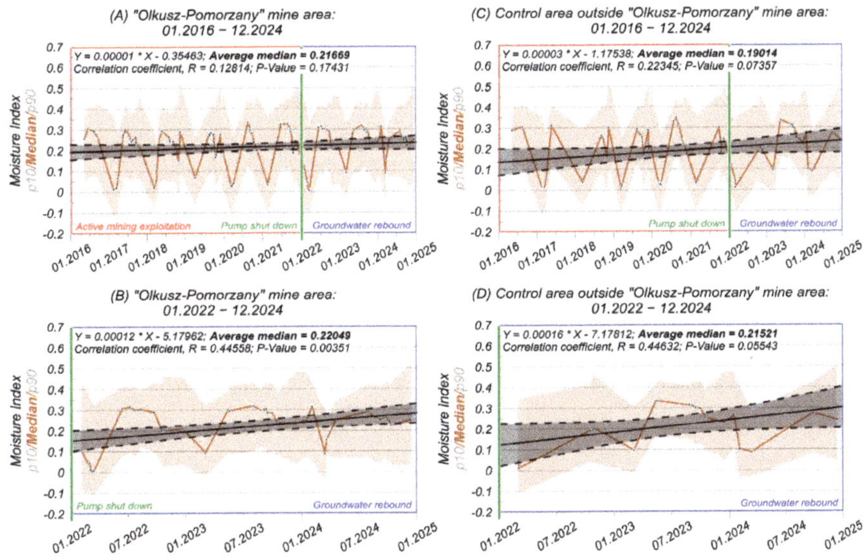

Figure 14. Sentinel-2-derived Moisture Index time series for (**A**,**B**) the "Olkusz–Pomorzany" mine area and (**C**,**D**) the control area outside the mine. (**A**,**C**) Entire study period (January 2016–December 2024); (**B**,**D**) period following groundwater rebound after mine closure and subsequent flooding (January 2022–December 2024). Linear trends (black solid lines) are shown with 95% confidence intervals (black dashed lines; grey areas).

For NDVI, the median value in the Olkusz–Pomorzany mine area during 2016–2024 was 0.45, slightly higher than in the control area (0.42). A weak decreasing trend was observed over the entire period, while in 2022–2024 the trend became slightly positive. In the control area, NDVI increased during both intervals (Figure 12). These results suggest an improvement in vegetation conditions after mine closure, most likely related to increasing soil moisture and rising groundwater levels.

NDWI values remained negative throughout the study period (around −0.34 in the mine area and −0.33 in the control area), which reflects the dominance of non-aquatic surfaces in the AOI. Linear trends were weak: slightly increasing for 2016–2024 and slightly decreasing for 2022–2024 in the mine area, and decreasing in the control area for both intervals (Figure 13). This can be explained by the fact that larger flooded areas only started to appear in the second half of 2024, and water levels continue to rise in 2025. NDWI analysis for the period ending in 2024 therefore does not fully capture these surface water changes. However, the limited increase in NDWI is consistent with the generally negative VWE indicating a climatic water deficit in the AOI (Figure 7), so that only in areas with strong groundwater rebound, as documented by Group 1 piezometers (Figure 9), surface water has started to appear.

The Moisture Index showed positive changes. Median values in the mine area increased from below 0.22 in 2016–2024 to slightly above 0.22 in 2022–2024, and from 0.19 to 0.22 in the control area. The strongest increase was observed in the mine area after flooding, which is consistent with enhanced soil and surface moisture conditions caused by groundwater rebound (Figure 14). This local increase in moisture contrasts with the overall negative VWE and regional water deficit (Figure 7), which suggests that mine flooding and groundwater rebound partly compensate for climate-driven drying in the central part of the AOI.

Based on these results, vegetation productivity and soil moisture in the AOI are strongly seasonal but also increasing in response to post-closure hydrogeological changes. The most visible changes occur in the mine area, overlapping with the zone of the strongest groundwater decline and ongoing land surface uplift. The increase in NDVI, NDWI, and Moisture Index is not very large, but it is noticeable compared to the control area. This suggests that groundwater rebound contributes to increased moisture and vegetation growth in the AOI, even though the region is generally subject to water deficits potentially linked to regional warming trends. The coincidence of higher NDVI and Moisture Index values with the zone of strongest groundwater rebound and uplift (Figures 8–11) indicates that the environmental response observed in the satellite indices is closely tied to the groundwater dynamics quantified in Sections 4.1 and 4.2.

To complement the time series analysis, spatial distributions of NDVI, NDWI, and Moisture Index were mapped for June 2016, 2021, 2022, and 2024, using cloud-free Sentinel-2 images acquired on 29 June 2016, 21 June 2021, 26 June 2022, and 15 June 2024 (Figures A1–A3 in Appendix A). These dates correspond to the beginning of data availability, the period immediately before mine flooding, the early flooding stage, and the end of the observation period. Only these four years were used, because of the large number of available images and since the full temporal evolution is already represented in the time series (Figures 12–14). The maps show that the largest changes in NDVI, NDWI, and Moisture Index occurred in the central part of the former mine, where groundwater rebound is strongest. NDWI and Moisture Index values clearly increased between 2022 and 2024, indicating rising soil moisture and the appearance of surface water (Figures A2 and A3). NDVI remained relatively stable over most of the AOI, with local decreases in areas where surface water replaced vegetation (Figure A1). Outside the mining zone, spatial patterns of all three indices remained stable, confirming that changes are concentrated in the central part of the former mine. Outside the mining zone, spatial patterns of all three indices remained stable, confirming that changes are concentrated in the central part of the former mine and spatially overlap with the uplift zone and the strongest rebound (Figure 8).

4.3.2. Land Cover Transformation Associated with Groundwater Rebound

For most of the observation period (2019–2023), the land cover structure in the AOI remained stable, with coniferous and deciduous forests, grassland vegetation, and anthropogenic areas dominating the landscape. The shift appeared only in 2024, when new surface water bodies formed in the central part of the former mining area, corresponding to the zone of the greatest historical groundwater drawdown and the present uplift of the land surface (Table 5, Figures 1B, 8 and 15).

Table 5. Land cover area for different land cover classes in the "Olkusz–Pomorzany" mine area, the control area outside the mine, and the surface water change area for the years 2019–2024.

Research Area	Land Cover Class	Land Cover Area [km^2]					
		2019	2020	2021	2022	2023	2024
AOI	Coniferous forests	174.94	176.13	180.26	182.43	183.68	183.03
	Deciduous forests	64.03	67.81	62.34	56.63	56.89	58.53
	Heathlands and shrublands	0.00	0.02	0.01	0.01	0.04	0.13
	Agricultural areas	35.45	33.21	36.60	38.35	36.87	40.45
	Grassland vegetation	136.31	134.26	130.7	129.49	129.64	128.2
	Natural areas without vegetation	8.86	8.73	8.45	12.95	12.23	8.79
	Wetlands	0.40	0.73	0.44	0.49	0.41	0.56
	Water bodies	0.31	0.31	0.28	0.46	0.59	1.06
	Anthropogenic areas	20.41	19.53	21.64	19.91	20.38	19.99
"Olkusz-Pomorzany" mine	Coniferous forests	36.05	36.13	37.33	37.38	38.63	37.54
	Deciduous forests	8.17	9.80	8.37	7.01	7.27	7.46
	Heathlands and shrublands	0.00	0.01	0.00	0.01	0.04	0.12
	Agricultural areas	6.31	5.66	6.02	6.70	5.22	6.37
	Grassland vegetation	24.89	23.65	23.10	22.80	22.95	22.98
	Natural areas without vegetation	3.29	3.27	3.31	4.64	3.92	3.54
	Wetlands	0.09	0.21	0.13	0.14	0.06	0.36
	Water bodies	0.08	0.07	0.07	0.19	0.32	0.71
	Anthropogenic areas	7.20	7.28	7.75	7.20	7.67	7.01
Control area outside the "Olkusz-Pomorzany" mine	Coniferous forests	138.89	140.00	142.93	145.05	145.05	145.49
	Deciduous forests	55.86	58.01	53.97	49.62	49.62	51.07
	Heathlands and shrublands	0.00	0.01	0.01	0.00	0.00	0.01
	Agricultural areas	29.14	27.55	30.58	31.65	31.65	34.08
	Grassland vegetation	111.42	110.61	107.60	106.69	106.69	105.22
	Natural areas without vegetation	5.57	5.46	5.14	8.31	8.31	5.25
	Wetlands	0.31	0.52	0.31	0.35	0.35	0.20
	Water bodies	0.23	0.24	0.21	0.27	0.27	0.35
	Anthropogenic areas	13.21	12.25	13.89	12.71	12.71	12.98

Table 5. *Cont.*

Research Area	Land Cover Class	Land Cover Area [km²]					
		2019	**2020**	**2021**	**2022**	**2023**	**2024**
Surface water changes	Coniferous forests	12.83	12.74	13.46	13.35	14.24	13.49
	Deciduous forests	2.02	2.79	2.20	1.64	1.58	1.62
	Heathlands and shrublands	0.00	0.01	0.00	0.01	0.04	0.11
	Agricultural areas	3.43	3.08	2.92	3.13	2.60	2.96
	Grassland vegetation	6.30	5.97	5.71	5.39	5.28	5.53
	Natural areas without vegetation	0.47	0.43	0.47	1.40	0.90	0.70
	Wetlands	0.00	0.01	0.01	0.01	0.01	0.27
	Water bodies	0.04	0.03	0.04	0.13	0.26	0.67
	Anthropogenic areas	4.41	4.44	4.70	4.44	4.60	4.14

The total area of open water bodies in the mine increased from 0.08 km^2 in 2019 to 0.71 km^2 in 2024. Most of this change occurred in 2024, in parallel with a rapid rise in groundwater levels recorded in the same period. The new water bodies developed mainly within former sand open-pit excavations and adjacent depressions, where the groundwater table has reached the land surface. Wetlands expanded from 0.09 to 0.36 km^2 in the mine area between 2019 and 2024, mainly around the edges of the newly formed water bodies. This indicates that progressively larger areas are becoming waterlogged. At the same time, grassland vegetation decreased slightly from 24.9 to 23.0 km^2, which is consistent with a change from dry to wetter surface conditions. Forest and agricultural areas showed only small year-to-year fluctuations, suggesting that the upland land cover types outside the flooding zone remained stable during the study period (Table 5, Figure 15).

In the control area, land cover remained practically unchanged, with only a small increase in surface water (from 0.23 to 0.35 km^2), most likely reflecting normal hydrometeorological variability rather than processes related to mine closure (Table 5, Figure 15). The surface water change area shows the spatial concentration of these transformations. In this area, water bodies expanded from 0.04 to 0.67 km^2. Sentinel-2 images confirm that no permanent surface water was visible in June 2023 (Figure 15Q), whereas by June 2024 water was present in several locations, including around the Dąbrówka shaft (Figure 15R). These changes closely follow the area of the former drainage center and the current uplift zone.

Overall, the observed land cover changes indicate that post-closure hydrogeological processes are driving surface water and wetland ecosystems change in the central part of the former mine. These transformations are concentrated in the zone of strongest groundwater rebound, while the surrounding landscape remains largely unaffected. The increase in soil moisture, the development of open water, and the appearance of wetland illustrate the early stages of landscape adjustment after the end of mining-induced aquifer system dewatering. These changes are still localized, but show a direction of environmental evolution as groundwater levels continue to rise and stabilize in the coming years, as already indicated by the ongoing rebound in the central part of the AOI and by the negative climatic water balance.

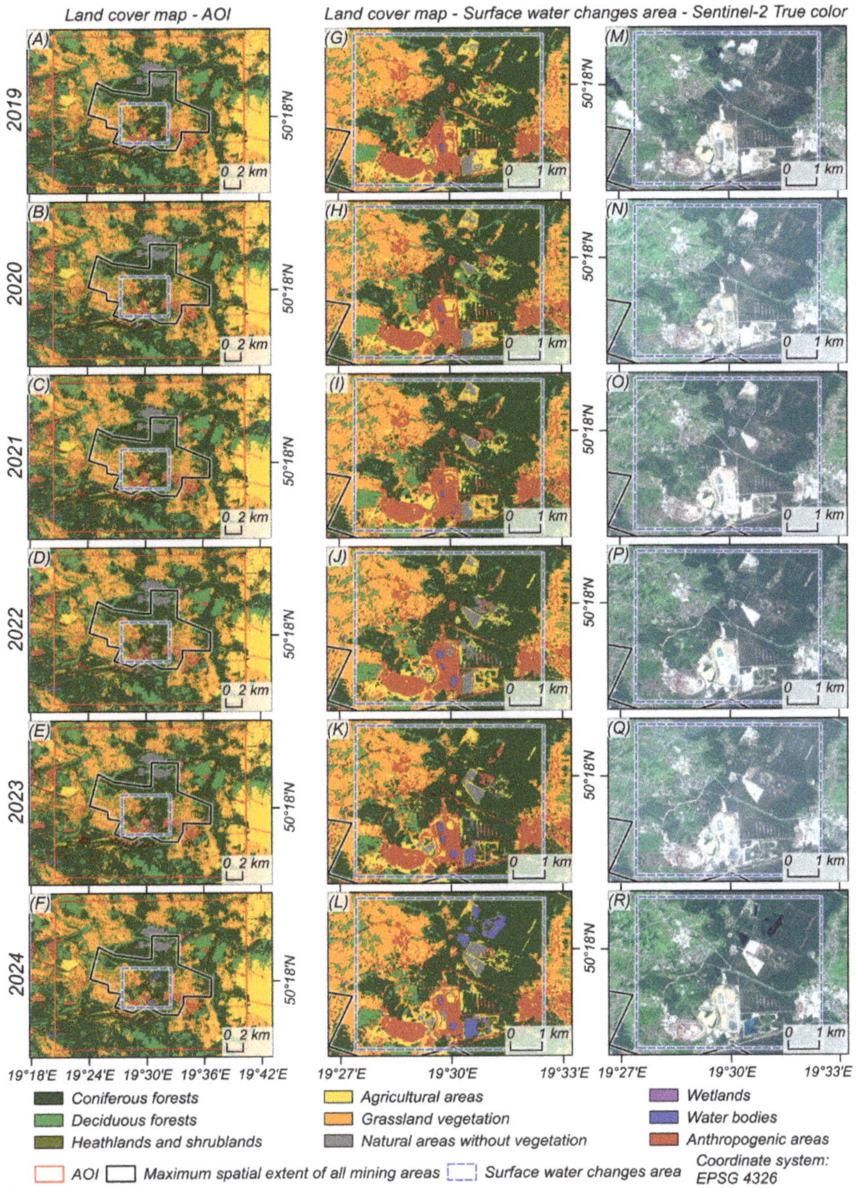

Figure 15. Land cover maps for the entire AOI (**A–F**), land cover maps for the surface water change area (**G–L**), and Sentinel-2 true color images of the surface water change area (**M–R**) for the years 2019 (**A,G,M**), 2020 (**B,H,N**), 2021 (**C,I,O**), 2022 (**D,J,P**), 2023 (**E,K,Q**), and 2024 (**F,L,R**). Data sources: Land Cover Map: Polish Space Agency [95]; Sentinel-2 true color imagery: Copernicus Browser [94].

5. Conclusions

This study integrates in situ meteorological data and groundwater head monitoring with remote sensing information, including EGMS-based vertical land motion, hydrospectral indices, and Sentinel-2-derived land cover datasets, to assess the environmental changes triggered by the flooding of the "Olkusz–Pomorzany" zinc and lead underground mine between 2021 and 2022, following several decades of mining activity. The results show a relationship between vertical land motion, groundwater rebound, and land cover transformation in a temperate climate context characterized by increasing water deficits and climatic pressure.

The analysis of vertical land deformation reveals minor but continuous subsidence, with a mean value of −0.8 mm/year across the AOI prior to mine closure and the development of a distinct uplift zone afterwards, with rates of about 1–2 mm/year. The spatial extent of this uplift corresponds to the former depression cone, especially in the area of the most intensive historical aquifer system drainage in the central part of the mine. The maximum groundwater rebound was recorded in the former shaft "Chrobry", reaching about +104 m between 2021 and 2024, accompanied by surface uplift of 3.6 mm/year. The onset of groundwater rise and related uplift was delayed by up to several months depending on the distance from the drainage center, indicating the non-linear propagation of hydraulic head recovery within the aquifer system.

Meteorological trends show gradual regional warming, particularly over the last decade, compared to the long-term 1981–2024 record. The mean monthly temperature increased from 8.83 °C (1981–2024) to 10.13 °C (2022–2024). Snowfall and snow cover duration declined, while the number of rainy days increased, despite the absence of a clear trend in total annual precipitation. These changes form a climatic background of increasing moisture deficits and persistently negative VWE, with mean values of −5.7 mm/month for 1981–2024 and −11.6 mm/month for 2022–2024. Against this climatic context, hydrospectral indices show a slight but consistent increase in vegetation activity and soil moisture in the former mine area after flooding, while conditions in the control area remained stable. This indicates that groundwater rebound can locally improve hydrological and ecological conditions, partly offsetting broader regional drying trends.

Land cover mapping between 2019 and 2024 shows that environmental transformations are primarily concentrated in the central part of the former mine. In this region, surface water area increased from 0.08 to 0.71 km^2 and wetland area increased from 0.09 to 0.36 km^2, mainly within former sand-pit excavations and adjacent depressions, where shallow groundwater levels intersect the land surface. Land cover outside the mine remained largely unchanged over the same period.

The results of this study confirm that post-closure hydrogeological processes are already reshaping the local environment, with the strongest transformations occurring in the areas most affected by past mining-induced drainage. The appearance of surface water and wetland areas marks the early stage of landscape adjustment following mine flooding (Figure 2, Supplementary Video). As groundwater levels continue to rise, further expansion of inundation zones and related ecological changes can be expected in the coming years, although the duration of these processes remains uncertain. Therefore, they should be systematically monitored and analyzed to better understand their development and to support science-based strategies for land use planning, infrastructure protection, and groundwater management in post-mining areas.

Certain limitations of the study should be noted. The analysis of vertical land movement was based on EGMS data, which approaches the detection threshold for millimeter-scale displacements. In addition, the EGMS point density is relatively low in the forested and vegetated areas that constitute a substantial part of the AOI, potentially reducing the spatial continuity of deformation results. Future research should therefore focus on a more detailed assessment of InSAR data under local AOI conditions to achieve the highest possible measurement point density, particularly in densely vegetated zones. Since increasing InSAR point density in such environments remains challenging, supplementary ground-based monitoring methods could be considered. Establishing a precise geodetic observation network, comprising leveling and GNSS measurement points, would strengthen vertical land motion monitoring and provide an independent validation dataset for InSAR observations. Furthermore, integrating InSAR observations with other remote sensing datasets (e.g., optical, thermal, and LiDAR) would improve the spatial resolution and reliability of deformation analyses, particularly in areas with limited radar coherence.

Groundwater monitoring was limited to a small number of piezometers within and outside the mine area. Expanding this network and combining the resulting observations with the development of a local-scale numerical groundwater flow model would enable more detailed and predictive assessments of groundwater rebound and its surface expression. Such modeling would also require a broader geological and hydrogeological characterization of the AOI, which remains highly complex due to the mining history and structural heterogeneity of the region, and currently is not thoroughly studied.

The analysis of hydrospectral indices was conducted up to the end of 2024, whereas major surface water changes became evident only in the second half of 2024 and continue into 2025. As more post-closure data become available, extended time series will allow for a more comprehensive analysis of the indices and a better characterization of the dynamics of surface water and vegetation changes.

Finally, integrating geodetic, InSAR, hydrogeological, and remote sensing datasets, supported by data fusion techniques and numerical modeling, would allow for a more precise quantification of the combined effects of climatic recharge and mining-related rebound on groundwater resources. Although technically demanding, such integrated and multi-source approaches represent a promising direction for future research.

Supplementary Materials: The following supporting information can be downloaded at https://zenodo.org/records/17466310 or https://doi.org/10.5281/zenodo.17466309 (accessed on 27 October 2025). Supplementary Video: Flooded Areas near Hutki Village Resulting from the Closure of the "Olkusz-Pomorzany" Mine, Poland.

Funding: This research was funded by the National Science Centre, Poland, Sonata grant titled "Novel algorithm of sinkhole precursors detection", no. 2021/43/D/ST10/02048. Additional support was provided by the Excellence Initiative-Research University program of AGH University of Krakow, Poland. It also forms part of the statutory research activities of the Land Subsidence and Hazard Mitigation Research Group, Faculty of Geo-Data Science, Geodesy and Environmental Engineering, AGH University of Krakow, Poland, grant no. 16.16.150.545.

Data Availability Statement: Meteorological data are available from the Polish Institute of Meteorology and Water Management—National Research Institute (www.imgw.pl). Vertical land surface movement data were obtained from the European Ground Motion Service (egms.land.copernicus.eu (accessed on 29 October 2025)), part of the Copernicus Programme of the European Space Agency. Groundwater head measurements are available from Motyka J. et al. [52]. Hydrospectral indices (NDVI, NDWI, Moisture Index) and Sentinel-2 true color images are accessible through the Copernicus Data Space Ecosystem Browser (dataspace.copernicus.eu/browser (accessed on 29 October 2025)). Land cover maps were obtained from the Polish Space Agency under the National Satellite Information System (nsisplatforma.polsa.gov.pl/portal (accessed on 29 October 2025)). All datasets used in this study are publicly available and were used exclusively for research purposes.

Acknowledgments: The author acknowledges the comments and suggestions provided by the anonymous reviewers, which improved the quality of this article. This work contributes to the UNESCO International Initiative on Land Subsidence (landsubsidence-unesco.org (accessed on 29 October 2025)).

Conflicts of Interest: The author declares no conflicts of interest. The funders had no role in the design of the study; in the collection, analysis, or interpretation of data; in the writing of the manuscript; or in the decision to publish the results.

Appendix A. Spatial Variability of Hydrospectral Indices Before and After the Mine Closure and Flooding

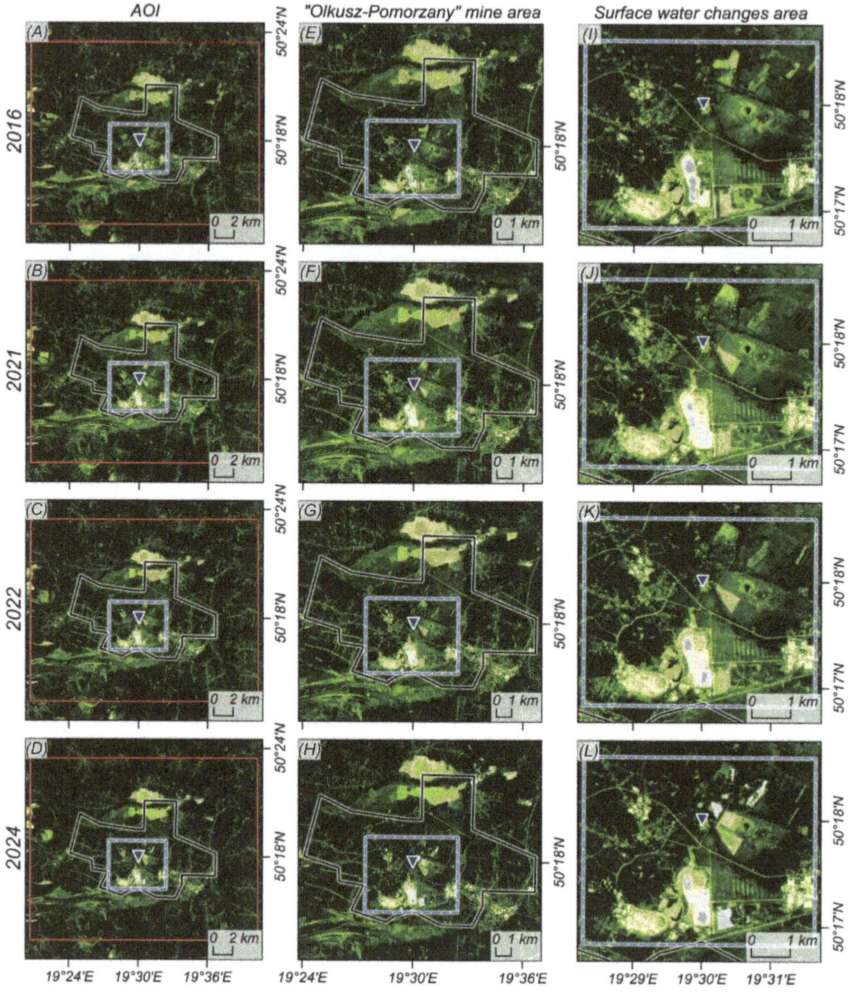

Figure A1. Spatial distribution of NDVI derived from Sentinel-2 imagery acquired in June of selected years (2016, 2021, 2022, and 2024) for the AOI (**A–D**), the "Olkusz–Pomorzany" mine area (**E–H**), and the surface water change area (**I–L**).

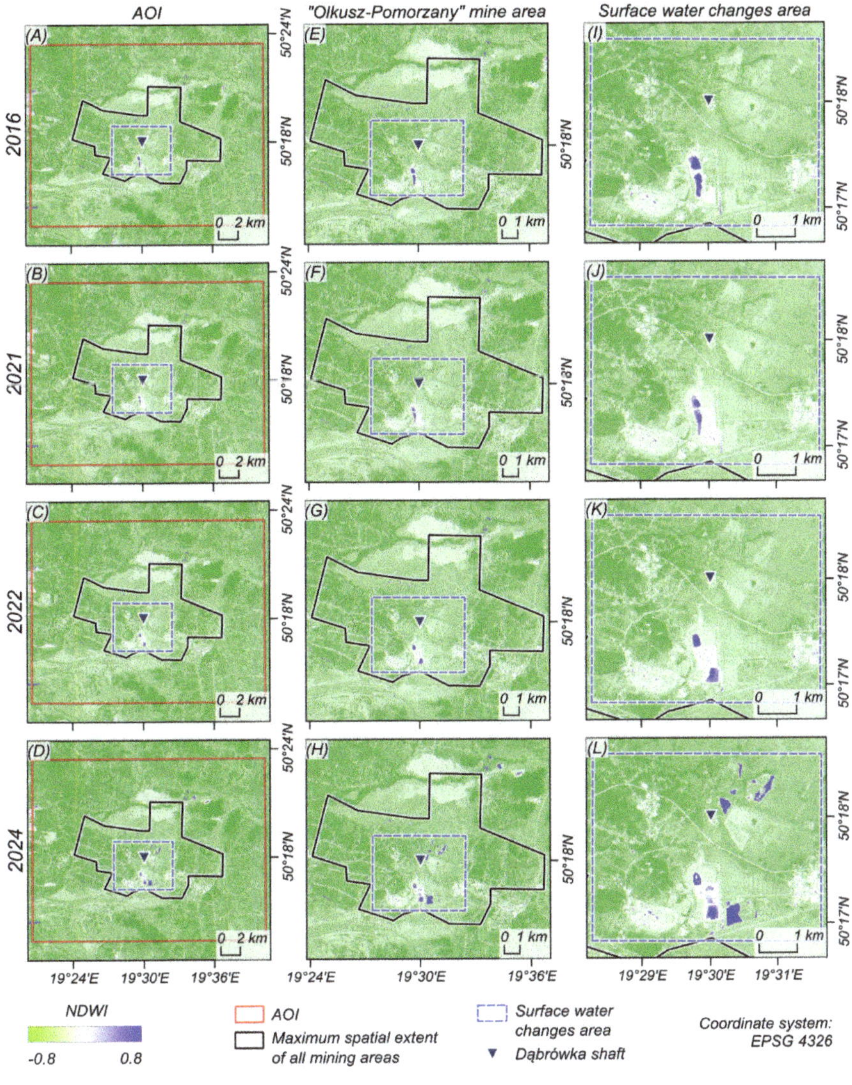

Figure A2. Spatial distribution of NDWI derived from Sentinel-2 imagery acquired in June of selected years (2016, 2021, 2022, and 2024) for the AOI (**A–D**), the "Olkusz–Pomorzany" mine area (**E–H**), and the surface water change area (**I–L**).

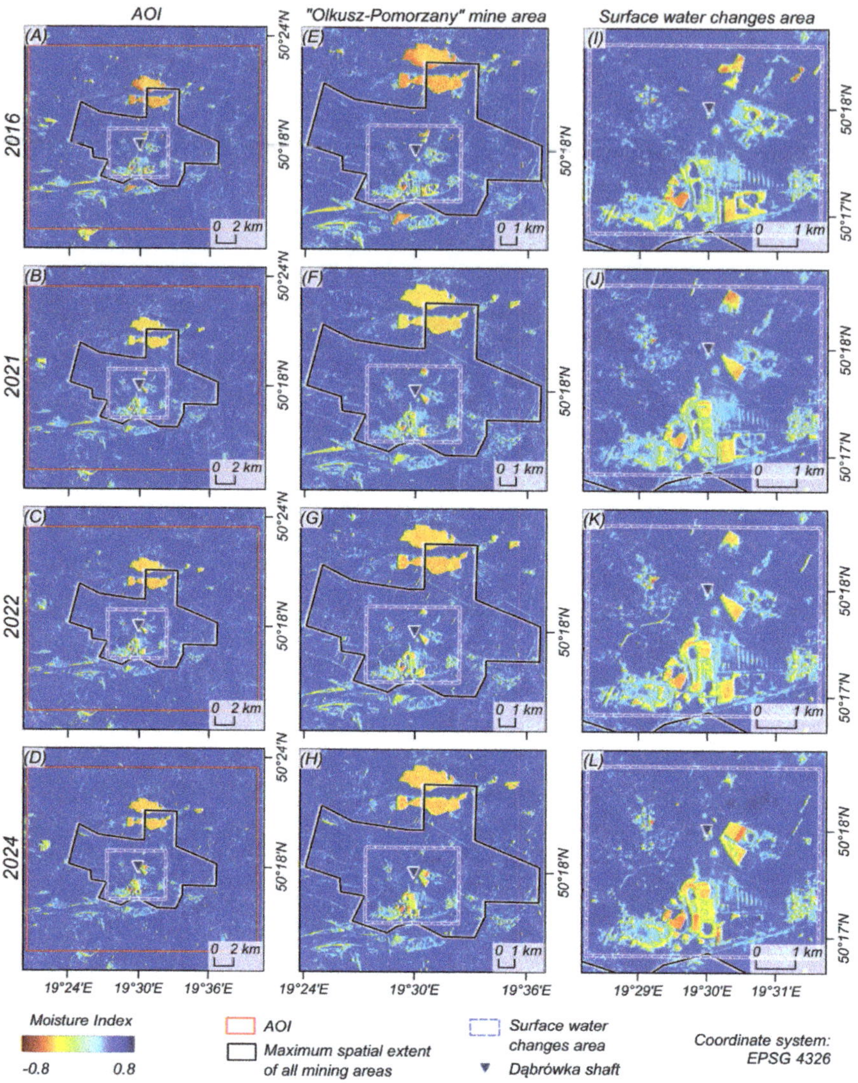

Figure A3. Spatial distribution of the Moisture Index derived from Sentinel-2 imagery acquired in June of selected years (2016, 2021, 2022, and 2024) for the AOI (**A–D**), the "Olkusz–Pomorzany" mine (**E–H**) area, and the surface water change area (**I–L**).

References

1. Oei, P.Y.; Brauers, H.; Herpich, P. Lessons from Germany's hard coal mining phase-out: Policies and transition from 1950 to 2018. *Clim. Policy* **2020**, *20*, 963–979. [CrossRef]
2. Pavloudakis, F.; Roumpos, C.; Spanidis, P.M. Planning the Closure of Surface Coal Mines Based on Circular Economy Principles. *Circ. Econ. Sustain.* **2024**, *4*, 75–96. [CrossRef]
3. Almeida, D.V.; Kolinjivadi, V.; Ferrando, T.; Roy, B.; Herrera, H.; Gonçalves, M.V.; Van Hecken, G. The 'Greening' of Empire: The European Green Deal as the EU first agenda. *Polit. Geogr.* **2023**, *105*, 102925. [CrossRef]
4. Egerer, J.; Grimm, V.; Lang, L.M.; Pfefferer, U. The coal phase-out in Germany and Central Western Europe under new framework conditions. *Energy Econ.* **2025**, *151*, 108886. [CrossRef]
5. Dvoracek, J. Problems associated with closing underground mines for economic reasons. *Górnictwo i Geoinżynieria* **2005**, *29*, 25–29.

6. Dvořáček, J.; Bauer, V.; Sousedíková, R.; Matušková, S.; Csikósová, A. Mine closure and resuming production options. *Resour. Policy* **2021**, *74*, 102323. [CrossRef]
7. Guzy, A.; Malinowska, A.A. State of the Art and Recent Advancements in the Modelling of Land Subsidence Induced by Groundwater Withdrawal. *Water* **2020**, *12*, 2051. [CrossRef]
8. Gambolati, G.; Teatini, P. Geomechanics of subsurface water withdrawal and injection. *Water Resour. Res.* **2015**, *51*, 3922–3955. [CrossRef]
9. Ma, R.; Gao, J.; Guan, C.H.; Zhang, B. Coal mine closure substantially increases terrestrial water storage in China. *Commun. Earth Environ.* **2024**, *5*, 418. [CrossRef]
10. Alexander, A.C.; Ndambuki, J.M. Impact of mine closure on groundwater resource: Experience from Westrand Basin-South Africa. *Phys. Chem. Earth Parts A/B/C* **2023**, *131*, 103432. [CrossRef]
11. Walczak, S.; Witkowski, W.T.; Stoch, T.; Guzy, A. Detecting sinkholes and land surface movement in post-mining regions using multi-source remote sensing data. *Remote Sens. Appl.* **2025**, *38*, 101560. [CrossRef]
12. Witkowski, W.T.; Guzy, A.; Lucka, M.; Kusztykiewicz, K. Study on Sinkhole Formation Mechanism in Abandoned Mines. *Int. Geosci. Remote Sens. Symp. (IGARSS)* **2023**, *2023*, 2462–2465. [CrossRef]
13. Kapica, J.; Jurasz, J.; Canales, F.A.; Bloomfield, H.; Guezgouz, M.; De Felice, M.; Kobus, Z. The potential impact of climate change on European renewable energy droughts. *Renew. Sustain. Energy Rev.* **2024**, *189*, 114011. [CrossRef]
14. Bevacqua, E.; Rakovec, O.; Schumacher, D.L.; Kumar, R.; Thober, S.; Samaniego, L.; Seneviratne, S.I.; Zscheischler, J. Direct and lagged climate change effects intensified the 2022 European drought. *Nat. Geosci.* **2024**, *17*, 1100–1107. [CrossRef]
15. Naumann, G.; Cammalleri, C.; Mentaschi, L.; Feyen, L. Increased economic drought impacts in Europe with anthropogenic warming. *Nat. Clim. Change* **2021**, *11*, 485–491. [CrossRef]
16. European Climate Risk Assessment—European Environment Agency. Available online: https://www.eea.europa.eu/publications/european-climate-risk-assessment (accessed on 16 October 2024).
17. Bianucci, P.; Sordo-Ward, A.; Lama-Pedrosa, B.; Garrote, L. How do environmental flows impact on water availability under climate change scenarios in European basins? *Sci. Total Environ.* **2024**, *911*, 168566. [CrossRef]
18. Kantor-Pietraga, I.; Krzysztofik, R.; Solarski, M. Planning Recreation around Water Bodies in Two Hard Coal Post-Mining Areas in Southern Poland. *Sustainability* **2023**, *15*, 10607. [CrossRef]
19. Cardoso, A.T.; Fan, F.M. A first evaluation of water resource conditions after an environmental reclamation effort at a former degraded coal mining area in Southern Brazil. *Environ. Monit. Assess.* **2021**, *193*, 632. [CrossRef] [PubMed]
20. Vervoort, A. Various phases in surface movements linked to deep coal longwall mining: From start-up till the period after closure. *Int. J. Coal Sci. Technol.* **2021**, *8*, 412–426. [CrossRef]
21. Matwij, W.; Gruszczyński, W.; Puniach, E.; Ćwiąkała, P. Determination of underground mining-induced displacement field using multi-temporal TLS point cloud registration. *Measurement* **2021**, *180*, 109482. [CrossRef]
22. Puniach, E.; Gruszczyński, W.; Ćwiąkała, P.; Matwij, W. Application of UAV-based orthomosaics for determination of horizontal displacement caused by underground mining. *ISPRS J. Photogramm. Remote Sens.* **2021**, *174*, 282–303. [CrossRef]
23. Zhang, Y.; Lian, X.; Ge, L.; Liu, X.; Du, Z.; Yang, W.; Wu, Y.; Hu, H.; Cai, Y. Surface Subsidence Monitoring Induced by Underground Coal Mining by Combining DInSAR and UAV Photogrammetry. *Remote Sens.* **2022**, *14*, 4711. [CrossRef]
24. Blachowski, J.; Ellefmo, S.L. Mining Ground Deformation Estimation Based on Pre-Processed InSAR Open Data—A Norwegian Case Study. *Minerals* **2023**, *13*, 328. [CrossRef]
25. Przyłucka, M.; Kowalski, Z.; Perski, Z. Twenty years of coal mining-induced subsidence in the Upper Silesia in Poland identified using InSAR. *Int. J. Coal Sci. Technol.* **2022**, *9*, 86. [CrossRef]
26. Guzy, A.; Witkowski, W.T. Land Subsidence Estimation for Aquifer Drainage Induced by Underground Mining. *Energies* **2021**, *14*, 4658. [CrossRef]
27. Szostak-Chrzanowski, A.; Chrzanowski, A.; Ortiz, E. Modeling of ground subsidence in oil fields. *Tech. Sc. Pap. Rep.* **2006**, *9*, 133–146.
28. Hejmanowski, R. Modeling of time dependent subsidence for coal and ore deposits. *Int. J. Coal Sci. Technol.* **2015**, *2*, 287–292. [CrossRef]
29. Hung, W.C.; Hwang, C.; Liou, J.C.; Lin, Y.S.; Yang, H.L. Modeling aquifer-system compaction and predicting land subsidence in central Taiwan. *Eng. Geol.* **2012**, *147–148*, 78–90. [CrossRef]
30. Vervoort, A. Impact of the Hydrogeological Conditions on the Calculated Surface Uplift above Abandoned and Flooded Coal Mines. *Geosciences* **2022**, *12*, 454. [CrossRef]
31. Misa, R.; Dudek, M.; Sroka, A.; Tajdus, K.; Mrochen, D. Contribution to the a priori assessment of the value of the caving zone expansion coefficient in the forecast of ground surface uplift caused by the flooding of closed coal mines in the Ruhr region/Germany. *Bull. Pol. Acad. Sci. Tech. Sci.* **2024**, *72*, e148611. [CrossRef]
32. González-Quirós, A.; Fernández-Álvarez, J.P. Conceptualization and finite element groundwater flow modeling of a flooded underground mine reservoir in the Asturian Coal Basin, Spain. *J. Hydrol.* **2019**, *578*, 124036. [CrossRef]

33. Vervoort, A. Challenge to Explain the Upward Surface Movement Above Abandoned Coal Mines. *Lect. Notes Civ. Eng.* **2022**, *228*, 155–165. [CrossRef]
34. Vervoort, A. Additional surface movements above coal longwalls in the post-mining phase: Case study based on remote sensing satellite data. *Geomech. Geophys. Geo-Energy Geo-Resour.* **2025**, *11*, 106. [CrossRef]
35. Guzy, A. Subsidence and Uplift in Active and Closed Lignite Mines: Impacts of Energy Transition and Climate Change. *Energies* **2024**, *17*, 5540. [CrossRef]
36. Blachowski, J.; Wajs, J.; Walerysiak, N.; Becker, M. Monitoring of Post-Mining Subsidence using Airborne and Terrestrial Laser Scanning Approach. *Arch. Min. Sci.* **2024**, *69*, 431–446. [CrossRef]
37. Kujawa, P.; Wajs, J.; Pleśniak, K. The approach to UAV image acquisition and processing for very shallow water mapping. *Int. J. Appl. Earth Obs. Geoinf.* **2025**, *141*, 104604. [CrossRef]
38. Guzy, A.; Piasecki, A.; Witkowski, W.T. Climate change and shallow aquifers—Unravelling local hydrogeological impacts and groundwater decline-induced subsidence. *Remote Sens. Environ.* **2025**, *321*, 114682. [CrossRef]
39. Guzy, A.; Malinowska, A.A. Assessment of the Impact of the Spatial Extent of Land Subsidence and Aquifer System Drainage Induced by Underground Mining. *Sustainability* **2020**, *12*, 7871. [CrossRef]
40. Chu, H.J.; Ali, M.Z.; Tatas; Burbey, T.J. Development of spatially varying groundwater-drawdown functions for land subsidence estimation. *J. Hydrol. Reg. Stud.* **2021**, *35*, 100808. [CrossRef]
41. Malinowska, A.A.; Witkowski, W.T.; Guzy, A.; Hejmanowski, R. Satellite-based monitoring and modeling of ground movements caused by water rebound. *Remote Sens.* **2020**, *12*, 1786. [CrossRef]
42. Misa, R.; Sroka, A.; Mrocheń, D. Evaluating Surface Stability for Sustainable Development Following Cessation of Mining Exploitation. *Sustainability* **2025**, *17*, 878. [CrossRef]
43. Kujawa, P.; Wajs, J.; Kasza, D.; Remondino, F. Monitoring a flooded open-cast mine with combining remote sensing techniques. The International Archives of the Photogrammetry. *Remote Sens. Spat. Inf. Sci.* **2025**, *XLVIII-2/W10-2025*, 155–161. [CrossRef]
44. Buczyńska, A.; Blachowski, J.; Bugajska-Jędraszek, N. Analysis of Post-Mining Vegetation Development Using Remote Sensing and Spatial Regression Approach: A Case Study of Former Babina Mine (Western Poland). *Remote Sens.* **2023**, *15*, 719. [CrossRef]
45. Smentek, A.; Kaczmarek, A.; Eksert, P.; Blachowski, J. Monitoring Surface Water Dynamics in Mining Areas Using Remote Sensing Indices: A Review and Cross-Case Analysis. *Water* **2025**, *17*, 2826. [CrossRef]
46. Chen, Q.; Vaudour, E.; Richer-de-Forges, A.C.; Arrouays, D. Spectral indices in remote sensing of soil: Definition, popularity, and issues. A critical overview. *Remote Sens. Environ.* **2025**, *329*, 114918. [CrossRef]
47. Yang, L.; Shi, L.; Li, J.; Kong, H.; Hua, D. Analysis of spatiotemporal changes and driving forces of ecological environment quality in Northwest China mining area. *Sci. Rep.* **2025**, *15*, 21273. [CrossRef]
48. Buczyńska, A.; Głąbicki, D.; Kopeć, A.; Modlińska, P. Understanding Surface Water Dynamics in Post-Mining Area Through Multi-Source Remote Sensing and Spatial Regression Analysis. *Remote Sens.* **2025**, *17*, 3218. [CrossRef]
49. Blachowski, J.; Dynowski, A.; Buczyńska, A.; Ellefmo, S.L.; Walerysiak, N. Integrated Spatiotemporal Analysis of Vegetation Condition in a Complex Post-Mining Area: Lignite Mine Case Study. *Remote Sens.* **2023**, *15*, 3067. [CrossRef]
50. Pawłuszek-Filipiak, K.; Wielgocka, N.; Rudziński, Ł. The Detectability of Post-Seismic Ground Displacement Using DInSAR and SBAS in Longwall Coal Mining: A Case Study in the Upper Silesian Coal Basin, Poland. *Remote Sens.* **2024**, *16*, 2533. [CrossRef]
51. Kaczmarek, A.; Monti, R.; Reguzzoni, M.; Blachowski, J. Assessment of Correlation Between InSAR Coherence and Multispectral Indices for Soil Moisture Monitoring. *Civ. Environ. Eng. Rep.* **2025**, *35*, 198–220. [CrossRef]
52. Motyka, J.; d'Obyrn, K.; Cień, D.; Wójcik, T.; Juśko, K.; Nguyen, B.T. Mine Water Rebound after The Closure of The Olkusz-Pomorzany Zn-Pb Ores Mine (Southern Poland)—Controlling Factors and Spatial Variability. *Pol. Miner. Eng. Soc.* **2025**, *1*, 261. [CrossRef]
53. Kos, J.; Wojciechowski, T.; Wójcik, A.; Zając, M.; Kamieniarz, S.; Przyłucka, M.; Perski, Z.; Karwacki, K.; Wódka, M.; Warmuz, B.; et al. Analytical Report on Land Deformations in the Area Affected by Zinc and Lead Ore Mining in the Olkusz Region (Raport z Prac Analitycznych o Deformacjach Terenu dla Rejonu Oddziaływania Eksploatacji Rud Cynku i Ołowiu w Rejonie Olkuskim). Warsaw, Poland. 2025. Available online: https://www.pgi.gov.pl/dokumenty-pig-pib-all/geozagrozenia/10967-raport-zapadliska-rejon-olkuski/file.html (accessed on 29 October 2025).
54. Ghazi, B.; Salehi, H.; Przybylak, R.; Pospieszyńska, A. Projection of climate change impact on the occurrence of drought events in Poland. *Sci. Rep.* **2025**, *15*, 1–14. [CrossRef]
55. Maciejowski, W. Natural environment types of the southern part of the Kraków-Częstochowa Upland. *Pr. Geogr.* **2025**, *2024*, 115–134. [CrossRef]
56. Kondracki, J. *Geografia Regionalna Polski*; Wydawnictwo Naukowe PWN: Warszawa, Poland, 1998.
57. National Geoportal of Poland. Available online: https://www.geoportal.gov.pl/ (accessed on 24 October 2025).
58. Gajdzik, B. The Polish Steel Industry in the Context of Industrial Transformation. In *Digital Transformation Towards Smart Steel Manufacturing: A Framework for Steel Enterprises in Poland*; Springer Nature: Cham, Switzerland, 2025; pp. 55–77. [CrossRef]

59. Mikulski, S.Z.; Strzelska-Smakowska, B. The prospective and hypothetical areas of zinc and/or lead ores in different types of deposits beyond the Upper Silesia Zn–Pb Ore District in Poland (Obszary perspektywiczne i hipotetyczne rud cynku i/lub ołowiu w różnych typach złóż poza Górnośląskim Okręgiem Rud Zn–Pb w Polsce). *Gospod. Surowcami Miner.Miner. Resour. Manag.* **2013**, *29*, 5–20. [CrossRef]
60. Strzelska-Smakowska, B. Złoża tlenowych rud cynku w okręgu śląsko-krakowskim. *Biul. Państwowego Inst. Geol.* **2010**, *439*, 371–374.
61. Habryn, R.; Buła, Z.; Nawrocki, J. The Kraków–Lubliniec Tectonic Zone in the Kraków Sector in the Light of Data from New Boreholes Trojanowice 2 and Cianowice 2 (Strefa tektoniczna Kraków–Lubliniec na odcinku krakowskim w świetle danych uzyskanych z nowych otworów badawczych Trojanowice 2 i Cianowice 2). *Biuletyn Państwowego Instytutu Geologicznego* **2014**, *459*, 45–60. Available online: https://geojournals.pgi.gov.pl/bp/article/view/29483 (accessed on 29 October 2025).
62. Mazur, S. An Outline of the Geology of Poland. In *Landscapes and Landforms of Poland*; Migoń, P., Jancewicz, K., Eds.; Springer International Publishing: Cham, Switzerland, 2024; pp. 3–17. [CrossRef]
63. Wilk, Z.; Motyka, J. Hydraulic contacts between aquifer levels in the Olkusz ore mining region (Kontakty między poziomami wodonośnymi w olkuskim rejonie kopalnictwa rud). *Ann. Soc. Geol. Pol.* **1977**, *47*, 115–143. Available online: https://geojournals.pgi.gov.pl/asgp/article/view/11575/10058 (accessed on 29 October 2025).
64. Motyka, J.; Pulido-Bosch, A.; Borczak, S.; Gisbert, J. Matrix hydrogeological properties of Devonian carbonate rocks of Olkusz (Southern Poland). *J. Hydrol.* **1998**, *211*, 140–150. [CrossRef]
65. Polish Geological Institute—National Research Institute. Available online: https://www.pgi.gov.pl/en/ (accessed on 24 October 2025).
66. OpenStreetMap contributors. *OpenStreetMap: Collaborative Project to Create a Free Editable Map of the World*; OpenStreetMap: Cambridge, UK, 2025.
67. Adamczyk, Z.; Motyka, J. Development of Water Inflows to Zinc and Lead Ore Mines in the Olkusz Region, Southern Poland (Rozwój Dopływów Wody Do Kopalń Rud Cynku i Ołowiu w Rejonie Olkusza). *Geol. Q.* **2000**, *48*, 171–175. Available online: https://geojournals.pgi.gov.pl/pg/article/view/15144/12925. (accessed on 29 October 2025).
68. Juśko, K.; Motyka, J.; d'Obyrn, K.; Adamczyk, Z. Construction of a numerical groundwater flow model in areas of intense mine drainage, as exemplified by the Olkusz Zinc and Lead Ore Mining Area in southwest Poland. *Geologos* **2018**, *24*, 237–244. [CrossRef]
69. Motyka, J.; Czop, M. Influence of Karst Phenomena on Water Inflow to Zn-Pb Mines in the Olkusz District (S Poland). In *Advances in Research in Karst Media*; Springer: Berlin/Heidelberg, Germany, 2010; pp. 449–454. [CrossRef]
70. Włodarz, B. Historic Ore Mine in Olkusz and Its Prospects (Zabytkowa Kopalnia Górnictwa Kruszcowego w Olkuszu—Perspektywy Uruchomienia). *Górnictwo I Geoinżynieria* **2006**, *30*, 131–141. Available online: http://yadda.icm.edu.pl/baztech/element/bwmeta1.element.baztech-article-AGH5-0018-0114 (accessed on 29 October 2025).
71. Paulo, A.; Wnuk, R. The Pomorzany Zn–Pb mine in Poland is 40 years old—When will its ore reserves become depleted? (Kopalnia rud cynku i ołowiu „Pomorzany skończyła 40 lat—Jak długo jeszcze wystarczy jej zasobów?). *Przegląd Geol.* **2015**, *63*, 1483–1490. Available online: https://geojournals.pgi.gov.pl/pg/article/view/27803/19507 (accessed on 29 October 2025).
72. Różkowski, K.; Różkowski, J.; Rahmonov, O. The Historical Aspect of the Impact of Zn and Pb Ore Mining and Land Use on Ecohydrological Changes in the Area of the Biała Przemsza Valley (Southern Poland). *Land* **2023**, *12*, 997. [CrossRef]
73. Cień, D.; Motyka, J.; d'Obyrn, K. Changes in the flow rate of the Sztoła River (Olkusz area) under mining drainage conditions (Zmiany natężenia przepływu w rzece Sztole (rejon olkuski) w warunkach drenażu górniczego). *Przegląd Geol.* **2022**, *70*, 761–769. [CrossRef]
74. Smentek, A.; Blachowski, J. Remote Sensing Analysis of Environmental Changes in a Post-Mining Area: A Case Study of the Olkusz Region—Preliminary Results. *Civ. Environ. Eng. Rep.* **2025**, *35*, 1–18. [CrossRef]
75. Motyka, J.; d'Obyrn, K. Reconstruction of the Groundwater Table in the Area of the Decommissioned Mines of ZGH „Bolesław" S.A.—What Do We Know about the Hydrological Situation? (Odbudowa ZwierciadłA Wody Podziemnej w Rejonie Likwidowanych Kopalń ZGH „Bolesław"—Co Wiemy o Sytuacji Hydrologicznej?). Bukowno, Poland. 2025. Available online: https://zghboleslaw.pl/images/lkkop/omow-dotychczasz-badan-monit2505.pdf (accessed on 29 October 2025).
76. Motyka, J.; Cień, D.; D'oByrn, K. The Use of Hydrochemical Zoning to Predict Water Hazards: The Example of the Olkusz-Pomorzany Zn-Pb Mine. *Mine Water Environ.* **2024**, *43*, 198–209. [CrossRef]
77. Cabala, J.; Teper, L.; Rutkowski, T. Rockmass deformations caused by Zinc and Lead ores mining in the Olkusz region (Southern Poland). *Acta Geodyn. Geomater.* **2004**, *1*, 47–58.
78. Pilecki, Z. The role of geophysical methods in the estimation of sinkhole threat in the post mining areas of shallow exploitation in the Upper Silesian Coal Basin, Poland (Rola metod geofizycznych w ocenie zagrożeń powstawania zapadlisk na terenach byłych obszarów górniczych o płytkiej eksploatacji w Górnośląskim Zagłębiu Węglowym, Polska). *Gospod. Surowcami Miner. Miner. Resour. Manag.* **2008**, *24*, 27–40. Available online: https://gsm.min-pan.krakow.pl/The-role-of-geophysical-methods-in-the-estimation-of-sinkhole-threat-in-the-post%2C96733%2C0%2C2.html (accessed on 29 October 2025).

79. Chaoshang, S.; Chaoyang, S.; Zhiming, Z.; Haixiao, L.; Zhenhua, L.; Feng, D.; Zhengzheng, C.; Pengtao, L.; Lin, L. Overburden failure characteristics and fracture evolution rule under repeated mining with multiple key strata control. *Sci. Rep.* **2025**, *15*, 28029. [CrossRef]
80. Lei, W.; Shang, Y.; Lu, X.; Zhang, D.; Gong, H.; Yi, W.; Zhou, H.; Yuchi, X.; Qin, B.; Du, X. Study on mining deformation and fracture response mechanism and rock pressure law of super thick composite key roof. *Sci. Rep.* **2025**, *15*, 19018. [CrossRef]
81. Kotyrba, A.; Kortas, Ł. Geophysical imprint of mining-induced rock mass deformation in the area of construction disaster in Bytom (Poland). *J. Sustain. Min.* **2023**, *22*, 169–184. [CrossRef]
82. Dudek, M.; Mrocheń, D.; Sroka, A.; Tajduś, K. Integrating the Finite Element Method with Python Scripting to Assess Mining Impacts on Surface Deformations. *Appl. Sci.* **2024**, *14*, 7797. [CrossRef]
83. Dudek, M.; Sroka, A.; Tajduś, K.; Misa, R.; Mrocheń, D. Assessment and Duration of the Surface Subsidence after the End of Mining Operations. *Energies* **2022**, *15*, 8711. [CrossRef]
84. Polish Institute of Meteorology and Water Management-National Research Institute. Available online: https://www.imgw.pl/ (accessed on 25 October 2025).
85. Allen, R.G. Using the FAO-56 dual crop coefficient method over an irrigated region as part of an evapotranspiration intercomparison study. *J. Hydrol.* **2000**, *229*, 27–41. [CrossRef]
86. He, R.; Zhang, L.; Chew, A.W.Z. Modeling and predicting rainfall time series using seasonal-trend decomposition and machine learning. *Knowl. Based Syst.* **2022**, *251*, 109125. [CrossRef]
87. Crosetto, M.; Solari, L.; Mróz, M.; Balasis-Levinsen, J.; Casagli, N.; Frei, M.; Oyen, A.; Moldestad, D.A.; Bateson, L.; Guerrieri, L.; et al. The Evolution of Wide-Area DInSAR: From Regional and National Services to the European Ground Motion Service. *Remote Sens.* **2020**, *12*, 2043. [CrossRef]
88. Crosetto, M.; Solari, L.; Balasis-Levinsen, J.; Bateson, L.; Casagli, N.; Frei, M.; Oyen, A.; Moldestad, D.A.; Mróz, M. Deformation Monitoring at European Scale: The Copernicus Ground Motion Service. *ISPRS—Int. Arch. Photogramm. Remote Sens. Spat. Inf. Sci.* **2021**, *XLIII-B3-2*, 141–146. [CrossRef]
89. Marmoni, G.M.; Antonielli, B.; Caprari, P.; Di Renzo, M.E.; Marini, R.; Mastrantoni, G.; Mazzanti, P.; Patelli, D.; Bozzano, F. Semi-automatic ranking of landslide candidate areas at a regional scale using EGMS InSAR data for territorial planning and risk management. *Landslides* **2025**, *22*, 4077–4095. [CrossRef]
90. Wood, J.; Whittaker, A.C.; Bell, R.E.; Coveney, S.M.; Kranis, H.D. Resolving Coseismic and Aseismic Normal Fault Slip Behaviors From InSAR Time Series of the European Ground Motion Service (EGMS). *Tectonics* **2025**, *44*, 2025TC008904. [CrossRef]
91. Cardello, G.L.; Barreca, G.; Monaco, C.; de Michele, M.; Antonioli, F. First comparison of subsidence/uplift rates between Copernicus European Ground Motion Service data and long-term MIS 5.5 geological record in Mediterranean regions. *Earth Sci. Rev.* **2025**, *265*, 105132. [CrossRef]
92. Crosetto, M.; Crippa, B.; Mróz, M.; Cuevas-González, M.; Shahbazi, S. Applications based on EGMS products: A review. *Remote Sens. Appl. Soc. Environ.* **2025**, *37*, 101452. [CrossRef]
93. Farías, C.A.; Sánchez, M.L.; Bonì, R.; Cigna, F. Statistical and Independent Component Analysis of Sentinel-1 InSAR Time Series to Assess Land Subsidence Trends. *Remote Sens.* **2024**, *16*, 4066. [CrossRef]
94. Imtiaz, F.; Farooque, A.A.; Randhawa, G.S.; Wang, X.; Esau, T.J.; Acharya, B.; Garmdareh, S.E.H. An inclusive approach to crop soil moisture estimation: Leveraging satellite thermal infrared bands and vegetation indices on Google Earth engine. *Agric. Water Manag.* **2024**, *306*, 109172. [CrossRef]
95. Wang, J.; Rich, P.M.; Price, K.P.; Kettle, W.D. Relations between NDVI and tree productivity in the central Great Plains. *Int. J. Remote Sens.* **2004**, *25*, 3127–3138. [CrossRef]
96. Gao, B.C. NDWI—A normalized difference water index for remote sensing of vegetation liquid water from space. *Remote Sens. Environ.* **1996**, *58*, 257–266. [CrossRef]
97. McFeeters, S.K. Using the Normalized Difference Water Index (NDWI) within a Geographic Information System to Detect Swimming Pools for Mosquito Abatement: A Practical Approach. *Remote Sens.* **2013**, *5*, 3544–3561. [CrossRef]
98. Abdulraheem, M.I.; Zhang, W.; Li, S.; Moshayedi, A.J.; Farooque, A.A.; Hu, J. Advancement of Remote Sensing for Soil Measurements and Applications: A Comprehensive Review. *Sustainability* **2023**, *15*, 15444. [CrossRef]
99. Copernicus Data Space Ecosystem. *Copernicus Data Space Ecosystem—Open Access to Copernicus Sentinel Satellite Data and Earth Observation Services*; European Space Agency (ESA): Paris, France, 2025.
100. Drusch, M.; Del Bello, U.; Carlier, S.; Colin, O.; Fernandez, V.; Gascon, F.; Hoersch, B.; Isola, C.; Laberinti, P.; Martimort, P.; et al. Sentinel-2: ESA's Optical High-Resolution Mission for GMES Operational Services. *Remote Sens. Environ.* **2012**, *120*, 25–36. [CrossRef]
101. Karlsen, S.R.; Stendardi, L.; Tømmervik, H.; Nilsen, L.; Arntzen, I.; Cooper, E.J. Time-Series of Cloud-Free Sentinel-2 NDVI Data Used in Mapping the Onset of Growth of Central Spitsbergen, Svalbard. *Remote Sens.* **2021**, *13*, 3031. [CrossRef]
102. Coluzzi, R.; Imbrenda, V.; Lanfredi, M.; Simoniello, T. A first assessment of the Sentinel-2 Level 1-C cloud mask product to support informed surface analyses. *Remote Sens. Environ.* **2018**, *217*, 426–443. [CrossRef]

103. Polish Space Agency (POLSA); National Satellite Information System (NSIS). National Satellite Information System (NSIS)—Polish National Satellite Data Platform and Geoportal. 2025. Available online: https://nsisplatforma.polsa.gov.pl/portal/ (accessed on 26 October 2025).
104. Malinowski, R.; Lewiński, S.; Rybicki, M.; Gromny, E.; Jenerowicz, M.; Krupiński, M.; Nowakowski, A.; Wojtkowski, C.; Krupiński, M.; Krätzschmar, E.; et al. Automated Production of a Land Cover/Use Map of Europe Based on Sentinel-2 Imagery. *Remote Sens.* **2020**, *12*, 3523. [CrossRef]
105. Ghazi, B.; Salehi, H.; Przybylak, R.; Pospieszyńska, A. Assessment of drought conditions under climate change scenarios in Central Europe (Poland) using the standardized precipitation index (SPI). *Clim. Serv.* **2025**, *39*, 100591. [CrossRef]
106. Tomczyk, A.M.; Bednorz, E.; Szyga-Pluta, K. Changes in Air Temperature and Snow Cover in Winter in Poland. *Atmosphere* **2021**, *12*, 68. [CrossRef]
107. Filho, W.L.; Dinis, M.A.P.; Nagy, G.J.; Fracassi, U.; Aina, Y.A. A ticket to where? Dwindling snow cover impacts the winter tourism sector as a consequence of climate change. *J. Environ. Manag.* **2024**, *356*, 120554. [CrossRef] [PubMed]
108. Liuzzo, L.; Viola, F.; Noto, L.V. Wind speed and temperature trends impacts on reference evapotranspiration in Southern Italy. *Theor. Appl. Clim.* **2014**, *123*, 43–62. [CrossRef]
109. Larsén, X.G.; Imberger, M.; Hannesdóttir, Á. The impact of Climate Change on extreme winds over northern Europe according to CMIP6. *Front. Energy Res.* **2024**, *12*, 1404791. [CrossRef]
110. EU. *Global Warming and Windstorm Impacts in the EU—JRC PESETA IV Project—Task 13*; EU Publications Office: Luxembourg, 2020. [CrossRef]
111. Pińskwar, I.; Choryński, A.; Kundzewicz, Z.W. Severe Drought in the Spring of 2020 in Poland—More of the Same? *Agronomy* **2020**, *10*, 1646. [CrossRef]
112. Vervoort, A. Uplift of the surface of the earth above abandoned coal mines. Part B: Framework to understand and explain uplift. *Int. J. Rock. Mech. Min. Sci. Géoméch. Abstr.* **2021**, *148*, 104947. [CrossRef]
113. Dudek, M.; Tajduś, K.; Misa, R.; Sroka, A. Predicting of land surface uplift caused by the flooding of underground coal mines—A case study. *Int. J. Rock. Mech. Min. Sci. Géoméch. Abstr.* **2020**, *132*, 104377. [CrossRef]
114. Vervoort, A. Uplift of the surface of the earth above abandoned coal mines. Part A: Analysis of satellite data related to the movement of the surface. *Int. J. Rock. Mech. Min. Sci. Géoméch. Abstr.* **2021**, *148*, 104896. [CrossRef]

water

MDPI

Article

Hydrogeochemical Evolution and Ecological Irrigation Evaluation of Mine Water in an Arid Coal Region: A Case Study from Northwest China

Hao Wang [1,2,3,4], Hongbo Shang [1,2,3,4], Tiantian Wang [2,3,4,*], Jiankun Xue [2,3,4,5], Xiaodong Wang [2,3,4,5], Zhenfang Zhou [2,3,4] and Qiangmin Wang [2,3,4]

1 China Coal Research Institute, Beijing 100013, China; wanghao@ccteg.com (H.W.); shanghongbo@ccteg.com (H.S.)
2 Xi'an Research Institute of China Coal Technology & Engineering, (Group), Corp, Xi'an 710077, China; xuejiankun@ccteg.com (J.X.); wangxiaodong@ccteg.com (X.W.); zhouzhenfang@ccteg.com (Z.Z.); wangqiangmin@ccteg.com (Q.W.)
3 Shaanxi Engineering Research Center of Mine Ecological Environment Protection and Restoration in the Middle of Yellow River Basin, Xi'an 710054, China
4 Shaanxi Key Laboratory of Prevention and Control Technology for Coal Mine Water Hazard, Xi'an 710077, China
5 State Key Laboratory of Disaster Prevention and Ecology Protection in Open Pit Coal Mines (Xi'an Research Institute), Xi'an 710077, China
* Correspondence: wangthpuedu@126.com; Tel.: +86-18291995907

Abstract

Investigating ecological irrigation risks associated with mine water utilization is of great significance for alleviating water resource shortages in arid mining regions of western China, thereby supporting efficient coal extraction and coordinated ecological development. In this study, a representative mining area in Xinjiang was investigated to reveal the evolution patterns of mine water quality under arid geo-environmental conditions in western China and to systematically assess environmental risks induced by ecological irrigation. Surface water, groundwater, and mine water samples were collected to study ion ratio coefficients, hydrochemical characteristics, and evolution processes. Based on this, a multi-index analysis was employed to evaluate ecological irrigation risks and establish corresponding risk control measures. The results show that the total dissolved solids (TDS) of mine water in the study area are all greater than 1000 mg/L. The evolution of mine water quality is mainly controlled by water–rock interaction and is affected by evaporation and concentration. The main ions Na^+, Cl^-, Ca^{2+}, and SO_4^{2-} originate from the dissolution of halite, gypsum, and anorthite. If the mine water is directly used for irrigation without treatment, the soluble sodium content, sodium adsorption ratio, salinity hazard, and magnesium adsorption ratio will exceed the limits, leading to the accumulation of Na^+ in the soil, affecting plant photosynthesis, and posing potential threats to the groundwater environment. Given the evolution process of mine water quality and the potential risks of direct use for irrigation, measures can be taken across three aspects: nanofiltration combined with reverse osmosis desalination, adoption of drip irrigation and intermittent irrigation technologies, and selection of drought-tolerant vegetation. These measures can reduce the salt content of mine water, decrease the salt accumulation in the soil layer, and lower the risk of groundwater pollution, thus reducing the environmental risks of ecological irrigation with mine water. The research will provide an important theoretical basis for the scientific utilization and management of mine water resources in arid areas by revealing the evolution law of mine water quality in arid areas and clarifying its ecological irrigation environmental risks.

Keywords: arid region; mine water; water quality evolution; ecological irrigation; environmental risk

1. Introduction

With coal resources steadily becoming depleted in the central and eastern parts of China, the coal mining center has rapidly shifted to the ecologically fragile arid areas in the west, where coal reserves account for 70% of the country's total and play a vital role in energy supply [1,2]. However, due to climatic factors, precipitation is scarce and evaporation is intense, and water resource and ecological environment problems are becoming increasingly serious [3]. Due to the superposition of sedimentation, structure, and man-made mining activities, the mine water produced in the coal mining process often contains a large amount of dust, coal dust suspended solids, and a certain amount of sulfate and heavy metals [4], and direct discharge not only wastes water resources, but also poses a threat to the soil and the surrounding ecological environment. According to statistics, for every 1 ton of raw coal mined, about 2.1 tons of mine water is produced [5]. Therefore, rationally treating and using mine water for ecological irrigation is of great significance in order to alleviate the shortage of water resources and to improve the ecological environment in the arid areas of western China.

The hydrochemical characteristics of mine water are the direct basis for determining whether the use of mine water for ecological irrigation poses risks to soil safety, vegetation growth, and the groundwater environment [6,7]. The main formation role and ion source of mine water quality can be determined by studying the water quality evolution process of mine water [8,9], which can provide a basis and guidance for the prevention and control of mine water ecological irrigation risks. Scholars have carried out a great number of studies on the chemical characteristics and water quality evolution of mine water irrigation [10,11]. The formation and evolution of mine water quality can be divided into three stages: natural equilibrium in the pre-mining period, intensive disturbance during mining, and rebalancing after pit closure [12–14]. Affected by coal mining activities, the sulfur-containing minerals in the coal and surrounding rock undergo a series of oxidation reactions to produce acidic substances, such as sulfate, and dissolve a variety of heavy metals, forming acidic mine water containing heavy metals [15]. However, the pH of the mine waters in the western mining area is mostly alkaline, and unlike the mining area in the central and eastern mining areas, the characteristics and evolution process of mine water quality need to be studied.

In terms of the environmental impact of mine water irrigation, the feasibility of coal mine drainage on farmland irrigation and its potential impact on soil performance were examined by analyzing key water quality parameters such as sodium adsorption ratio, electrical conductivity (EC), sodium percentage (Na%) and Kelly's index (KI) [16–18]. This provided theoretical support for mine water ecological irrigation [19]. The formation and evolution of neutral mine water are less affected by coal mining disturbances [20]. However, the pH of the mine waters in the western mining area is mostly alkaline, and unlike the mining area in the central and eastern mining areas, the characteristics and evolution process of mine water quality need to be studied. These studies mainly focused on the analysis of mine water hydrochemistry and ecological irrigation suitability. However, systematic analysis and research on the effects of ecological irrigation of mine water on soil, vegetation, and groundwater are lacking. Therefore, it is essential to systematically study the environmental risks of mine water ecological irrigation based on an understanding of the characteristics and evolution process of mine water in arid areas.

Yushuquan mine is located on the southern foothills of the Tianshan Mountains, and the northern edge of the Tarim Basin. It is a typical arid climate zone, with scarce rainfall and scarce surface water and groundwater resources. The mine water is a potential water resource for ecological water such as vegetation irrigation. Therefore, we study the hydrochemical characteristics and formation of mine water, and conduct an evaluation of the applicability of mine water for irrigation. This research aims to (1) investigate the hydrochemical characteristics and formation of mine water using mathematical statistics and ionic scale factors, and (2) analyze the impact of mine water irrigation on soil safety, vegetation growth, and the groundwater environment by combining factors such as soluble sodium content (*SSP*), sodium adsorption ratio (*SAR*), permeability index (*PI*), heavy metal pollution index (*HPI*), and the Nemero index (*F*).

2. Materials and Methods

2.1. Study Area

Yushuquan mine is situated at the southern foothills of the Tianshan Mountains, along the northern boundary of the Tarim Basin. This region experiences a typical temperate continental arid climate, characterized by long periods of drought, limited rainfall, and a fragile ecosystem. The surface bedrock is exposed in the mining area, with relatively flat terrain in the center, and in the south and north the surface is largely covered by red clinker formed by the natural oxidation of the coal seams, where a number of alluvial gullies have developed. The main coal resources are deposited in the tectonic basins in this region, and its hydrogeological structures are mainly affected by folding and faulting tectonics. In addition, the wind-eroded landforms in the region provide favorable conditions for the conversion of surface rainfall into groundwater flow, and the main aquifers include Quaternary pore water and Jurassic fissure water.

The strata in the mining area include Triassic Huangshanjie Formation (T_3hs), Jurassic Tariqike Formation (J_1t), Jurassic Ahe Formation (J_1a), and Quaternary (Q), from oldest to newest (Figure 1). Based on the water content and water-resisting property of these strata, the sequence can be divided into five aquifers, one permeable stratum and four aquicludes. There are 13 coal seams in the well field, which are mainly distributed in the Tarichik Formation, of which there are four recoverable coal seams with stable conditions, including no.5 coal, no.7 coal, no.8 coal and no.10 coal. Currently, the mine water of the no.10 major minable seams mainly comes from direct water-filled water sources (no.10 coal roof confined aquifer) and indirect water-filled water sources (clinker phreatic aquifer and the Quaternary phreatic aquifer).

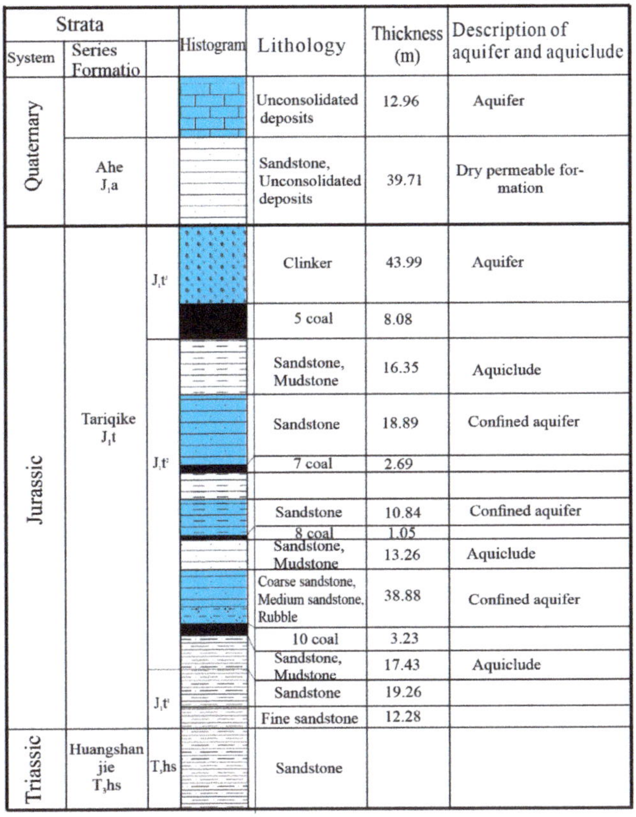

Figure 1. Comprehensive hydrogeological column map of the study area.

2.2. Sample Collection and Analysis

To investigate the water quality evolution process of mine water and to quantitatively evaluate the environmental risks of using mine water for ecological irrigation, a total of 23 water samples were collected from different water-filled water sources in the mine, different locations within the underground coal mine, and surface gullies. The sampling included three surface water samples collected from the western and eastern alluvial gullies. Six quaternary groundwater samples, two clinker phreatic water samples, and six samples of no.10 coal roof groundwater were collected during the hydrogeological exploration phase (Figure 2). Six mine water samples were collected from different locations in the no.10 coal seam underground mine at a depth of 208 m. Mine water sampling points include the coal mining face, the return airway of the mine, and the central water tank. Prior to water sampling, in situ measurements of pH and total dissolved solids (TDS) were conducted using a portable multi-parameter monitoring device (YSI, Yellow Springs, OH, USA; HACH, Loveland, CO, USA) with a precision of ±0.2 (pH) and ±1% (TDS), respectively. Then, the sampling bottles were flushed 2~3 times with the water to be sampled, and the water samples were sealed, labeled and transported to the laboratory for testing. The concentrations of cations, including K^+, Na^+, Ca^{2+}, and Mg^{2+}, were tested using ICP-AES (Thermo, Waltham, MA, USA); Concentrations of major anions F^-, Cl^-, NO_3^-, and SO_4^{2-} were determined with ion chromatography (ICS2000, Dionex, Sunnyvale, CA, USA). The concentration of HCO_3^- was determined through chemical titration. Each water sample was tested three times and averaged, and the standard deviation was controlled within 10%. The ionic charge imbalance between the main cations and anions was less than 5%. The concentrations of heavy metals, including FeMn, Cu, Zn, Hg, As, Cd, and Pb were tested using ICP-AES (Thermo), with detection accuracies of 5%. Cr^{6+} was measured as chromate according to a modified version of U.S. EPA Method 218.6.38.46. This method is based on anion exchange chromatography using a Thermo Scientific Dionex IonPac AS7 (Thermo, Massachusetts, USA) column at Duke University with a method detection limit (MDL) for chromate of 0.004 μg/L and a reporting limit of 0.012 μg/L.

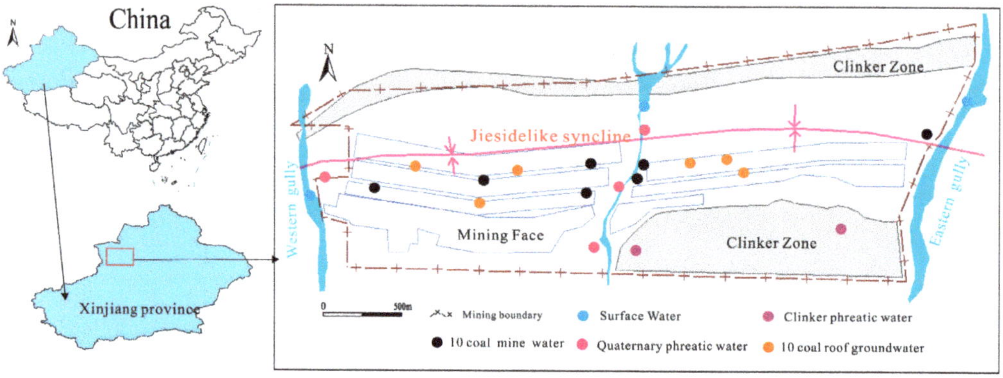

Figure 2. Distribution of sampling points.

3. Results and Discussion

3.1. Hydrochemistry of Mine Water

Analyzing the chemical characteristics of mine water and its water-filled water sources is the premise for studying the evolution of mine water quality, and it is also the basis for understanding the environmental risks caused by ecological irrigation of mine water. Detection of the concentrations of routine ions and physicochemical indicators in different types of water samples from the Yushuquan mine and the statistical analysis results are shown in Table 1. Waters of different types in the Yushuquan mine are characterized by an average pH value of 7.78, classified as weakly alkaline. Their TDS ranged from 985.42 to 3051.66 mg/L, with most values exceeding 1000 mg/L, classifying the water as high mineralization. In addition, the TDS variation ranges for surface water, Quaternary groundwater, clinker phreatic water, and the no.10 coal roof groundwater were 1658.43–2158.77, 985.42–1984.22, 1468.09–1854.34, and 1428.27–2748.88 mg/L, respectively. The TDS variation range for mine water was 1507.27–3051.66 mg/L, which is significantly higher than that of surface water and groundwater (Figure 3a). The overall increase in the TDS of mine water is due to the influence of mining activities, water-conducting fissure zones communicating different aquifers, and enhancing groundwater mixing, and superimposes the secondary water–rock effect in the goaf [21]. The major anions in the water samples were Cl^- and $SO_4{}^{2-}$ and the cations were predominantly Na^+ and Ca^{2+}. The coefficient of variation for surface water, Quaternary groundwater, clinker phreatic water, confined water from the no.10 coal seam roof, and mine water from the no.10 coal seam is all less than 1, indicating that the spatial variability of each indicator within the same water sample type is relatively small.

Table 1. Results of conventional ion detection of study area.

Sample Types	Samples Number	Statistical Analyses	Na^+ (mg/L)	Ca^{2+} (mg/L)	Mg^{2+} (mg/L)	Cl^- (mg/L)	SO_4^{2-} (mg/L)	HCO_3^- (mg/L)	pH	TDS (mg/L)	Temperature (°C)
Surface water	3	minimum	275.47	125.65	46.30	409.92	518.49	116.25	7.70	1658.43	31.2
		maximum	395.58	145.53	82.94	505.30	969.69	226.08	7.87	2158.77	35.5
		mean	335.53	135.59	64.62	457.61	744.09	189.47	7.79	1842.22	33.4
		variance	60.05	9.94	18.32	47.69	225.60	63.41	0.09	275.33	47.2
		CV	0.02	0.02	0.07	0.02	0.02	0.04	0.04	0.01	0.21
Quaternary groundwater	6	minimum	197.60	77.94	41.76	255.20	123.54	151.09	7.82	985.42	30.4
		maximum	469.45	233.30	92.36	737.33	510.00	290.90	8.40	1984.22	35.4
		mean	293.53	121.05	61.45	461.65	272.87	253.83	8.04	1349.53	32.8
		variance	105.66	57.66	18.20	211.35	134.13	53.75	0.25	395.37	86.5
		CV	0.04	0.06	0.07	0.03	0.04	0.03	0.06	0.01	0.28
Clinker phreatic water	2	minimum	357.68	98.89	61.58	411.00	400.58	251.63	7.15	1468.09	30.6
		maximum	502.75	147.13	91.13	735.53	533.08	392.82	7.84	1854.34	35.2
		mean	430.22	123.01	76.36	573.27	466.83	322.23	7.50	1661.22	32.9
		variance	102.58	34.11	20.90	229.48	93.69	99.84	0.49	273.12	92.3
		CV	0.02	0.05	0.06	0.03	0.02	0.03	0.09	0.01	0.29
10 coal roof groundwater	6	minimum	276.72	87.02	45.16	554.29	191.46	117.85	7.10	1428.27	30.8
		maximum	937.13	169.46	154.04	919.41	775.87	415.48	8.13	2748.88	34.8
		mean	526.56	135.25	78.42	697.23	475.58	302.20	7.72	1908.78	32.7
		variance	242.42	27.87	39.07	144.50	299.37	99.98	0.40	505.22	96.8
		CV	0.03	0.04	0.08	0.02	0.04	0.03	0.08	0.01	0.30
10 coal mine water	6	minimum	326.53	136.16	63.05	633.29	201.05	175.85	7.73	1507.27	31.2
		maximum	920.13	208.20	167.90	902.41	845.69	428.30	9.40	3051.66	35.2
		mean	522.19	171.59	109.16	772.16	428.83	329.80	8.46	2035.66	32.4
		variance	212.03	27.79	38.26	99.73	287.16	88.47	0.67	573.62	76.7
		CV	0.03	0.03	0.06	0.01	0.04	0.03	0.10	0.01	0.27

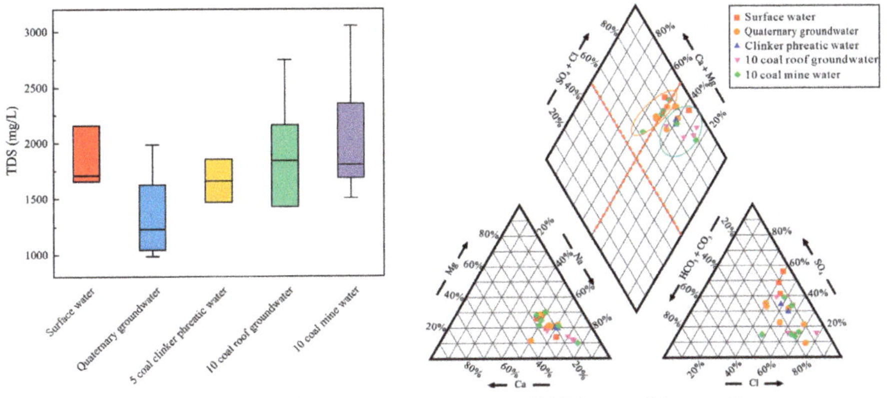

(**a**) TDS concentration box diagram (**b**) Piper trilinear diagram

Figure 3. TDS concentration box diagram and Piper trilinear diagram, (**a**) TDS concentration box diagram, (**b**) Piper trilinear diagram.

Based on the results of the ion concentration detection, a Piper trilinear diagram (Figure 3b) was plotted to analyze the water chemistry types and major ion compositions of surface water, groundwater, and mine water in the Yushuquan mine [22]. According to the piper plots, in the anion triangle, each type of water samples is close to SO_4^{2-} and Cl^- terminal elements, and the anion is mainly dominated by SO_4^{2-} or Cl^-. In the cation triangle, the distribution of different types of water samples is close to the Na^+ terminal element, with a small portion close to the Ca^{2+} terminal element. The main hydrochemical types of water samples in the Yushuquan mine are Cl-Na and SO_4-Ca. The basic water chemistry characteristics of the no.10 coal mine water are most similar to the no.10 coal roof groundwater, suggesting that the direct source of the no.10 coal mine water is the no.10 coal roof groundwater.

3.2. Hydrogeochemical Evolution of Mine Water

The Gibbs diagram can reveal the controlling factors of major ion evolution in surface water and phreatic water by using the relationship between the ion concentration ratio $Na^+/(Na^+ + Ca^{2+})$ and TDS [23]. The hydrochemistry mechanisms can be classified into three types: evaporation, encompassing water–rock interaction, and atmospheric precipitation. In recent years, more and more scholars have used the Gibbs diagram to study the evolution process of major ions in surface water and groundwater. The Gibbs map of surface water, groundwater, and mine water in the Yushuquan mine is shown in Figure 4. The TDS of water samples from the Yushuquan mine are greater than 1000 mg/L, with $Na^+/(Na^+ + Ca^{2+})$ values from 0.55 to 0.95, and $Cl^-/(Cl^- + HCO_3^-)$ from 0.45 to 0.95. Different types of water samples all fall within the scope of rock weathering and evaporation concentration, indicating that rock weathering and evaporation concentration is the controlling mechanism of the evolution of groundwater and mine water. In addition, the confined groundwater of the no.10 coal seam roof is close to the no.10 coal mine water in the Gibbs diagram, which further proves that the main source of mine water in the evolution of mine water is the no.10 coal roof groundwater.

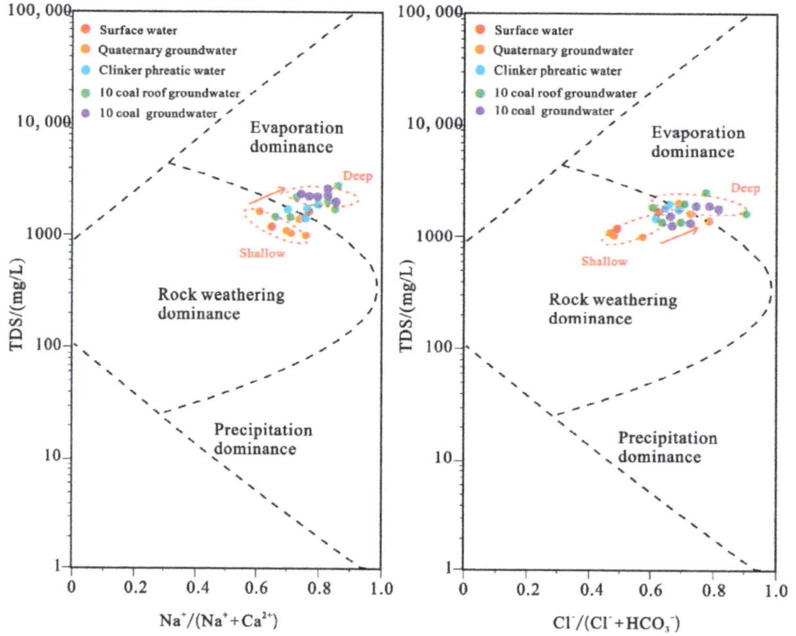

Figure 4. Gibbs chart of water samples from study area.

From the water general chemistry, it can be observed that the TDS of no.10 coal roof groundwater and no.10 coal mine water is greater than 1000 mg/L. The hydrochemical parameter correlation diagram between no.10 coal roof groundwater and major ions in no.10 coal mine water further reveals the evolution of mine water and the sources of major ions (Figure 5). The hydrochemical parameter correlation diagram illustrates that the TDS concentrations exhibited strong correlations with Na^+ (R = 0.92), SO_4^{2-} (R = 0.87), Cl^- (R = 0.75), and Ca^{2+} (R = 0.48), indicating that the contribution to the mine water TDS is, in descending order, from Na^+, SO_4^{2-}, Cl^-, and Ca^{2+}. The ionic correlation of the mine water was similar to that of the no.10 coal roof groundwater, as the no.10 coal roof groundwater is the direct filling source of the mine water, and the correlation coefficients of TDS with Na^+,

SO_4^{2-}, and Cl^- were 0.89, 0.84, and 0.73, respectively. Furthermore, it can be seen from the correlation analysis diagram of each ion in mine water that the correlation between Na^+ and Cl^-, and Ca^{2+} and SO_4^{2-} are high, and the correlation coefficient is greater than 0.5, indicating that Na and Cl^-, Ca^{2+} and SO_4^{2-} have a similar or the same source.

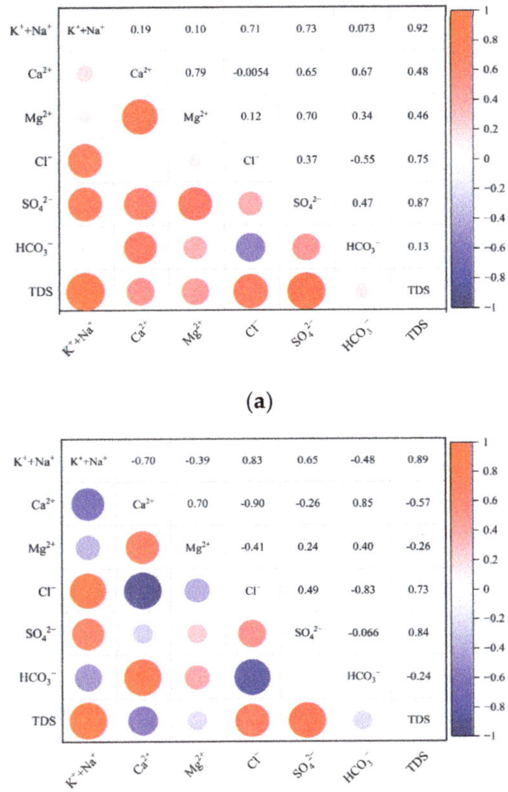

Figure 5. Correlation analysis diagram of no.10 coal roof groundwater and no.10 coal mine water. (**a**) Correlation analysis diagram of no.10 coal roof groundwater; (**b**) Correlation analysis diagram of no.10 coal mine water.

The relationship between Na^+ and Cl^- (Figure 6a) and the relationship between Ca^{2+} and SO_4^{2-} (Figure 6b) can be further analyzed for sources of major ions. The concentrations of Na^+, K^+, and Cl^- in groundwater are generally relatively stable, so the ratio of $\gamma(Na^+)/\gamma(Cl^-)$ (γ means concentration of ions) can be used to identify the sources of Na^+ and Cl^- in groundwater. When the ratio of $\gamma(Na^+)/\gamma(Cl^-)$ is near 1, it indicates that the dissolution of rock salt minerals is occurring; when the ratio of $\gamma(Na^+)/\gamma(Cl^-)$ is greater than 1, it suggests that the dissolution of silicate minerals is taking place. The no.10 coal roof groundwater and mine water are mainly distributed below the ratio line of $\gamma(Na^+)/\gamma(Cl^-)$, indicating that for Na^+ and Cl^-, in addition to the dissolution of rock salt, there was also a reverse cation adsorption process taking place.

$$NaCl \rightarrow Na^+ + Cl^- \quad (1)$$

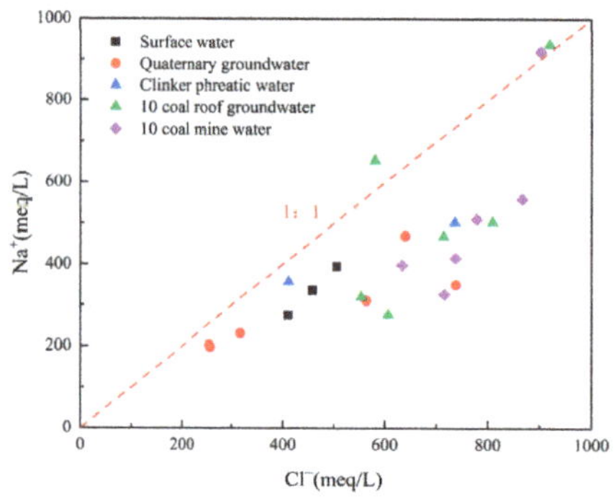

(a)

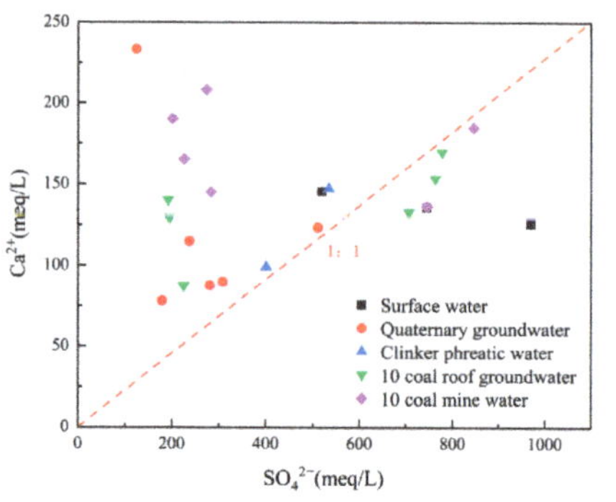

(b)

Figure 6. Main ion relationship diagram of mine water, (**a**) Na^+ vs. Cl^- ion relationship diagram, (**b**) Ca^{2+} vs. $SO_4{}^{2-}$ ion relationship diagram.

If the ratio of $\gamma(Ca^{2+})/\gamma(SO_4{}^{2-})$ is 1, this indicates that Ca^{2+} and $SO_4{}^{2-}$ only originate from the dissolution of gypsum, without involvement of any other reactions (Figure 6b). The mine water samples are mainly distributed in the upper-left region of the 1:1 line, indicating that there are other sources of Ca^{2+}, such as the dissolution of carbonate rocks or silicate rocks. Meanwhile, the correlation coefficient between Ca^{2+} and $HCO_3{}^-$ is 0.85, suggesting that Ca^{2+} not only originates from the dissolution of sulfate rocks (gypsum) but also from the dissolution of calcium feldspar:

$$CaSO_4 \cdot 2H_2O \rightarrow Ca^{2+} + SO_4^{2-} + 2H_2O \tag{2}$$

$$2CaAl_2Si_2O_8(\text{calcium feldspar}) + 4CO_2 + 6H_2O \rightarrow 2Al_2Si_2O_5(OH)_4 + 2Ca^{2+} + 4HCO_3^- \tag{3}$$

3.3. *Ecological Irrigation Risk Evaluation*

From the analysis of the water quality evolution of mine water, it can be observed that its initial sources are mainly atmospheric precipitation and surface water, while being influenced by multiple factors, including original geological deposition characteristics, water–rock interactions, and disturbances caused by coal mining activities. From multiple perspectives, the environmental risks of mine water ecological irrigation are quantitatively characterized and discussed, including the impact of untreated mine water for ecological irrigation on soil safety, plant growth safety, and groundwater safety.

3.3.1. Risks to Soil Safety Induced by Mine Water Ecological Irrigation

As an unconventional water resource for irrigation, the soil safety risks caused by mine water are quantified using four indicators: soluble sodium percentage, sodium adsorption ratio, residual sodium carbonate, and alkalinity hazard and permeability index. These indicators help characterize the soil salinization and permeability issues caused by mine water. Alkalinity hazard and permeability index.

(1) Soluble Sodium Percentage (*SSP*)

Na^+ in water can exchange with Ca^{2+} and Mg^{2+} on clay particles, thereby hindering the movement of soil moisture [24]. When the Na^+ concentration in mine water is high and the $SSP > 60\%$, as demonstrated by the red line in the figure, this water source is not suitable for irrigation [25]. The formula of *SSP* is as follows:

$$SSP = \frac{Na^+ + K^+}{Na^+ + K^+ + Ca^{2+} + Mg^{2+}} \times 100\% \tag{4}$$

where the concentrations of Na^+, K^+, Ca^{2+}, and Mg^{2+} are all measured in meq/L.

(2) Sodium Adsorption Ratio (*SAR*)

The sodium adsorption ratio (*SAR*) indicates the relative activity of Na+ in soil exchange reactions and is used to assess the degree of soil alkalization caused by irrigation water [26]. According to the standard requirements of "Urban Sewage Reuse for Green Space Irrigation Water Quality", the degree of soil alkalization, $SAR \leq 9$, will not cause soil alkalization [27]. The formula of *SAR* is as follows:

$$SAR = \frac{Na^+}{\sqrt{\frac{Ca^{2+} + Mg^{2+}}{2}}} \quad (5)$$

where the concentrations of Na^+, Ca^{2+}, and Mg^{2+} are all measured in meq/L.

(3) Residual Sodium Carbonate (*RSC*)

Residual sodium carbonate (*RSC*) is one of the key indicators to identify the degree of alkalinity damage of irrigation water on soil. When RSC exceeds 2.5 mmol/L, the water source is not suitable for irrigation. Conversely, when *RSC* is less than 1.25 mmol/L, the water source is suitable for irrigation [28,29].

$$RSC = HCO_3^- + 2CO_3^- - 2\left(Ca^{2+} + Mg^{2+}\right) \quad (6)$$

(4) Permeability Index (*PI*)

The permeability index (*PI*) is used to assess the impact of water on soil permeability. It will not affect the soil's permeability if the *PI* > 70%. However, it will impact the soil's permeability if the *PI* < 25% [30]. The formula is as follows:

$$PI = \frac{Na^+ + \sqrt{HCO_3^-}}{Na^+ + Ca^{2+} + Mg^{2+}} \quad (7)$$

where the concentrations of Na^+, Ca^{2+}, Mg^{2+}, and HCO_3^- are all measured in meq/L.

The soluble sodium percentage (*SSP*), sodium adsorption ratio (*SAR*), residual sodium carbonate (*RSC*), and permeability index (*PI*) for the no.10 coal roof groundwater and the no.10 coal mine water were calculated, respectively (Figure 7).

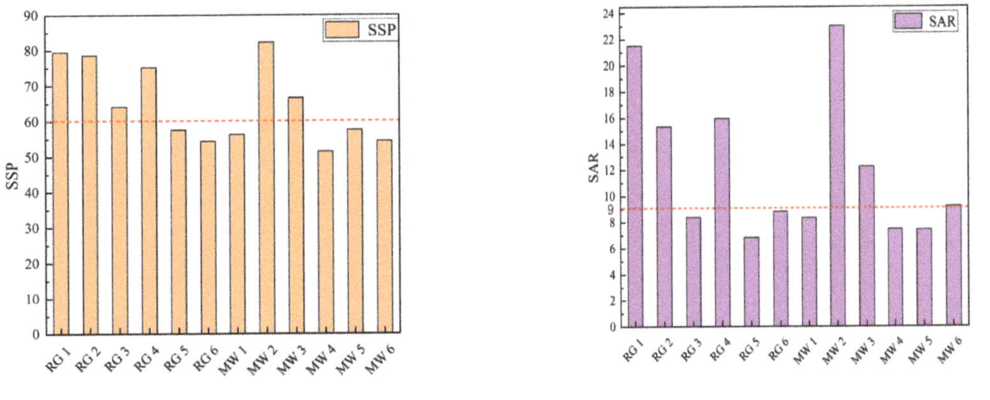

(**a**) (**b**)

Figure 7. *Cont.*

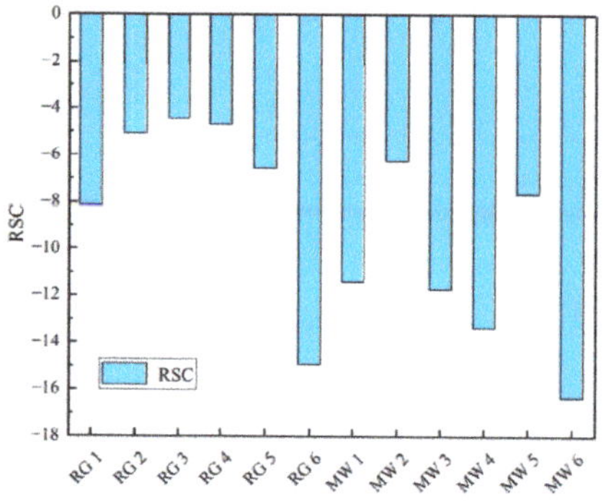

(c)

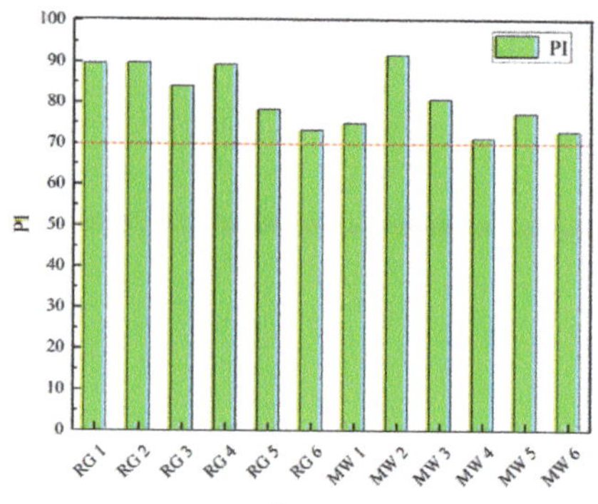

(d)

Figure 7. Soil safety risk assessment map. (**a**) Soluble sodium percentage (*SSP*) bar chart; (**b**) Sodium adsorption ratio (*SAR*) bar chart; (**c**) Residual sodium carbonate (*RSC*) bar chart; (**d**) permeability index (*PI*) bar chart.

Among the 12 mine water samples, six samples had an *SSP* greater than 60%, with an exceedance rate of 50%; five samples had an *SAR* greater than 9, with an exceedance rate of 41.67%; all 12 samples had *RSC* values less than 0, with no exceedance; and all 12 mine water samples had a *PI* greater than 70%, which will not affect the permeability of the soil.

In the comprehensive analysis, the residual sodium carbonate and soil permeability index in the soil meet the standards, and the *SSP* and *SAR* appeared to be exceeded. This caused the accumulation of Na^+ in the soil, and long-term accumulation would induce soil salinization when directly irrigating with untreated mine water.

3.3.2. Risks to Plant Growth Induced by Mine Water Ecological Irrigation

As an unconventional water resource for irrigation, three indicators (salinity hazard, magnesium adsorption ratio, and heavy metal pollution index) are used to quantitatively characterize the safety risks to plant growth caused by mine water.

(1) Salinity Hazard (*S*)

Salinity hazard (*S*) indicates the maximum amount of Cl^- and ${SO_4}^{2-}$ salts that may occur in the water. Most plant seedlings will be inhibited when $S > 10$ mmol/L [31]. The formula is as follows:

$$\begin{cases} Na^+ > Cl^- + 2SO_4^{2-}; & S = Cl^- + 2SO_4^{2-} \\ Na^+ < Cl^- + 2SO_4^{2-}; & S = Na^+ \end{cases} \quad (8)$$

where the concentrations of Na^+, Cl^-, and ${SO_4}^{2-}$ are all measured in meq/L.

(2) Magnesium Adsorption Ratio (K_{Mg})

The magnesium adsorption ratio (*MAR*) represents the amount of magnesium adsorbed by the soil, which is toxic to plants when $K_{Mg} > 50\%$ [32]. The formula is as follows:

$$K_{Mg} = \frac{Mg^{2+}}{Ca^{2+} + Mg^{2+}} \times 100\% \quad (9)$$

where the concentrations of Mg^{2+}, and Ca^{2+} are all measured in meq/L.

(3) Heavy Metal Pollution Index (*HPI*)

The total amount of heavy metals in water reflects the toxicity to plants, indicating that the level of heavy metal contamination in the water body has exceeded the maximum level tolerated by plants [33], when the $HPI > 100$ [34]. The formula is as follows:

$$HPI = \frac{\sum_{i=1}^{n} q_i w_i}{\sum_{i=1}^{n} w_i} \tag{10}$$

$$q_i = \frac{|m_i - I_i|}{(S_i - I_i)} \times 100 \tag{11}$$

where $w_i = k/S_i$: w_i is the weight of the heavy metal indicator, k is the constant of proportionality, taken as 1; q_i is the quality grade index of the heavy metal indicator; m_i is the actual concentration of heavy metals in the water (mg/L); S_i is the maximum limit value of the heavy metal indicator; and I_i is the ideal value of the heavy metal indicator. S_i and I_i are selected from the Class III groundwater limits in the "Groundwater Quality Standards" [35], which are applicable to agricultural water and centralized drinking water (mg/L). The Class III groundwater standard specifies the following limit values for Cu, Hg, As, Cd, and Pb: 1 mg/L, 0.001 mg/L, 0.01 mg/L, 0.005 mg/L, and 0.01 mg/L, respectively.

The salinity hazard (S), magnesium adsorption ratio (K_{Mg}), and heavy metal pollution index (HPI) for the no.10 coal roof groundwater and the no.10 coal mine water were calculated, respectively (Figure 8).

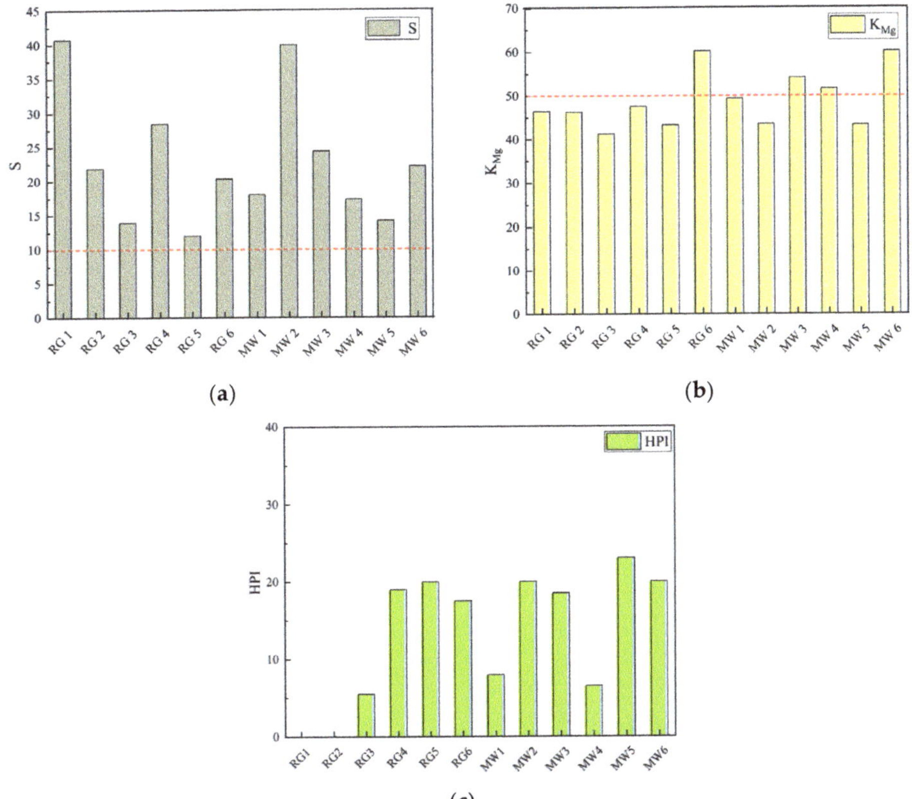

Figure 8. Plant growth risk assessment map. (**a**) Salinity hazard bar chart; (**b**) Magnesium adsorption ratio bar chart; (**c**) HPI bar chart.

The salinity hazard (S) of the water sample is greater than 10, which can easily cause disturbances in plant growth, and the high content of sulphate and chloride ions inhibited a number of physiological processes in the plant, resulting in yellowing and wilting of the leaves, slow growth, or cessation of growth [36]. Furthermore, four samples of the 12 mine water samples exceeded the magnesium adsorption ratio, which would result in increased magnesium levels in the soil leading to symptoms of burns or necrosis in plant leaves. The heavy metal indexes of Cu, Zn, Hg, As, Cd, Cr^{6+}, and Pb in the 12 mine water samples were all below the detection limit, and only Fe and Mn were detected. Meanwhile, their *HPI* value ranged from 5.5 to 23, which is far lower than the critical value of 100, and the accumulation of heavy metals does not pose a threat to plants (Table 2).

Table 2. Heavy Metal Pollution Index (*HPI*) Evaluation Index Test Results (mg/L).

Serial Number	Sample Types	Sample Name	Fe	Mn	Cu	Zn	Hg	As	Cd	Cr^{6+}	Pb	*HPI*
1	10 coal roof groundwater	10 coal roof groundwater 1	<0.08	<0.05	<0.001	<0.005	<0.00005	0.002	<0.0005	<0.005	<0.001	/
2		10 coal roof groundwater 2	<0.08	<0.05	<0.001	<0.005	<0.00005	<0.001	<0.0005	<0.005	<0.001	/
3		10 coal roof groundwater 3	<0.08	0.10	<0.001	<0.005	<0.00005	0.005	<0.0005	<0.005	<0.001	5.50
4		10 coal roof groundwater 4	0.13	0.06	<0.001	<0.005	<0.00005	0.005	<0.0005	<0.005	<0.001	19.00
5		10 coal roof groundwater 5	0.11	0.08	<0.001	<0.005	<0.00005	0.004	<0.0005	<0.005	<0.001	20.00
6		10 coal roof groundwater 6	0.13	0.09	<0.001	<0.005	<0.00005	0.001	<0.0005	<0.005	<0.001	17.50
7	10 coal mine water	10 coal mine water 1	0.22	<0.05	<0.001	<0.005	<0.00005	<0.001	<0.0005	<0.005	<0.001	8.00
8		10 coal mine water 2	0.10	<0.05	<0.001	<0.005	<0.00005	0.008	<0.0005	<0.005	<0.001	20.00
9		10 coal mine water 3	0.15	0.03	<0.001	<0.005	<0.00005	<0.001	<0.0005	<0.005	<0.001	18.50
10		10 coal mine water 4	0.24	0.09	<0.001	<0.005	<0.00005	0.002	<0.0005	<0.005	<0.001	6.50
11		10 coal mine water 5	0.09	0.06	<0.001	<0.005	<0.00005	0.002	<0.0005	<0.005	<0.001	23.00
12		10 coal mine water 6	0.12	0.06	<0.001	<0.005	<0.00005	0.002	<0.0005	<0.005	<0.001	20.00

3.3.3. Potential Risks to the Groundwater Induced by Mine Water Ecological Irrigation

As an unconventional water resource for irrigation, groundwater quality standards are used to quantitatively characterize the potential risks to the groundwater environment caused by mine water. The Nemerow Index is calculated based on the individual pollution indices of water samples to obtain a comprehensive score. It takes into account both the average and maximum values of the pollution indices for each detected parameter, with a focus on the impact and influence of the largest pollution items on the quality of the water environment. The formula is as follows:

$$F = \sqrt{\frac{I_{imax}^2 + I_{iave}^2}{2}} \tag{12}$$

where F is the Nemerow index; I_{iave} is the arithmetic mean of all individual pollution indices; I_{imax} is the maximum value among all individual pollution indices.

The Nemerow index determines the degree of mine water pollution and thus the potential impact on groundwater. When $F < 1$, the water quality is clean; $1 \leq F \leq 2$, mildly polluted; $2 < F \leq 3$, polluted; $3 < F \leq 5$, heavily polluted; $F > 5$, severely polluted. Among the 12 mine water samples, four samples have F values greater than 2, indicating mild pollution, while eight samples have F values less than 1, indicating clean water quality (Figure 9). In summary, the direct irrigation of untreated mine water poses a slight potential threat to the groundwater environment.

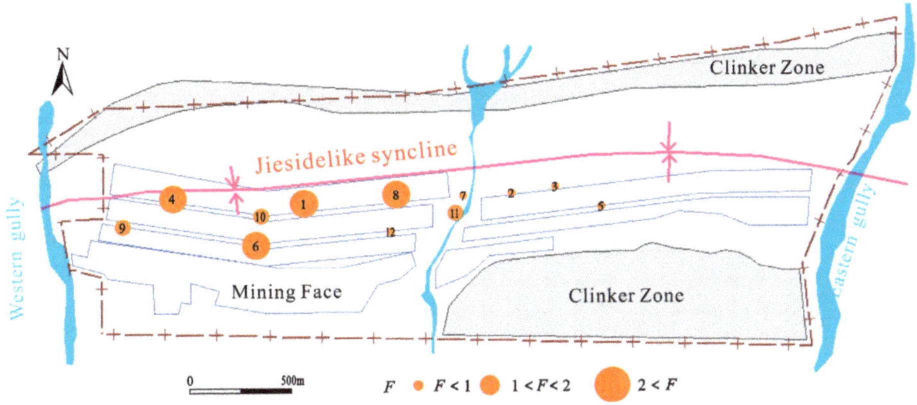

Figure 9. Distribution map of mine water Nemerow index.

3.4. Management and Preventive Measures for Using Mine Water

Risk prevention and control measures for ecological irrigation with mine water should be taken from three aspects: mine water treatment, irrigation technology, and vegetation selection. These measures are aimed at addressing the high salinity characteristics, evolutionary process, and environmental risks caused by direct irrigation of mine water in the arid regions of the western area.

In terms of mine water treatment, the high levels of TDS and major ions (Na^+, Ca^{2+}, Mg^{2+}, $SO_4{}^{2-}$) in mine water can cause soil salinization when directly irrigated. This not only threatens soil safety and affects the growth of plants but is also a potential threat to the groundwater environment. Therefore, desalination treatment of mine water is necessary. It is recommended to choose a combination of nanofiltration and reverse osmosis processes. The nanofiltration process can be used as the primary treatment stage to effectively remove sulfate ions from the mine water, reducing the load on the reverse osmosis membrane. The reverse osmosis process can further desalinate the water, removing sodium ions, and ultimately ensuring that the water quality meets the requirements for ecological irrigation.

In terms of irrigation technology, drip irrigation and micro-sprinkler irrigation systems can be used to deliver mine water directly to the plant roots, reducing water evaporation and salt accumulation on the soil surface. Furthermore, the use of mine water and groundwater intermittent irrigation techniques allows sufficient time for the soil to absorb and discharge excess salts through phased irrigation. The cyclical irrigation of mine water and groundwater provides sufficient buffer time for the soil to reduce the salt concentration in the surface soil, reducing damage to plants and potential harm to groundwater.

In terms of the selection of salt-tolerant vegetation, salt-tolerant and drought-tolerant vegetation can be selected for planting, such as Salicornia europaea and Carex scabrifolia, etc., according to the special climate of arid areas and the water quality characteristics of mine water. Their salt tolerance thresholds are 29,250 mg/L and 20,475 mg/L, respectively, which are 29.25 and 20.475 times higher than that of common plants (threshold 1000 mg/L). As a result, these plants can effectively absorb and stabilize the salts in the soil. In addition, soil structure can be improved through mixed cropping and crop rotation, reducing salt accumulation and increasing overall ecological stability and productivity.

4. Conclusions

In the Yushuquan mine, the mine water primarily originates from the coal roof groundwater, with an average pH value of 8.46, indicating weak alkalinity. The total dissolved solids (TDS) are greater than 1000 mg/L, with the main cations being Na^+ and Ca^{2+}, and the main anions being SO_4^{2-} and Cl^-. The evolution of the mine water quality is primarily influenced by water–rock interactions and evaporation concentration processes. The Na^+ and Cl^- in the mine water mainly come from the dissolution of salt rocks, while Ca^{2+} and SO_4^{2-} primarily originate from the dissolution of minerals such as gypsum and anhydrite.

The comprehensive evaluation of the untreated mine water indicates some pollution, and its direct use for ecological irrigation threatens the groundwater environment. The soluble sodium content (*SSP*) and sodium adsorption ratio (*SAR*) of the mine water exceed the recommended limits, leading to the accumulation of Na^+ in the soil and inducing soil salinization and alkalization in the long term. At the same time, the salinity hazard (S) and magnesium adsorption ratio (K_{Mg}) exceed the recommended limits, which can negatively impact plant photosynthesis and threaten plant growth. In this regard, the three methods of desalination of mine water, improvement of irrigation technology, and selection of salt-tolerant plants should be adopted to reduce the TDS of mine water, reduce the accumulation of soil salts, and select drought- and salt-tolerant vegetation, thus reducing the environmental risks of mine water ecological irrigation in arid areas.

Author Contributions: H.W.: Writing—original draft, Data curation, Funding acquisition. H.S.: Water sampling and, Formal analysis. T.W.: Writing—review & editing. J.X.: Writing—review & editing, Methodology. X.W.: Writing—review & editing. Z.Z.: Data collection and, Formal analysis. Q.W.: Writing—review & editing, Funding acquisition. All authors have read and agreed to the published version of the manuscript.

Funding: The research was financially supported by the Innovation Capability Support Program of Shaanxi (2025JC-JCQN-012, 2024RS-CXTD-44, 2025ZC-KJXX-138) and the National Key R&D Program of China (2023YFC3012105). The authors gratefully acknowledge the peer reviewers and editors for their valuable comments.

Data Availability Statement: The original contributions presented in this study are included in the article. Further inquiries can be directed to the corresponding author.

Conflicts of Interest: Authors Hao Wang, Hongbo Shang, Tiantian Wang, Jiankun Xue, Xiaodong Wang, Zhenfang Zhou and Qiangmin Wang were employed by Xi'an Research Institute of China Coal Technology & Engineering (Group). All authors declare that the research was conducted in the absence of any commercial or financial relationships that could be construed as a potential conflict of interest.

References

1. Zhang, D.S.; Li, W.P.; Lai, X.P.; Fan, G.W.; Liu, W.Q. Development on the basic theory of water protection during coal mining in northwest of China. *J. China Coal Soc.* **2017**, *42*, 36–43.
2. Wang, S.M.; Shen, Y.J.; Sun, Q.; Hou, E.K. Scientific issues of coal detraction mining geological assurance and their technology expectations in ecologically fragile mining areas of Western China. *J. Min. Strat. Control Eng.* **2020**, *2*, 043531.
3. Fan, L.M.; Ma, X.D.; Ji, R.J. Progress in engineering practice of water-preserved coal mining inwestern eco-environment frangible area. *J. China Coal Soc.* **2015**, *40*, 1711–1717.
4. Yang, J.; Wang, H.; Wang, T.T.; Wang, Q.; Liu, J. Removal law of typical pollution components during underground storage of mine water: Taking Mindong No.1 Mine Inner Mongolia as an example. *J. China Coal Soc.* **2020**, *45*, 2918–2925.
5. Gu, D.Z.; Li, J.F.; Cao, Z.G.; Wu, B.Y.; Jiang, B.Y.; Yang, Y.; Yang, J.; Chen, Y. Technology and engineering development strategy of water protection and utilization of coal mine in China. *J. China Coal Soc.* **2021**, *46*, 3079–3089.
6. Guo, Y.N. Spatial distribution characteristics of suita-bility of mine water for irrigation in the Shendong mining area. *J. Min. Sci. Technol.* **2024**, *9*, 561–572.

7. Merritt, P.; Power, C. Assessing the long-term evolution of mine water quality in abandoned underground mine workings using first-flush based models. *Sci. Total Environ.* **2022**, *846*, 157390. [CrossRef]
8. Zhang, J.; Chen, L.W.; Chen, Y.F.; Ge, R.T.; Ma, L.; Zhou, K.D.; Shi, X.P. Discrimination of water-inrush source and evolution analysis of hydrochemical environment under mining in Renlou coal mine, Anhui Province, China. *Environ. Earth Sci.* **2020**, *79*, 61. [CrossRef]
9. Chen, L.W.; Xie, W.P.; Feng, X.Q.; Zhang, N.; Yin, X. Formation of hydrochemical composition and spatio-temporal evolution mechanism under mining-induced disturbance in the Linhuan coal-mining district. *Arab. J. Geosci.* **2017**, *10*, 57. [CrossRef]
10. Gui, H.R.; Song, X.M.; Lin, M.L. Water-inrush mechanism research mining above karst confined aquifer and applications in North China coalmines. *Arab. J. Geosci.* **2017**, *10*, 180. [CrossRef]
11. Wang, T.T.; Zhao, W.; Wang, Z.H.; Zhou, Z.; Yang, J.; Xu, F.; Xue, J.; Li, G. Occurrence, Main Source and Health Risks of Fluorine in Mine Water. *Expo Health* **2025**, *17*, 279–292. [CrossRef]
12. Sun, Y.J.; Zhang, L.; Xu, Z.M.; Chen, G.; Zhao, X.; Li, X.; Gao, Y.; Zhang, S.; Zhu, L. Mutifield action mechanism and research progress of coal mine water quality formation and evolution. *J. China Coal Soc.* **2022**, *47*, 423–437.
13. Roisenberg, C.; Loubet, M.; Formoso, M.L.; Berger, G.; Munoz, M.; Dani, N. Tracing the Origin and Evolution of Geochemical Characteristics of Waters from the Candiota Coal Mine Area (Southern Brazil): Part I. *Mine Water Environ.* **2016**, *35*, 29–43. [CrossRef]
14. Zhang, Y.J.; Cui, B.Y.; Wang, Y.N.; Zhang, S.; Feng, G.; Zhang, Z. Evolution Law of Shallow Water in Multi-Face Mining Based on Partition Characteristics of Catastrophe Theory. *Fractal Fract.* **2023**, *7*, 779. [CrossRef]
15. Johnson, D.B.; Hallberg, K.B. Acid mine drainage remediation options: A review. *Sci. Total Environ.* **2005**, *338*, 3–14. [CrossRef]
16. Batarseh, M.; Imreizeeq, E.; Tilev, S.; Al Alaween, M.; Suleiman, W.; Al Remeithi, A.M.; Al Tamimi, M.K.; Al Alawneh, M. Assessment of groundwater quality for irrigation in the arid regions using irrigation water quality index (IWQI) and GIS-Zoning maps: Case study from Abu Dhabi Emirate, UAE. *Groundw. Sustain. Dev.* **2021**, *14*, 100611. [CrossRef]
17. Yıldız, S.; Karakuş, C.B. Estimation of irrigation water quality index with development of an optimum model: A case study. *Environ. Dev. Sustain.* **2020**, *22*, 4771–4786. [CrossRef]
18. Shi, J.F.; Wu, P.; Zhang, R.X.; Li, X.X.; Cha, X.F. Water quality characteristics and irrigation suitability evaluation under the influence of coal mine drainage–a case study of the river watershed in Zhijin, Guizhou. *Water Sav. Irrig.* **2016**, *11*, 71–76.
19. Ni, S.H.; Wang, H.L. Risk assessment and control of farmland irrigation with coal mine water. *Coal Econ. Res.* **2021**, *41*, 48–53.
20. Chen, S.; Gui, H.R. Hydrogeochemical characteristics of groundwater in the limestone aquifers of the Taiyuan Group and its geological significance in the Suxian mining area. *Hydrogeol. Eng. Geol.* **2016**, *43*, 33–41.
21. Wang, T.T.; Yang, J.; Jin, D.W.; Li, G.; Zhou, Z.; Xue, J.; Shang, H. The hydrogeochemical characteristics and formation mechanism of high-fluoride mine water. *J. Clean. Prod.* **2023**, *430*, 139671. [CrossRef]
22. Piper, A.M. A graphical procedure in the geochemical interpretation of wateranalysis. Eos Trans. *Am. Geophys. Union* **1944**, *25*, 914–928.
23. Gibbs, R.J. Mechanisms controlling world water chemistry. *Science* **1970**, *170*, 795–840. [CrossRef]
24. Anbazhagan, S.; Jothibasu, A. Modeling water quality index to assess shallow groundwater quality for sustainable utilization in Southern India. *Int. J. Adv. Geosci.* **2014**, *2*, 122. [CrossRef]
25. Mao, M.; Zhu, X.Q. Chemical characteristics of groundwater in Xuanhua basin and assessment of irrigation applicability. *J. Arid Land Resour. Environ.* **2020**, *34*, 142–149.
26. *GB/T 25499-2010*; Water Quality Standard of Green Land Irrigation for Urban Sewage Reclamation. General Administration of Quality Supervision, Inspection and Quarantine of the People's Republic of China, Standards Press of China: Beijing, China, 2010.
27. Jandu, A.; Malik, A.; Dhull, S.B. Fluoride and nitrate in groundwater of rural habitations of semiarid region of northern Rajasthan, India: A hydrogeochemical, multivariate statistical, and human health risk assessment perspective. *Environ. Geochem. Health* **2021**, *43*, 3997–4026. [CrossRef] [PubMed]
28. Xie, F. *The Groundwater Quality Assessment and Response of Different Irrigation Water Sources to Groundwater in Jinghuiqu Irrigation District*; Northwest A&F University: Xianyang, China, 2016; pp. 32–36.
29. Yidana, M.S. Groundwater classification using multivariate statistical methods: SouthernGhana. *J. Afr. Earth Sci.* **2010**, *57*, 455–469. [CrossRef]
30. Raju, N.J.; Shukla, U.K.; Ram, P. Hydrogeochemistry for the assessment of groundwater quality in Varanasi:a fast-urbanizing center in Uttar Pradesh, India. *Environ. Monit. Assess.* **2011**, *173*, 279–300. [CrossRef]
31. Nag, S.K.; Das, S. Assessment of groundwater quality from Bankura I and II Blocks, Bankura District, West Bengal, India. *Appl. Water Sci.* **2017**, *7*, 2787–2802. [CrossRef]
32. Amrani, S.; Hinaje, S.; El Fartati, M.; Gharmane, Y.; Yaagoub, D. Assessment of groundwater quality for drinking and irrigation in the Timahdite-Almis Guigou Area (Middle Atlas, Morocco). *Appl. Water Sci.* **2022**, *12*, 82. [CrossRef]
33. Wang, T.T.; Jin, D.W.; Yang, J. Heavy metal pollution characteristics and source analysis of water drainage from a mine in Inner Mongolia. *Coal Geol. Explor.* **2021**, *49*, 45–51.

34. Singha, S.; Pasupuleti, S.; Singha, S.S.; Kumar, S. Effectiveness of groundwater heavy metal pollution indices studies by deep-learning. *J. Contam. Hydrol.* **2020**, *235*, 103718. [CrossRef] [PubMed]
35. *GB/T 14848-2017*; Standard for Groundwater Quality. National Standard of the People's Republic of China; General Administration of Quality Supervision, Inspection and Quarantine of the People's Republic of China, Standards Press of China: Beijing, China, 2017.
36. Gao, M.X. *Effects of Sulfur Supply on Cadmium up Take and Accumulation in Rice Seedlings*; Northwest A&F University: Xianyang, China, 2009.

water

MDPI

Article

Numerical Simulation Study of Rainfall-Induced Saturated–Unsaturated Landslide Instability and Failure

Zhuolin Wu [1], Gang Yang [1], Wen Li [1], Xiangling Chen [1], Fei Liu [2,*] and Yong Zheng [3]

[1] Chongqing Survey and Design Institute of Water Resources, Electric Power and Architecture Co., Ltd., Chongqing 401120, China; 15923047773@163.com (Z.W.); 13658387758@163.com (G.Y.); liwen880827@163.com (W.L.); 17623003057@163.com (X.C.)
[2] School of Civil and Transportation Engineering, Henan University of Urban Construction, Pingdingshan 467000, China
[3] College of River and Ocean Engineering, Chongqing Jiaotong University, Chongqing 400074, China; zhengyong@cqjtu.edu.cn
* Correspondence: liufei1991c@163.com

Abstract

Rainfall infiltration is a key factor affecting the stability of the slope. To study the impact of rainfall on the instability mechanism and stability of slopes, this paper employs numerical simulation to establish a rainfall infiltration slope model and conducts a saturated–unsaturated slope flow and solid coupling numerical analysis. By combining the strength reduction method with the calculation of slope stability under rainfall infiltration, the safety factor of the slope is obtained. A comprehensive analysis is conducted from the perspectives of the seepage field, displacement field and other factors to examine the impact of heavy rainfall patterns and rainfall intensities on the instability mechanism and stability of the slope. The results indicate that heavy rainfall causes the transient saturation zone within the landslide body to continuously move upward, forming a continuous sliding surface inside the slope, which may lead to instability and sliding of the soil in the upper part of the slope toe. The heavy rainfall patterns significantly affect the temporal and spatial evolution of pore water pressure, displacement and safety factors of the slope. Pore water pressure and displacement show a positive correlation with the rainfall intensity at various times during heavy rainfall events. The pre-peak rainfall pattern causes the largest decrease in the safety factor of the slope, and the slope failure occurs earlier, which is the most detrimental to the stability of the slope. The rainfall intensity is inversely proportional to the safety factor. As the rainfall intensity increases, the decrease in the slope's safety factor becomes more significant, and the time required for slope instability is also shortened. The results of this study provide a scientific basis for analyzing rainfall-induced slope instability and failure.

Keywords: rainfall infiltration; slope; strength reduction method; safety factor

1. Introduction

Landslides are the second most significant natural disaster after earthquakes, causing hundreds of billions of dollars in economic losses and serious casualties worldwide each year [1,2]. The occurrence of landslides is closely related to natural factors and human activities, with rainfall being one of the major influencing factors [3,4]. Rainfall, especially heavy rainfall, can cause the slope materials to (1) increase in moisture content and decrease in matric suction [5,6]; (2) experience a reduction in negative pore water pressure (PWP) and effective stress; (3) present an increase in self-weight and a decrease in shear strength [7,8]. Furthermore, continuous intense rainfall can cause rapid increases in water head within fractures, leading to greater driving forces for slope failure and thereby triggering slope instability [9,10]. Figure 1 shows the relationship between the monthly occurrence frequency of landslides and the average monthly rainfall in the Chongqing region of China. The monthly landslide occurrence frequency is positively correlated with the average monthly rainfall, indicating that the rainwater infiltration effect is a significant trigger for landslides in this area. Therefore, studying the stability of saturated–unsaturated slopes under intense rainfall infiltration has significant practical importance.

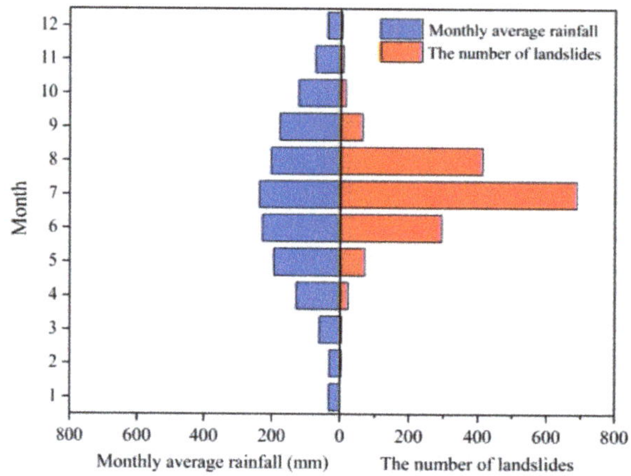

Figure 1. The relationship between the number of landslides and the average monthly precipitation in the Chongqing area.

Considering the significant impact of rainfall infiltration on landslide stability, researchers both domestically and internationally have conducted extensive studies on this issue. These studies primarily focus on the relationships between rainfall and the strength of landslide materials, pore water pressure and landslide failure mechanisms. Sun et al. [11] conducted a simulated analysis of landslide stability changes under rainfall using Geo-studio 2018 in conjunction with indoor model experiments. They discovered that landslide stability is closely related to the depth of rainfall infiltration. Kafle et al. [12] conducted a transient hydrological analysis to examine the effects of rainfall and reservoir water level fluctuations on the characteristics of slope masses. They found that rainfall increases pore water pressure and reduces shear strength, making it the primary cause of ongoing deformation in slopes, while reservoir water level fluctuations are the main cause of sudden displacements in slopes. Sun et al. [13] conducted numerical simulations based on field monitoring results, concluding that changes in pore water pressure at the sliding surface caused by rainfall infiltration are key factors in landslide initiation. By monitoring relevant data on rainfall-induced instability of fill slopes, they identified that the reduction in matrix suction, which leads to decreased shear strength at the sliding surface, is a critical factor for landslide triggering. Reid et al. [14] investigated the relationship between rainfall amounts and the probability of slope instability. They found that when monthly rainfall reaches or exceeds 300 mm and heavy rainfall exceeds 250 mm, the probability of landslide instability is significantly higher. Rahardjo et al. [15] found that slope stability and pore water pressure are highly sensitive to prior rainfall, with prolonged effects. Regmi et al. [16] conducted model tests on the failure characteristics of gently inclined slopes, revealing a certain pattern between rainfall intensity, the initiation time of sliding and the location of the slip at the base of the slope. Collins and Znidarcic [17], through numerous rainfall-induced landslide cases and theoretical analysis, identified two types of failure modes for rainfall-induced landslides. One mode occurs when the infiltration rate of the slope's soil and rock is high, leading to rapid development of pore water pressure and quick advancement of the wetting front. In this case, the soil and rock are more likely to reach saturation quickly. Under rainfall conditions, this type of slope experiences shallow instability, as permeability typically plays a dominant role at this time. The other mode occurs when the infiltration rate of the slope's soil and rock is low, resulting in a slow development of pore water pressure and a gradual advancement of the wetting front. In this case, the soil and rock are less likely to reach saturation. Under rainfall conditions, this type of landslide experiences deeper instability and failure, as the reduction in matrix suction leads to a decrease in the shear strength of the slip surface. Tohari et al. [18] confirmed this through laboratory tests, demonstrating that during the rainfall infiltration process, the unstable area of the slope gradually expands, and instability can be induced without the soil and rock reaching saturation. Xie et al. [19–21] found that the slope deformation in active rotating landslides is mainly attributed to the sliding or rotation of the landslide mass along the sliding surface via experiments and field tests. The pre-failure tilt behavior in rain-induced rotating landslides can be predicted through slope tilt measurement. In recent years, deep learning has been widely applied in the field of slope seepage [22,23]. By using intelligent algorithms to analyze and process rainfall-induced landslide monitoring data, effective predictions of instability and failure have been achieved. This also provides new insights for the stability analysis of rainfall-induced slopes.

As mentioned above, it is clear that previous scholars have recognized the importance of rainfall infiltration for slope stability. Rainfall not only raises the groundwater level, creating a transient saturation zone in the soil and rock above the groundwater table, but it also increases the pore water pressure in the corresponding areas. Due to the influence of factors such as the initial moisture content of the soil and rock, infiltration capacity, type of rainfall, duration of rainfall and rainfall intensity, the process of rainfall infiltration in saturated–unsaturated slopes is highly complex and uncertain. Therefore, revealing the comprehensive and accurate impact of rainfall infiltration on slope instability and failure is a topic worthy of in-depth research. This study takes a typical landslide in the Three Gorges Reservoir area of southwest China as an example to establish a numerical simulation model. It considers the interaction between unstable rainfall infiltration and non-linear deformation to conduct a fluid–solid coupling analysis of saturated–unsaturated slopes. By analyzing the effects of different types and intensities of rainfall on slope instability and failure from multiple perspectives, including the seepage field, displacement field and safety factor (*SF*), this study aims to reveal the intrinsic patterns of slope instability and failure under rainfall infiltration.

2. Numerical Methodology

2.1. Geometry and Boundary Conditions

The established slope model is shown in Figure 2, where the length of the slope crest (BA) is 20 m; the slope angle θ is 40°; the height of the right slope (FA) is 30 m; the height of the left slope (ED) is 10 m; and the length of the slope base (EF) is 55 m. The landslide model is discretized into 4-node Lagrangian elements with degrees of freedom for displacement and PWP (C3D8P) for coupled flow analysis. The slope is meshed with an element size of 0.5 m, resulting in a total of 25,124 quadrilateral elements in the model area. The constitutive model for the soil is based on ideal elastoplasticity, and the yield criterion follows the Mohr–Coulomb strength criterion. The calculations determine the stress, displacement and plastic strain for all nodes within the model. Rainwater infiltration occurs perpendicular to the slope crest surface (i.e., the BA surface) by applying surface pore fluid in ABAQUS, with an infiltration rate equal to the rainfall intensity q. For the slope surface (i.e., the BC surface), the corresponding infiltration rate is equal to $q\cos\theta$. It should be noted that no other external loads are applied to the slope. It should be noted that a landslide with multiple-material domains or complex geometries may be more realistic, but this would also increase the complexity of finite element simulations. Herein, a simplified finite element model is used for the interpretation of results. Similar simplifications of the numerical model can also be found in Gu et al. [24], Cho et al. [25] and Cheng et al. [26].

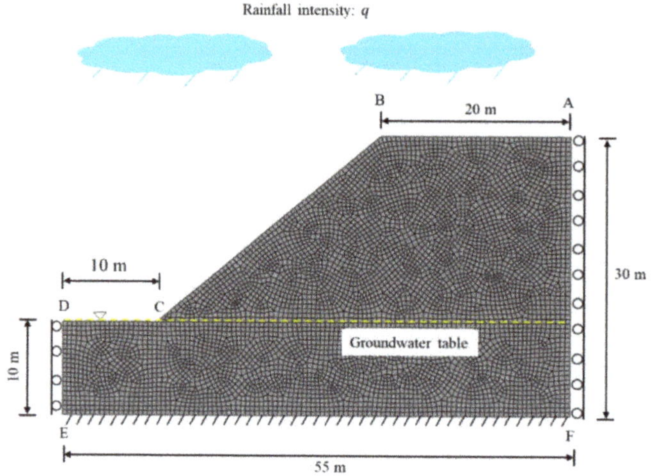

Figure 2. Numerical model.

2.2. Analysis Step

The current fluid–solid coupling numerical analysis of rainfall-induced slope stability is conducted. This analysis consists of three stages: the steady-state analysis step, the transient analysis step and the static analysis step. In the steady-state analysis step, the fluid velocity and volume do not change over time, allowing for the balance of self-weight stresses in the slope before rainfall, which establishes the initial state of the slope. In the transient analysis step, the variations in PWP, displacement and saturation concerning rainfall duration can be obtained, providing the slope conditions at different rainfall times. Subsequently, a general static analysis step is used to conduct a stability analysis of the slope after rainfall, during which the strength reduction method (SRM) is applied to calculate the *SF* of the slope.

2.3. Model Parameters

The slope is assumed to be homogeneous with identical properties throughout the entire numerical domain. The parameters are summarized in Table 1. It is important to note that, there has been no experimental research on the properties of unsaturated soil in landslides within the Three Gorges Reservoir area. In this context, the soil properties of the unsaturated slope are based on the studies conducted by Cho et al. [24] and Gu et al. [25].

To reflect the influence of the stress in the slope soil on the infiltration process, material properties are assigned to the numerical model. The seepage parameters of the slope soil under intermittent rainfall are calculated to assess the impact of the infiltration process on the principal stresses over time in the elements. The pore pressure function is given by Equation (1):

$$k_w = \frac{a_w k_{ws}}{a_w + [b_w(u_a - u_w)]^{c_w}} \quad (1)$$

where k_w is the permeability coefficient of the soil; k_{ws} is the permeability coefficient of the soil when it is saturated, with a value of 5.0×10^{-6} m/s; u_a and u_w represent the air pressure and water pressure in the soil, respectively. Since the slope surface is in contact with the atmosphere, u_a is set to 0. a_w, b_w and c_w are the soil coefficients, with values of 1000, 0.01 and 1.70, respectively.

Table 1. Soil parameters of the unsaturated slope.

Soil Parameters	Values
Effective cohesion c' (kPa)	15
Effective friction angle θ' (°)	30
Young's modulus E (MPa)	100
Unit weight γ (kN/m^3)	19.5
Poisson's ratio v	0.3
Saturated hydraulic conductivity ks (m/s)	2.8×10^{-6}

The relationship between the saturation of the soil material and the matric suction is given by Equation (2):

$$s_r = s_i + \frac{(s_n - s_i)a_s}{a_s + [b_s(u_a - u_w)]^{c_s}} \quad (2)$$

where s_r is the saturation; s_i is the residual saturation, with a value of 0.08; s_n is the maximum saturation, and its value is 1; a_s, b_s and c_s are material coefficients, with values of 1, 5.0×10^{-5} and 3.5, respectively.

By using Equations (1) and (2), the change in the permeability coefficient with saturation and the moisture retention curve can be calculated, as shown in Figures 3 and 4. It is clear that soil saturation increases with the increase in the permeability coefficient, while the matric suction of the soil increases as the saturation decreases.

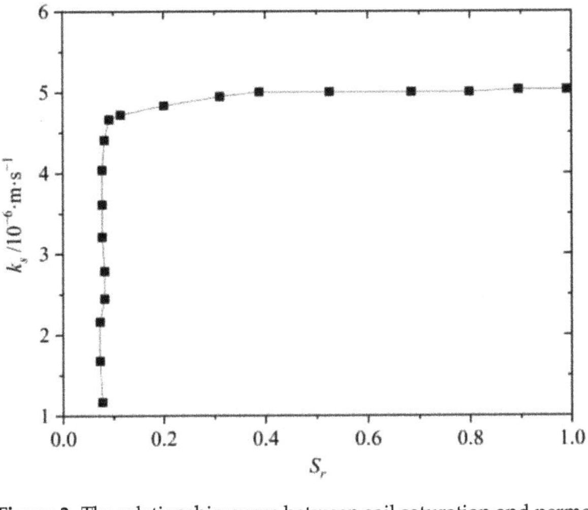

Figure 3. The relationship curve between soil saturation and permeability coefficient.

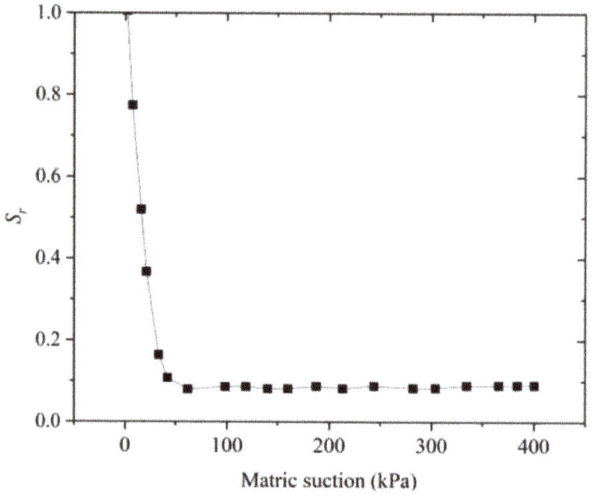

Figure 4. The relationship curve between matric suction and soil saturation.

2.4. Simulation Scheme

This study primarily investigates the effects of rainfall patterns, rainfall intensity and other parameters on the seepage and stability of landslides. The specific numerical analysis scheme is presented in Table 2. Four different rainfall patterns are selected: uniform, triangular, pre-peak and post-peak, as shown in Figure 5. The uniform pattern maintains a constant intensity throughout the rainfall duration (see Figure 5a). The triangular pattern features an intensity that initially increases over time, reaching a maximum during the mid-rainfall period, after which the intensity gradually decreases (see Figure 5b). The pre-peak rainfall pattern has an intensity that gradually decreases over time (see Figure 5c), while the post-peak pattern exhibits an intensity that gradually increases (see Figure 5d). To better reveal the impact of rainfall patterns on the mechanisms of slope instability and stability, the cumulative rainfall for the four patterns is equal for rainfall.

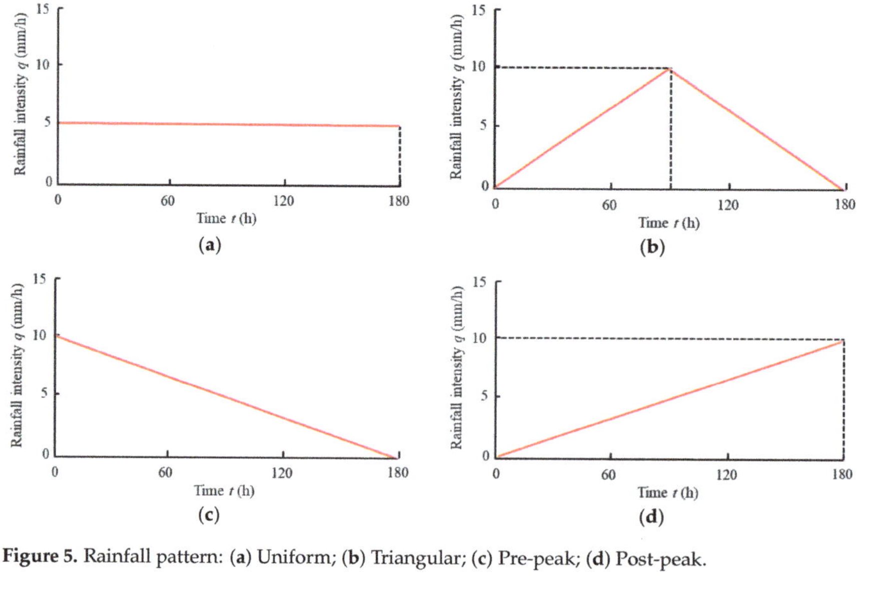

Figure 5. Rainfall pattern: **(a)** Uniform; **(b)** Triangular; **(c)** Pre-peak; **(d)** Post-peak.

Table 2. Numerical analysis scheme.

Scheme	Rainfall Intensity q (mm/h)	Rainfall Duration t (h)	Rainfall Patterns	Influencing Factors
Scheme 1	10	90	Uniform	Rainfall
Scheme 2	10	90	Uniform, triangular, pre-peak, post-peak	Rainfall patterns
Scheme 3	10, 5, 2	90	Uniform	Rainfall intensity

According to local rainfall data and meteorological rainfall classification standards [27,28], the average annual number of rainy days in the region over the past 10 years is 288 days, with an average annual precipitation of 914.6 mm. Most of the rainfall occurs between June and September. In the past 5 years, several instances of intense short-duration rainfall in the area have exceeded 40 mm/h, with 24 h rainfall reaching up to 250 mm. The longest duration of intermittent rainfall can last up to 10 days. Therefore, to study the effects of rainfall intensity on slope seepage and stability, three common rainfall levels are selected: moderate rain, heavy rain and torrential rain, with the corresponding q being 2 mm/h, 5 mm/h and 10 mm/h, respectively. The rainfall pattern is uniform, and the rainfall duration time t is 90 h.

3. Results and Discussion

3.1. Analysis of Slope Stability Under Rainfall

To investigate the extent and degree of the impact of heavy rainfall on slope instability and failure, a fluid–solid coupling analysis of the entire process of rainfall infiltration in the slope is conducted. The rainfall pattern is uniform, with the q considered under extreme conditions set at 10 mm/h, lasting for 90 h, resulting in a total precipitation of 900 mm by the end of the rainfall.

Figure 6a shows the distribution of pore water pressure (PWP) in the slope before rainfall (i.e., $t = 0$ h). The dashed line represents the wetting front, which is the groundwater level under steady-state conditions and serves as the boundary between the saturated and unsaturated zones of the slope soil. The PWP values above the dashed line are negative, indicating the presence of an unsaturated zone, which is beneficial for slope stability. As the height of the slope decreases, the magnitude of PWP increases linearly.

Figure 6b shows the PWP distribution map of the landslide after 90 h of rainfall. It can be seen in this figure that the difference in PWP in the unsaturated zone above the groundwater level before and after rainfall is significant. The variation in PWP is greatest in the soil at the top of the slope, followed by the slope face, while the variation is smallest in the soil at the bottom of the slope. Rainfall can cause a temporary increase in negative PWP in the originally unsaturated zone of the slope, leading to a closed phenomenon where the pressure head contour lines from the top of the slope move inward from high (zero) to low and back to high. This occurs because the shallow soil in the landslide responds first to the rainfall, while the deeper soil responds more slowly. Additionally, after 90 h of rainfall, the wetting front within the landslide body continuously rises toward the surface, with the wetting front at the toe of the slope being the first to rise slowly. This indicates that the transient saturated zone will continuously move upward, initially affecting the toe of the slope. Therefore, when the slope becomes unstable, the toe is the first to experience failure.

Figure 7 shows the distribution of accumulated equivalent plastic strain (PEEQ) before and after rainfall. The location and magnitude of PEEQ serve as important criteria for assessing slope instability and failure. In the figure, it can be seen that after rainfall infiltration, the areas of increased shear strain in the slope shift compared to before the rainfall. At this point, a continuous slip surface has formed, with the lower part of the slip surface connecting to the toe of the slope and the upper part reaching the crest. The PEEQ occurs on the upper surface of the slope toe, indicating that the soil in the upper part of the slope toe may experience instability and sliding under the influence of rainfall infiltration. As rainfall fully infiltrates, the surface layer of the soil gradually approaches saturation, and the values of PEEQ increase sharply, indicating that local failure may have occurred under these conditions.

In practical engineering, the *SF* is commonly used to evaluate slope stability [20,21]. This paper employs the SRM using Abaqus finite element analysis to calculate the *SF* of the slope. The procedure for determining the *SF* via SRM can be outlined in the following steps:

(1) Define the field variable, typically the strength reduction factor (SRF).
(2) Specify the material properties that change with the field variables.
(3) Establish the boundary conditions, and achieve numerical equilibrium.
(4) Adjust and increase the field variable (i.e., SRF) until the numerical calculation fails to converge.

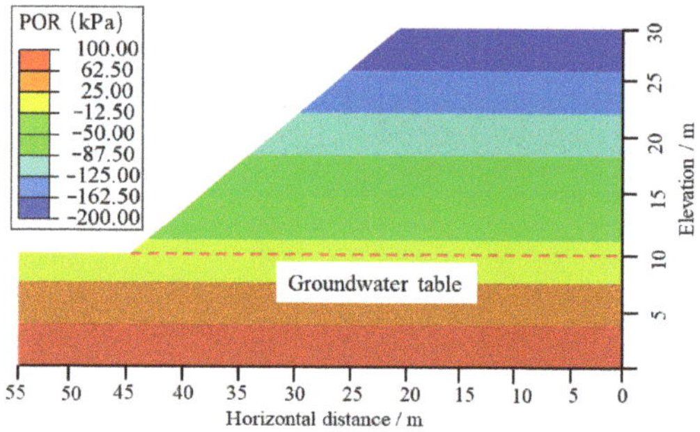

(a)

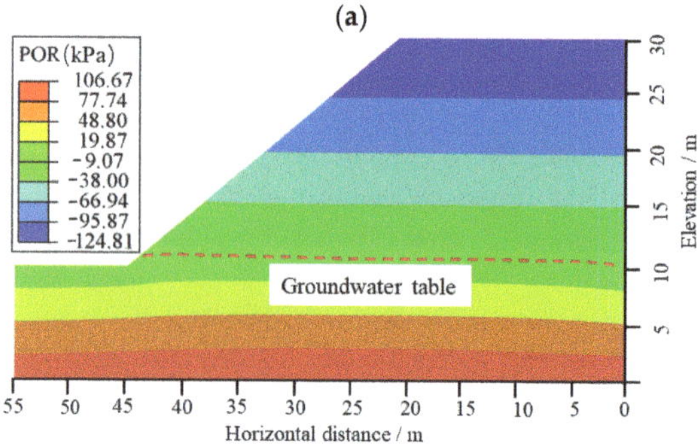

(b)

Figure 6. PWP zone distribution of landslide for (**a**) t = 0 h; (**b**) t = 90 h.

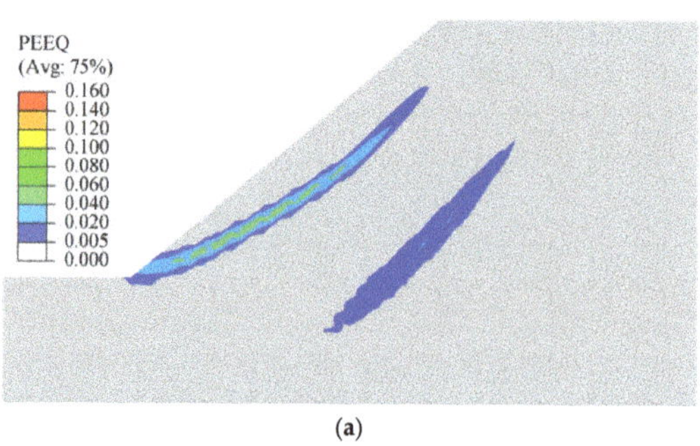

(a)

Figure 7. *Cont.*

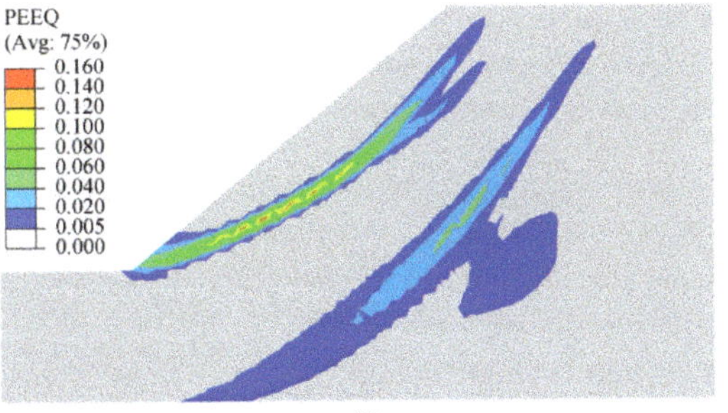

(**b**)

Figure 7. Plastic zone distribution of landslide for (**a**) $t = 0$ h; (**b**) $t = 90$ h.

The SRM for calculating the *SF* of the ABAQUS has been utilized by many scholars and experts for landslide stability analysis.

Figure 8 shows the variation in the slope safety factor with rainfall duration. It can be observed in the figure that during prolonged continuous rainfall, the slope safety factor gradually decreases with the duration of rainfall. The *SF* of the slope before rainfall is 1.23, indicating a stable condition. Under heavy rainfall intensity ($q = 10$ mm/h), the *SF* of the slope gradually decreases with the duration of rainfall. The variation in the *SF* is more significant during the early phase of rainfall, while the changes are smaller in the later phase. This is because during the early phase of rainfall, the water content in the soil is low. As rainfall infiltrates, there is a significant change in the water content of the slope, leading to a substantial increase in PWP. This greatly reduces the soil's matric suction, resulting in a significant decrease in the soil's shear strength, which causes the *SF* to decrease rapidly. After a period of continuous rainfall, once the water content in the upper soil layer of the slope increases to a certain level, it can no longer accommodate additional rainfall infiltration. The permeability coefficient of the slope soil also decreases accordingly, and at this point, the *SF* no longer decreases significantly.

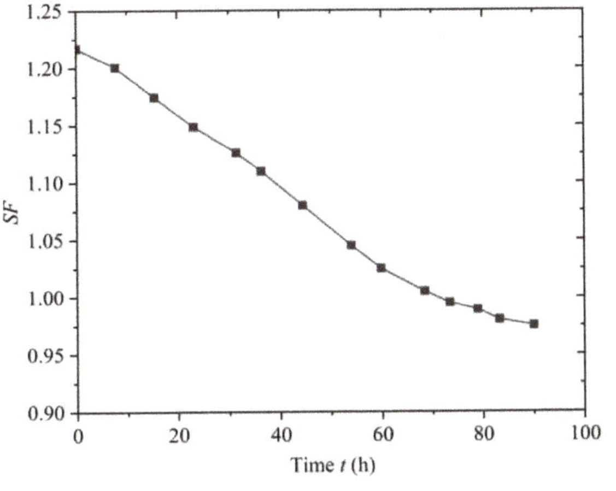

Figure 8. The relationship curve between the *SF* of the slope and the duration of rainfall.

3.2. The Impact of Rainfall Patterns on Slope Instability and Failure

Figure 9 illustrates the distribution of PWP within the slope after 90 h of rainfall under different rainfall patterns. It can be observed that, with the same total rainfall amount and duration, the distribution of PWP within the slope varies with different rainfall patterns, showing differences in PWP changes. Based on the variations in the contour lines of PWP, it is evident that the pre-peak rainfall has a relatively small impact on the PWP within the slope, while the post-peak rainfall has a significantly larger effect on the PWP of the slope soil.

To further investigate the influence of rainfall patterns on PWP distribution in the slope, Figure 10 presents the temporal and spatial evolution of PWP at the slope-free face (i.e., the BC face) above the groundwater level under four different rainfall patterns. In Figure 10, it is evident that rainfall patterns have a significant impact on the distribution of PWP at the slope-free face. After 24 h of rainfall, the order of influence of the four rainfall patterns is as follows: pre-peak > uniform > triangular > post-peak. This is primarily because pre-peak represents a sudden heavy rainfall, which increases the PWP at the slope-free face due to its rapid intensity, while post-peak has a lower initial rainfall intensity, resulting in the smallest impact on the PWP at the slope-free face. After 60 h of rainfall, the influence changes to triangular > uniform > post-peak > pre-peak. This is mainly because the rainfall intensity of triangular sharply increases during this period, becoming significantly greater than the other three types of rainfall. As a result, the PWP at the slope-free face is higher than that corresponding to the other rainfall types at the same time. After the rainfall ends (i.e., after 90 h), the order of influence is post-peak > uniform > triangular > pre-peak. This is primarily because the rainfall intensity of post-peak continues to increase over time, leading to a rise in PWP at the slope-free face. In contrast, pre-peak experiences a decrease in rainfall intensity, allowing PWP to dissipate over time, making it the lowest among all rainfall patterns. Overall, under all rainfall conditions, the variation in PWP at the slope-free face aligns closely with elevation, with higher PWP corresponding to rainfall patterns with greater intensity at the same time.

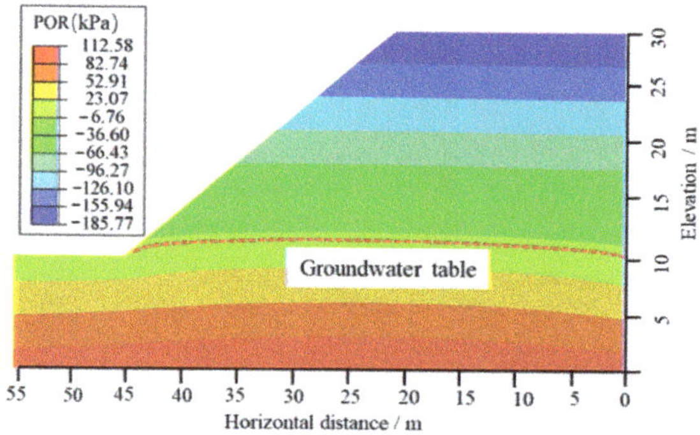

(a)

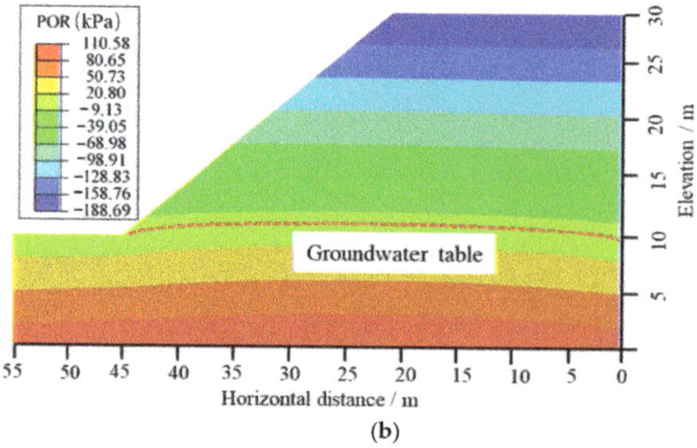

(b)

Figure 9. *Cont.*

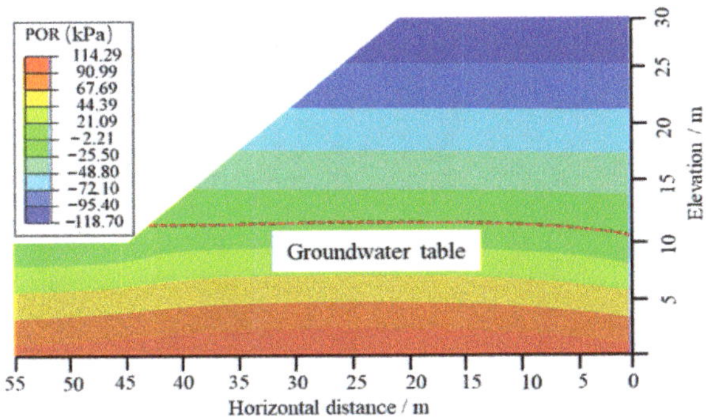

(c)

Figure 9. PWP zone distribution of landslide for (**a**) Triangular; (**b**) Pre-peak; (**c**) Post-peak rainfall.

Figure 11 shows the time-series curves of maximum horizontal deformation $U_{1,\ max}$ for the landslide under different rainfall patterns. It is important to note that the $U_{1,\ max}$ refers to the highest deformation value at various locations within the landslide at a specific moment of rainfall. In this figure, it can be seen that, for the uniform and post-peak rainfall patterns, the $U_{1,\ max}$ increases with the duration of rainfall. In contrast, for the triangular and pre-peak rainfall patterns, the $U_{1,\ max}$ initially increases and then decreases as rainfall time progresses. Due to the differing concentrated rainfall periods of the four rainfall patterns, the timing of the maximum values of their respective $U_{1,\ max}$ also varies. For the uniform intensity and post-peak rainfall patterns, since their rainfall intensity does not decrease over time, the $U_{1,\ max}$ for both occurs at the end of the rainfall, reaching 25 mm and 36 mm, respectively. The $U_{1,\ max}$ for the triangular rainfall pattern is 27 mm, occurring at 58 h of rainfall. The $U_{1,\ max}$ for the pre-peak rainfall pattern is 36 mm, occurring at 38 h of rainfall. Subsequently, due to the decrease in rainfall for these two patterns, the amount of infiltrated water within the slope decreases, combined with drainage from the slope, resulting in a slight rebound in horizontal displacement.

Figure 12 presents the slope displacement cloud map at the time of maximum horizontal displacements under four rainfall patterns. In the figure, it can be observed that when the maximum horizontal displacements are reached for all four patterns, the horizontal displacements are concentrated at the bottom of the slope. It can be inferred that during the slope failure process, deformation will first occur at the bottom of the slope, leading to a loss of support in the upper and middle sections. Under the effects of rainfall and self-weight, sliding will occur downward, resulting in overall instability and failure of the slope, which exhibits distinct characteristics of traction failure. Additionally, the pre-peak rainfall reaches the maximum horizontal displacement earlier than the other three rainfall patterns, and its maximum horizontal displacement is also the highest among the four. Therefore, under the same total rainfall and duration, the pre-peak rainfall is the most unfavorable for slope stability.

Figure 13 shows the variation in the *SF* of the slope over time under different rainfall patterns. In this figure, it can be seen that, with the same total rainfall and duration, different rainfall patterns significantly affect the timing of slope failure. Both the uniform and post-peak rainfall patterns exhibit a trend where the *SF* decreases over time. The *SF* of the slope under uniform rainfall shows a relatively stable decreasing trend overall. At 73 h into the rainfall, the safety factor drops from 1.23 to 0.995 and then continues to decline as the rainfall persists. Under trailing peak rainfall, the *SF* of the slope drops from 1.23 to 1.07 at 67 h into the rainfall. At this point, the slope remains stable, but the safety margin is low. Following this, as the rainfall intensity gradually increases, the *SF* decreases rapidly.

By 83 h into the rainfall, the *SF* falls to 0.995, and the slope begins to destabilize and fail. The *SF* continues to decrease with the ongoing rainfall, reaching a minimum value of 0.965 by the end of rainfall.

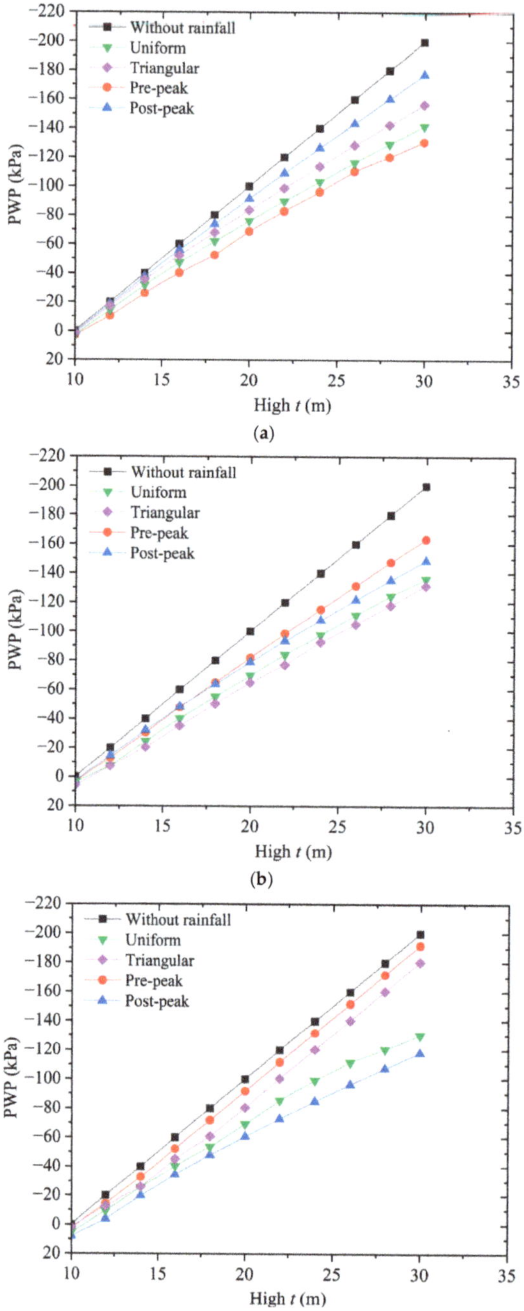

(c)

Figure 10. PWP distribution on the slope-free face for different rainfall patterns at (**a**) t = 24 h; (**b**) t = 60 h; (**c**) t = 90 h.

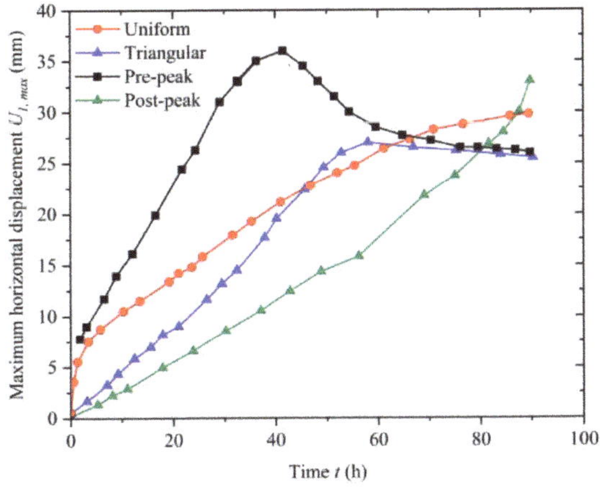

Figure 11. Curves showing the relationship between maximum horizontal deformation of the slope and rainfall duration under different rainfall patterns.

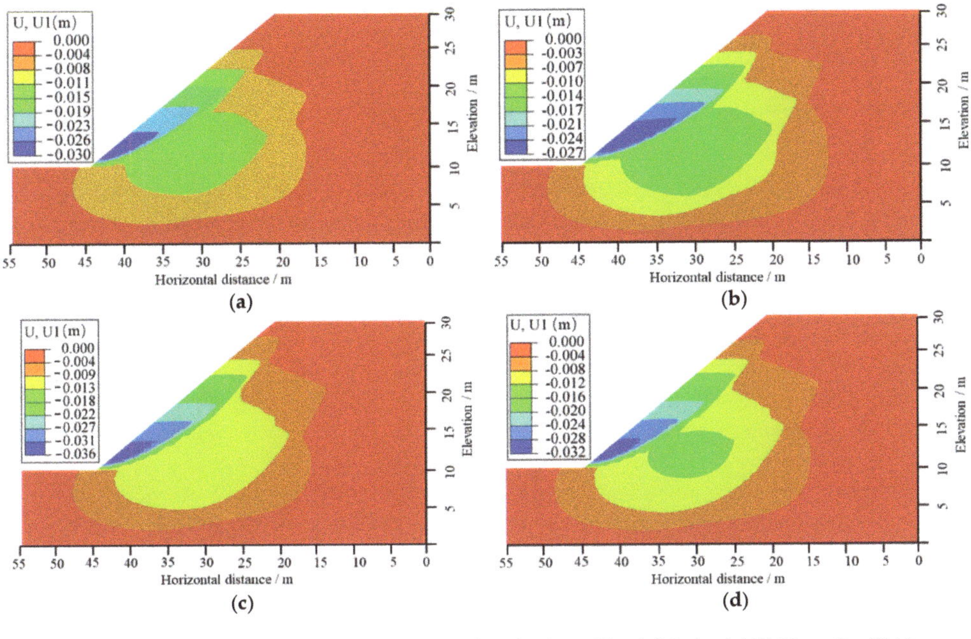

Figure 12. Horizontal soil displacement zone distribution of landslide for (**a**) Uniform (t = 90 h); (**b**) Triangular (t = 58 h); (**c**) Pre-peak (t = 36 h); (**d**) Post-peak (t = 180 h).

Both triangular and pre-peak rainfall patterns show a trend where the *SF* initially decreases and then increases over time. However, due to differences in the concentration of rainfall, the times at which the slopes begin to destabilize and fail vary between the two. For the triangular rainfall, the *SF* falls below 1 at 52 h, indicating that the slope begins to destabilize. By 70 h, the *SF* reaches a minimum value of 0.95. For the pre-peak rainfall, most of the rainfall is concentrated in the first 30 h, causing the *SF* to significantly decrease by the 5th hour. It drops to 0.998 by 22 h, indicating that the slope begins to destabilize and fail. After this point, the *SF* continues to decline, reaching a minimum value of 0.930 at 30 h into the rainfall. Subsequently, as the rainfall intensity decreases, the *SF* starts to rise slowly. Therefore, under conditions of the same total rainfall and duration, the pre-peak rainfall results in the earliest occurrence of slope failure, and the minimum *SF* is the lowest among the four rainfall types. This indicates that pre-peak rainfall is the least favorable for slope stability.

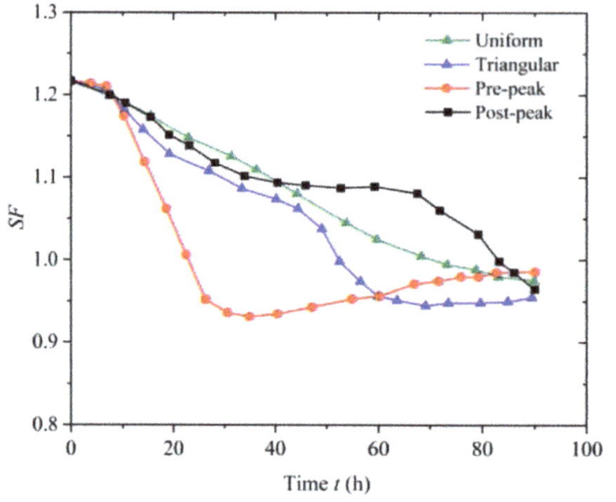

Figure 13. Variation in *SF* over time under different rainfall patterns.

3.3. The Impact of Rainfall Intensity on Slope Instability and Failure

Figure 14 shows the distribution of PWP in the landslide during a 90 h rainfall event for both heavy and moderate rainfall. In Figures 6 and 14, it can be seen that, under the same rainfall duration, different levels of rainfall have varying effects on the PWP of slope soil. As the rainfall intensity changes, the PWP in the slope body continuously varies. At the top of the slope, in the initial state, the PWP value of the surface soil is between −198 kPa and −00 kPa. After 90 h of rainfall, when the rainfall intensity reaches a heavy downpour, the PWP value of the surface soil at the top of the slope is between −123 kPa and −124 kPa. When the rainfall intensity is heavy rain, the PWP value of the surface soil at the top of the slope is between −154 kPa and −155 kPa. When the rainfall intensity is moderate rain, the PWP value is between −180 kPa and −181 kPa. It can be concluded that, within the same rainfall duration, the higher the rainfall intensity, the greater the impact on the PWP of the surface soil at the top of the slope. As the rainfall intensity increases, the PWP of the surface soil also rises, indicating a corresponding decrease in matric suction. This trend is also observed in the PWP of the soil on the slope surface and at the base. Additionally, with the increase in rainfall intensity, the wetting front continuously rises from its initial position toward the surface of the slope, indicating that the transient saturated zone is moving upward progressively.

Figure 15 illustrates the temporal and spatial evolution of PWP above the groundwater level at the slope face under three different rainfall intensities. In the figure, it is evident that as the rainfall intensity increases, the changes in PWP on the slope face soil are more significant, with more rainwater infiltrating these areas, thereby increasing the likelihood of slope instability. Additionally, Figure 15 also shows that, with greater rainfall intensity, the depth of influence on the slope face soil becomes deeper.

Figure 16 illustrates the trend of $U_{1,\,max}$ of the slope over time under different rainfall intensities. In the figure, it can be observed that the $U_{1,\,max}$ of the slope increases with the duration of rainfall for all three rainfall intensities. In the initial phase of rainfall, the differences in $U_{1,\,max}$ among the three intensities are minimal. However, after 40 h of rainfall, the $U_{1,\ max}$ significantly increases with higher rainfall intensity. As the rainfall intensity increases, the amount of water infiltration also rises, resulting in a larger saturated zone at the sliding surface and toe of the slope by the end of rainfall. This leads to the soil absorbing water, softening and losing shear strength, making it more susceptible to failure. At the end of rainfall, the $U_{1,\ max}$ of the slope under rainfall intensities of q = 10 mm/h, 5 mm/h and 2 mm/h is 25 mm, 19 mm and 17 mm, respectively. It is evident that the $U_{1,\ max}$ at a rainfall intensity of q = 10 mm/h is significantly greater than that at q = 5 mm/h and q = 2 mm/h, indicating a correspondingly more severe degree of slope damage.

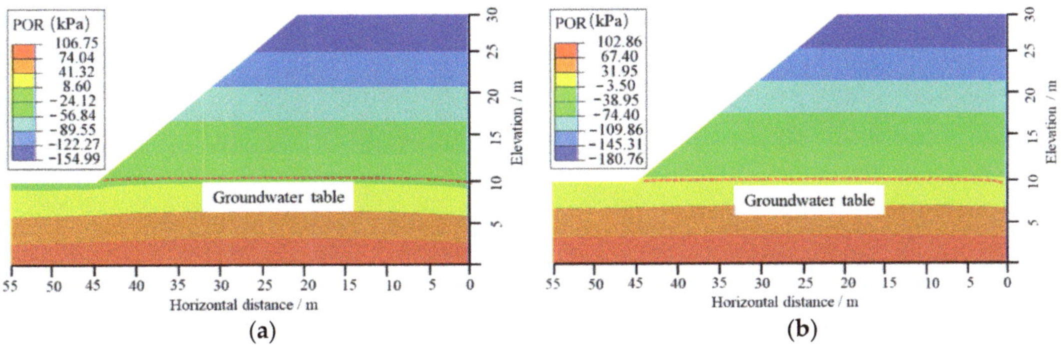

Figure 14. PWP zone distribution of landslide for (**a**) Heavy rainfall; (**b**) Moderate rainfall.

Figure 17 presents the displacement contour maps for the $U_{1,\ max}$ timings under rainfall intensities q = 10 mm/d, 5 mm/d and 2 mm/d. In Figures 12a and 17, it is observed that when the $U_{1,\ max}$ is reached for all three intensities, the displacements are concentrated at the base of the slope. The slope exhibits distinct characteristics of tensile failure. During the infiltration process, a significant amount of moisture transfers to the toe of the slope, increasing pore water pressure and decreasing shear strength at this location. This makes the toe more susceptible to failure compared to other parts of the slope, subsequently affecting the overall stability of the slope. As rainfall intensity increases, the horizontal displacement at the toe of the slope also increases, further elevating the risk of instability and failure.

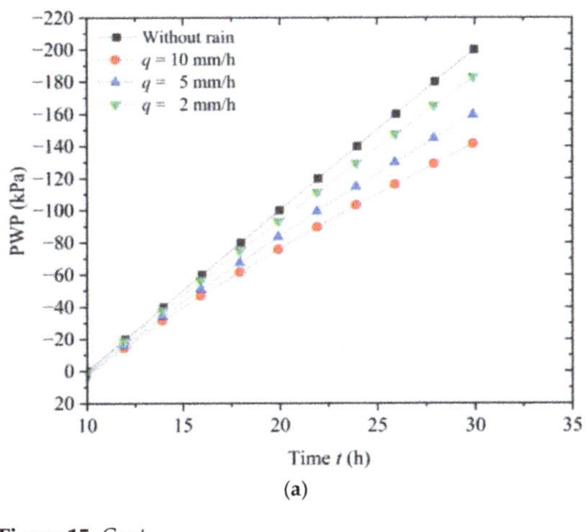

Figure 15. *Cont.*

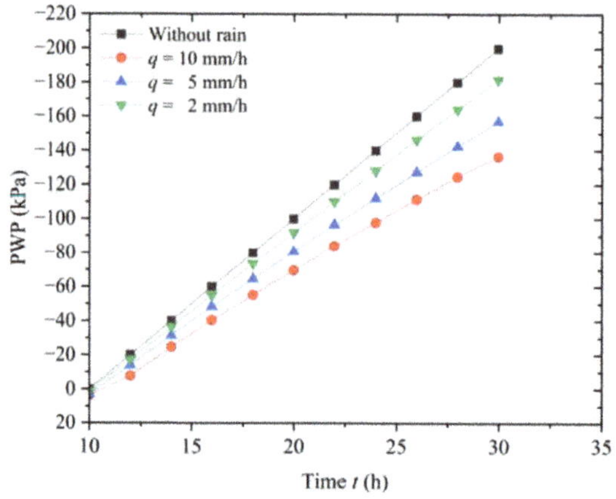

(b)

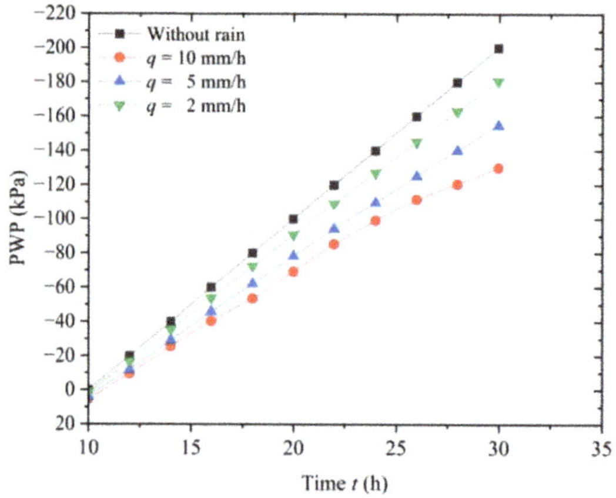

(c)

Figure 15. PWP distribution on the slope-free face of different rainfall intensity at (**a**) t = 24 h; (**b**) t = 60 h; (**c**) t = 90 h.

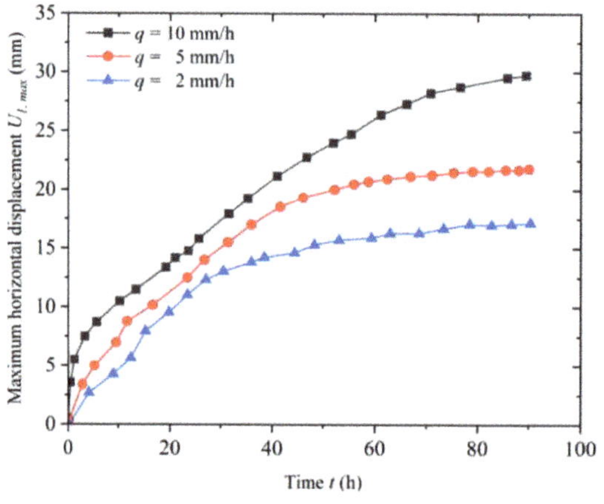

Figure 16. Curves showing the relationship between maximum horizontal deformation of the slope and rainfall duration under different rainfall intensities.

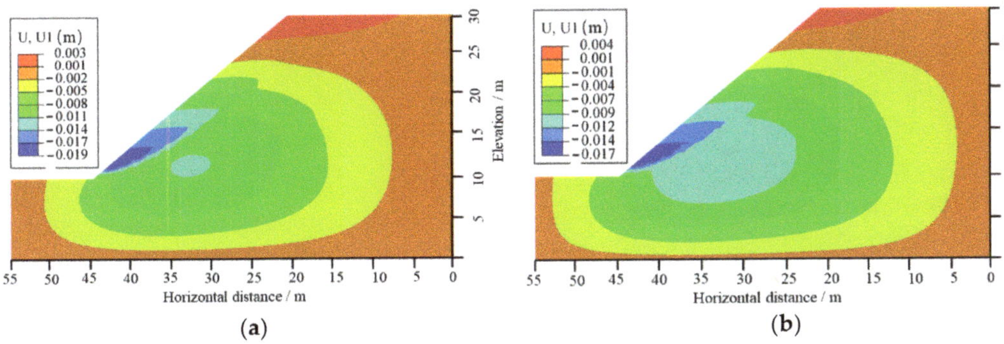

Figure 17. Horizontal soil displacement zone distribution of landslide for (**a**) q = 5 mm/h; (**b**) q = 2 mm/h.

Figure 18 shows the PEEQ contour maps at 90 h under different rainfall intensities. In Figures 6a and 18, it can be observed that, while maintaining the same rainfall duration of 90 h, the infiltration effects of different rainfall intensities lead to varying phenomena in the PEEQ zones. At a rainfall intensity of 2 mm/h, the plastic strain zone of the slope is primarily concentrated in a smaller area. At 5 mm/h, the PEEQ zone extends to a certain degree compared to the previous intensity. In contrast, at a higher rainfall intensity of 10 mm/h, the PEEQ zone further extends and presents two distinct areas that are approaching connection.

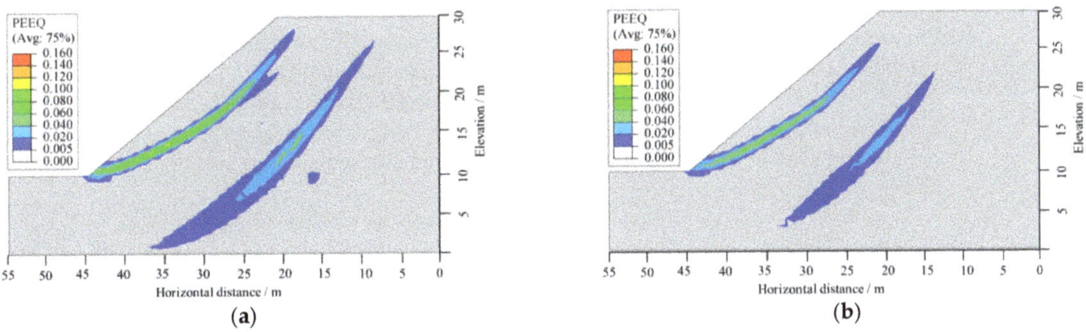

Figure 18. Plastic zone distribution of landslide for: (**a**) q = 5 mm/h; (**b**) q = 2 mm/h.

Figure 19 shows the variation in the *SF* with rainfall duration under different rainfall intensities. It can be observed that, for all three rainfall intensities, the *SF* of the slope decreases as the rainfall duration increases, and the higher the rainfall intensity, the lower the *SF*. With rainfall intensities q = 5 mm/h and 2 mm/h, the *SF* of the slope reaches minimum values of 1.07 and 1.04, respectively, indicating that while the slope does not fail, the safety margin is not high. At a rainfall intensity of q = 10 mm/d, instability begins to occur after 73 h of rainfall. Additionally, the stability of the slope is worse at the end of rainfall, with the safety factor being the lowest among the three intensities at 0.97. Therefore, the greater the rainfall intensity, the poorer the stability of the slope at the end of rainfall.

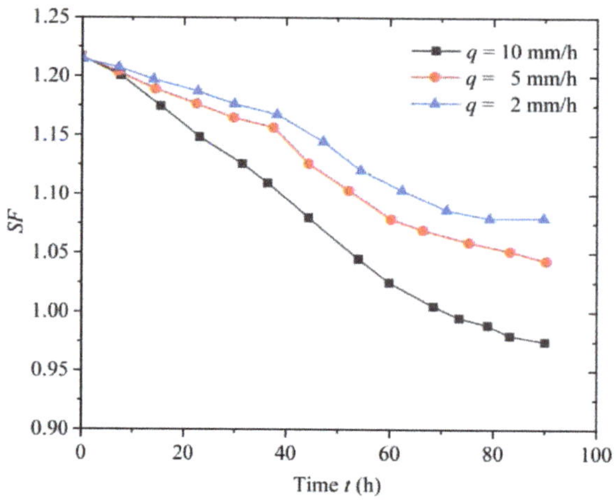

Figure 19. Variation in *SF* over time under different rainfall intensities.

4. Conclusions

In this paper, numerical simulations were performed to examine the instability and failure of saturated–unsaturated landslides triggered by rainfall. The impacts of rainfall, rainfall patterns and rainfall intensity on pore water pressure, displacement and the safety factor of the slope were thoroughly analyzed. The main findings are summarized as follows:

(1) Heavy rainfall can cause a temporary increase in negative PWP in the original unsaturated zone of the slope, leading to a closed phenomenon where the PWP isohyets extend from the slope crest to the interior. This process also elevates the free water surface. Meanwhile, under the infiltration of heavy rainfall, a through-going slip surface forming from the slope toe to the crest may develop within the slope, which is likely to trigger slope instability and sliding.

(2) The influence of heavy rainfall patterns on the spatio-temporal evolution of slope PWP, displacement and *SF* is remarkably significant. The dynamic evolution of PWP is closely related to the rainfall intensity at various times, showing a direct proportional relationship. Moreover, for the same time point, rainfall patterns with higher intensities exhibit more pronounced deformation responses, particularly at the toe of the slope, where the horizontal deformation response of the soil and rock is most evident. With a constant total rainfall amount, the pre-peak rainfall pattern resulted in the greatest decrease in the SF of the slope and the earliest occurrence of failure. This indicates that the pre-peak rainfall pattern is most detrimental to slope stability. Therefore, monitoring and preventive measures should be enhanced for slopes under pre-peak rainfall pattern conditions.

(3) For a uniform rainfall pattern, when the rainfall duration is the same, a higher rainfall intensity results in more water infiltrating into the slope. This leads to greater changes in PWP and maximum displacement, making the slope more susceptible to instability and failure. As the rainfall intensity increases, the reduction in the *SF* of the slope becomes more significant, and the time required for slope failure decreases. Therefore, in practical engineering, it is essential to enhance the forecasting of heavy rainfall and strengthen the monitoring and reinforcement of at-risk slopes.

Author Contributions: Z.W.: Methodology, Software, Investigation, Validation, Data Curation, Funding Acquisition; G.Y.: Methodology, Supervision, Resources, Writing—Review and Editing; W.L.: Writing—Review and Editing, Software, Funding Acquisition; X.C.: Writing—Review and Editing; F.L.: Writing—Review and Editing, Funding Acquisition; Y.Z.: Writing—Review and Editing. All authors have read and agreed to the published version of the manuscript.

Funding: This work is financially supported by the National Natural Science Foundation of China (No. 52108304), the Natural Science Foundation of Chongqing (Grant No. 2023NSCQ-MSX3411), the Scientific and Technological Research Program of Chongqing Municipal Education Commission (Grant No. KJQN202301307), the Fundamental Research Funds for the Central Universities (Grant No. 300102265506) and the Henan Province Science and Technology Research Project (Grant No. 252102320028).

Data Availability Statement: The data presented in this study are available from the corresponding author upon request.

Conflicts of Interest: Authors Zhuolin Wu, Gan Yang, Wen Li and Xiangling Chen were employed by the Chongqing Survey and Design Institute of Water Resources, Electric Power and Architecture Co., Ltd. The funders had no role in the design of the study; in the collection, analyses, or interpretation of data; in the writing of the manuscript; or in the decision to publish the results. The remaining authors declare that the research was conducted in the absence of any commercial or financial relationships that could be construed as a potential conflict of interest.

References

1. Liu, Y.; Cheng, P.; Liu, E.; Hu, J.; Cheng, Y.; Su, L. Seismic response and dynamic failure mode of a class of bedding rock slope subjected to freeze–thaw cycles. *Rock Mech. Rock Eng.* **2025**, *58*, 3289–3307. [CrossRef]
2. Chen, R.H.; Chen, H.P.; Chen, K.S.; Zhung, H.B. Simulation of a slope failure induced by rainfall infiltration. *Environ. Geol.* **2009**, *58*, 943–952. [CrossRef]
3. Rosly, M.H.; Mohamad, H.M.; Bolong, N.; Harith, N.S.H. An overview: Relationship of geological condition and rainfall with landslide events at East Malaysia. *Trends Sci.* **2022**, *19*, 3464. [CrossRef]
4. Ijaz, N.; Ye, W.; Rehman, Z.; Dai, F.; Ijaz, Z. Numerical study on stability of lignosulphonate-based stabilized surficial layer of unsaturated expansive soil slope considering hydro-mechanical effect. *Transp. Geotech.* **2022**, *32*, 100697. [CrossRef]
5. Qi, S.; Vanapalli, S.K. Hydro-mechanical coupling effect on surficial layer stability of unsaturated expansive soil slopes. *Comput. Geotech.* **2015**, *70*, 68–82. [CrossRef]
6. Yang, K.H.; Uzuoka, R.; Lin, G.L.; Nakai, Y. Coupled hydro-mechanical analysis of two unstable unsaturated slopes subject to rainfall infiltration. *Eng. Geol.* **2017**, *216*, 13–30. [CrossRef]
7. Ran, Q.; Wang, F.; Gao, J. Modelling effects of rainfall patterns on runoff generation and soil erosion processes on slopes. *Water* **2019**, *11*, 2221. [CrossRef]
8. Tian, D.F.; Zheng, H.; Liu, D.F. A 2D integrated FEM model for surface water–groundwater flow of slopes under rainfall condition. *Landslides* **2017**, *14*, 577–593. [CrossRef]
9. Zhang, J.; Hu, F.; Zhang, Q.; Wang, J.; Deng, W.; Zhang, L.; Shao, X. Evaluation of the Effects of Rainfall Infiltration Boundaries on the Stability of Unsaturated Soil Slopes Using the Particle Flow Code. *Water* **2024**, *16*, 3704. [CrossRef]
10. Luo, S.; Huang, D.; Peng, J.; Aierken, A.; Li, Z.; Kuang, X.B.; Tomás, R. Performance and application of a novel drainage anti-slide pile on accumulation landslide with a chair-like deposit-bedrock interface in the Three Gorges Reservoir area, China. *Comput. Geotech.* **2023**, *155*, 105199. [CrossRef]
11. Sun, Y.; Yang, K.; Hu, R.; Wang, G.; Lv, J. Model test and numerical simulation of slope instability process induced by rainfall. *Water* **2022**, *14*, 3997. [CrossRef]
12. Kafle, L.; Xu, W.; Zeng, S.; Nagel, T. A numerical investigation of slope stability influenced by the combined effects of reservoir water level fluctuations and precipitation: A case study of the Bianjiazhai landslide in China. *Eng. Geol.* **2022**, *297*, 106508. [CrossRef]
13. Sun, P.; Wang, H.; Wang, G.; Li, R.; Zhang, Z.; Huo, X. Field model experiments and numerical analysis of rainfall-induced shallow loess landslides. *Eng. Geol.* **2021**, *295*, 106411. [CrossRef]
14. Reid, M.E.; Lahusen, R.G. Real-time monitoring of active landslides along highway 50, El Dorado County. *Calif. Geol.* **1998**, *51*, 17–20.
15. Rahardjo, H.; Nio, A.S.; Leong, E.C.; Song, N.Y. Effects of groundwater table position and soil properties on stability of slope during rainfall. *J. Geotech. Geoenviron. Eng.* **2010**, *136*, 1555–1564. [CrossRef]

16. Regmi r k jung k Nakagawa, H.; Kang, J. Study on mechanism of retrogressive slope failure using artificial rainfall. *Catena* **2014**, *122*, 27–41. [CrossRef]
17. Collins, B.D.; Znidarcic, D. Stability analyses of rainfall induced landslides. *J. Geotech. Geoenviron. Eng.* **2004**, *130*, 362–372. [CrossRef]
18. Tohari, A.; Nishigaki, M.; Komatsu, M. Laboratory rainfall-induced slope failure with moisture content measurement. *J. Geotech. Geoenviron. Eng.* **2007**, *133*, 575–587. [CrossRef]
19. Xie, J.; Uchimura, T.; Wang, G.; Selvarajah, H.; Maqsood, Z.; Shen, Q.; Mei, G.; Qiao, S. Predicting the sliding behavior of rotational landslides based on the tilting measurement of the slope surface. *Eng. Geol.* **2020**, *269*, 105554. [CrossRef]
20. Xie, J.; Uchimura, T.; Wang, G.; Shen, Q.; Maqsood, Z.; Xie, C.; Liu, J.; Lei, W.; Tao, S.; Chen, P.; et al. A new prediction method for the occurrence of landslides based on the time history of tilting of the slope surface. *Landslides* **2020**, *17*, 301–312. [CrossRef]
21. Xie, J.; Uchimura, T.; Huang, C.; Maqsood, Z.; Tian, J. Experimental study on the relationship between the velocity of surface movements and tilting rate in pre-failure stage of rainfall-induced landslides. *Sensors* **2021**, *21*, 5988. [CrossRef]
22. Yang, H.Q.; Shi, C.; Zhang, L. Ensemble learning of soil–water characteristic curve for unsaturated seepage using physics-informed neural networks. *Soils Found.* **2025**, *65*, 101556. [CrossRef]
23. Lan, P.; Su, J.; Zhu, S.; Huang, J.; Zhang, S. Reconstructing unsaturated infiltration behavior with sparse data via physics-informed deep learning. *Comput. Geotech.* **2024**, *168*, 106162. [CrossRef]
24. Gu, X.; Wang, L.; Ou, Q.; Zhang, W. Efficient stochastic analysis of unsaturated slopes subjected to various rainfall intensities and patterns. *Geosci. Front.* **2023**, *14*, 101490. [CrossRef]
25. Cho, S.E.; Lee, S.R. Instability of unsaturated soil slopes due to infiltration. *Comput. Geotech.* **2001**, *28*, 185–208. [CrossRef]
26. Cheng, P.; Liu, Y.; Hu, J.; Fang, K.; Hu, Y. Experimental study on dynamic response and failure mode of bedding rock slope with cracks under earthquake. *Bull. Eng. Geol. Environ.* **2025**, *84*, 81. [CrossRef]
27. Li, B.; Gao, Y.; Yin, Y.; Wan, J.; He, K.; Wu, W.; Zhang, H. Rainstorm-induced large-scale landslides in Northeastern Chongqing, China, August 31 to September 2, 2014. *Bull. Eng. Geol. Environ.* **2022**, *81*, 271. [CrossRef]
28. Yang, G.; Wu, Z.; Zhang, L.; Yang, G.; Wu, Z.; Zhang, L.; Hou, J.; Tong, S.; Liu, F.; Zheng, Y. A Study on the Deformation Mechanism of a Landslide Reinforced with an Anti-Slip Pile Under the Effect of Reservoir Water Level Decline. *Water* **2025**, *17*, 1390. [CrossRef]

water

MDPI

Article

Numerical Simulation of Fracture Failure Propagation in Water-Saturated Sandstone with Pore Defects Under Non-Uniform Loading Effects

Gang Liu [1,2,*], Yonglong Zan [1], Dongwei Wang [1], Shengxuan Wang [1], Zhitao Yang [2], Yao Zeng [1], Guoqing Wei [1] and Xiang Shi [1]

1 Heilongjiang Ground Pressure and Gas Control in Deep Mining Key Laboratory, Heilongjiang University of Science and Technology, Harbin 150022, China; 17353693207@163.com (Y.Z.); 18297919986@163.com (D.W.); w13140438798@163.com (S.W.); 15261825386@163.com (Y.Z.); w18853152380@163.com (G.W.); 15254932807@163.com (X.S.)

2 Baotailong New Material Co., Ltd., Qitaihe 154604, China; w120559779@163.com

* Correspondence: liugang@usth.edu.cn; Tel.: +86-189-4463-0110

Abstract: The instability of mine roadways is significantly influenced by the coupled effects of groundwater seepage and non-uniform loading. These interactions often induce localized plastic deformation and progressive failure, particularly in the roof and sidewall regions. Seepage elevates pore water pressure and deteriorates the mechanical properties of the rock mass, while non-uniform loading leads to stress concentration. The combined effect facilitates the propagation of microcracks and the formation of shear zones, ultimately resulting in localized instability. This initial damage disrupts the mechanical equilibrium and can evolve into severe geohazards, including roof collapse, water inrush, and rockburst. Therefore, understanding the damage and failure mechanisms of mine roadways at the mesoscale, under the combined influence of stress heterogeneity and hydraulic weakening, is of critical importance based on laboratory experiments and numerical simulations. However, the large scale of in situ roadway structures imposes significant constraints on full-scale physical modeling due to limitations in laboratory space and loading capacity. To address these challenges, a straight-wall circular arch roadway was adopted as the geometric prototype, with a total height of 4 m (2 m for the straight wall and 2 m for the arch), a base width of 4 m, and an arch radius of 2 m. Scaled physical models were fabricated based on geometric similarity principles, using defect-bearing sandstone specimens with dimensions of 100 mm × 30 mm × 100 mm (length × width × height) and pore-type defects measuring 40 mm × 20 mm × 20 mm (base × wall height × arch radius), to replicate the stress distribution and deformation behavior of the prototype. Uniaxial compression tests on water-saturated sandstone specimens were performed using a TAW-2000 electro-hydraulic servo testing system. The failure process was continuously monitored through acoustic emission (AE) techniques and static strain acquisition systems. Concurrently, FLAC^{3D} 6.0 numerical simulations were employed to analyze the evolution of internal stress fields and the spatial distribution of plastic zones in saturated sandstone containing pore defects. Experimental results indicate that under non-uniform loading, the stress–strain curves of saturated sandstone with pore-type defects typically exhibit four distinct deformation stages. The extent of crack initiation, propagation, and coalescence is strongly correlated with the magnitude and heterogeneity of localized stress concentrations. AE parameters, including ringing counts and peak frequencies, reveal pronounced spatial partitioning. The internal stress field exhibits an overall banded pattern, with localized variations induced by stress anisotropy. Numerical simulation results further show that shear failure zones tend to cluster regionally, while tensile failure zones are more evenly distributed. Additionally, the stress field configuration at the specimen crown significantly

 https://doi.org/10.3390/w17121725

influences the dispersion characteristics of the stress–strain response. These findings offer valuable theoretical insights and practical guidance for surrounding rock control, early warning systems, and reinforcement strategies in water-infiltrated mine roadways subjected to non-uniform loading conditions.

Keywords: water-saturated sandstone; fracture failure; defect-bearing sandstone; non-uniform loading; FLAC^{3D}

1. Introduction

In geotechnical engineering and underground space development, rock mass instability is a critical issue that poses a significant threat to engineering safety. Recent studies have demonstrated that rock mass failure often does not occur under uniform loading conditions; instead, it is triggered by the combined effects of non-uniform loading and multiple physical factors, such as groundwater and chemical processes [1–3]. In tunnel engineering, excavation disturbances or structural asymmetry can lead to substantial stress redistribution within the rock mass, resulting in localized zones of high-stress concentration. Concurrently, groundwater infiltrates through fractures, exacerbating the softening of joint surfaces, increasing pore water pressure, and degrading cementation structures. This ultimately promotes the degradation of rock mass strength and the propagation of fractures, potentially leading to sudden structural instability.

In practical geotechnical engineering applications such as tunnel excavation and underground mining, rock masses are frequently subjected to non-uniform loading conditions, which can induce localized stress concentrations, crack initiation, and progressive failure. Wang et al. and Yu et al. [4,5] investigated these phenomena by designing trapezoidal rock specimens and employing asymmetric loading devices to examine the failure characteristics of cavity-bearing rocks. Zhao and Ge et al. [6,7] further explored the coupled effects of eccentric external loading and internal voids, proposing damage constitutive models to characterize stress redistribution and fracture initiation mechanisms. These studies provide essential theoretical guidance for understanding rock failure behavior under complex loading paths and offer a solid basis for the experimental design and analysis of fracture evolution in saturated sandstone under non-uniform loading in this study.

The interaction between water and rock plays a crucial role in influencing the mechanical behavior of rock masses, particularly in groundwater-rich geological environments [8]. Li, Duan, Zhang, and Zhou et al. [9–12] systematically examined how variations in water content, pore pressure, and fracture network characteristics affect rock strength, energy dissipation, and crack propagation through uniaxial compression, direct shear tests, and infrared thermography. Thörn and Cardona et al. [13–15] focused on the nonlinear relationship between fracture aperture behavior and fluid flow, emphasizing that fracture permeability is highly sensitive to microstructural changes during closure and reopening cycles. He et al. [16] further demonstrated that water saturation and crack tip intersection angles significantly influence the mechanical response and fracture trajectories in sandstone. These findings deepen the understanding of hydro-mechanical coupling mechanisms and provide crucial support for this study's investigation into water-induced damage and fracture behavior in saturated rock.

Geometric characteristics of rock masses, such as fracture inclination, cavity shape, and spatial configuration, are key factors governing fracture initiation and propagation. Luo and Chen [17,18] employed PFC2D with the parallel bond model to simulate the nonlinear effects of various fracture inclinations and loading rates on rock failure modes. Further research by Luo et al. [19] indicated that a fracture dip angle of 15° results in the most severe stress concentration at the crack tip and shows a strong nonlinear relationship between peak strength and loading rate. Tian [20] used fluid–structure coupling simulations to reveal a permeability "jump phenomenon" during crack propagation, supporting Cardona's [14] model of abrupt hydraulic transitions in natural fracture systems. Zhang and Zhu [21,22] integrated acoustic emission monitoring with energy analysis to highlight the dominant role of fracture geometry in controlling damage evolution and failure mechanisms. These insights provide a solid theoretical foundation for this study's focus on the influence of cavity geometry on stress localization and fracture development under coupled hydro-mechanical conditions.

In summary, substantial progress has been made in the study of non-uniform loading, hydro-mechanical coupling, and fracture geometry effects, which has laid a solid foundation for understanding the damage evolution behavior of rock masses. However, existing research often focuses on the influence of single factors, lacking a systematic understanding of the coupling mechanisms between non-uniform loading and saturated environments. In particular, under engineering scenarios such as roadway excavation and deep groundwater infiltration, the multidimensional interactions among crack propagation, stress distribution, fluid flow, and structural features remain to be fully clarified. Therefore, investigating the mechanical behavior and fracture evolution of saturated porous rock specimens under non-uniform loading conditions is essential to provide theoretical support for engineering stability assessment and disaster prevention (Figure 1).

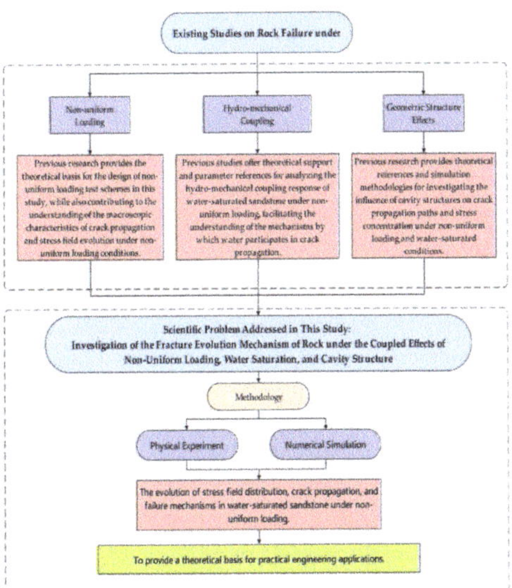

Figure 1. Theoretical and Methodological Support from Previous Research for This Study.

2. Geological Control Factors of Mining Tunnel Stability Under Groundwater Erosion

In underground mining engineering, long-term groundwater seepage and erosion are among the principal factors governing the structural evolution and failure mechanisms of deep rock masses. The impact of groundwater on the stability of mining tunnels is comprehensively controlled by various geological conditions, including rock mass structure, lithological characteristics, the degree of weathering, and the spatial distribution of tectonic fracture zones. For instance, Bayat et al. [23] investigated the Nowsud Tunnel and reported that in karst regions, the coupling between fault zones and groundwater activity readily forms water inrush pathways, significantly compromising tunnel stability. Coli [24] highlighted that groundwater seepage through tectonic fracture zones often exhibits transient flow behavior, with hydraulic properties playing a critical role in the evolution of surrounding rock stress fields. Hack [25] emphasized that rock weathering not only reduces shear strength but also accelerates hydration and dissolution processes, which in turn hasten rock mass degradation. Moreover, Xing et al. [26], based on studies of an underground mine in the United States, demonstrated that stress concentration and seepage interaction in structurally complex zones are key mechanisms triggering instability and failure.

(1) The structural characteristics of the rock mass control the groundwater seepage path and the integrity of the surrounding rock structure. Discontinuous features such as joints, fractures, bedding planes, and faults are commonly present, forming high-permeability flow channels. These structural discontinuities not only reduce the overall strength of the rock mass but also provide active interfaces for water–rock interactions, significantly diminishing shear strength, which may result in structural displacements and slippage.

(2) Lithology and the physical–mechanical properties of the rock mass directly determine the response to water erosion. Weakly cemented and highly porous rocks, such as sandstone, mudstone, and shale, are highly susceptible to softening, swelling, and deterioration under the influence of water, leading to rapid loss of strength and stiffness. In contrast, dense and erosion-resistant rocks, such as granite and quartzite, exhibit greater stability under water exposure and retain superior mechanical properties.

(3) Both the degree of weathering and the intensity of tectonic fracturing are critical geological controls on the erosive impact of groundwater. Highly weathered zones are typically characterized by high porosity, low cementation, and a loose structural framework, making them highly susceptible to groundwater infiltration and associated deterioration. In contrast, tectonic fracture zones—marked by extensive rock fragmentation and enhanced permeability—often act as primary channels for groundwater accumulation and flow and are frequently correlated with increased risks of surrounding rock instability.

Although these geological factors provide pathways for groundwater seepage and weaken the mechanical properties of the surrounding rock, they are typically latent triggers. The primary cause of tunnel instability and failure is often the combined effect of groundwater erosion and a non-uniform stress field. Mining-induced disturbances alter the in situ stress distribution, creating localized zones of high-stress concentration. When these high-stress areas are further affected by water-induced softening, the surrounding rock undergoes dual degradation: structural weakening and loss of load-bearing capacity, which leads to the initiation of the failure process characterized by "stress release—crack propagation—structural collapse". Therefore, the stress–water coupling effect not only integrates geological conditions but also serves as the primary driving mechanism for the transition of the surrounding rock from stable to unstable. Thus, studying the evolution of rock damage and failure under stress–water coupling is critical for improving our understanding of tunnel instability and failure modes under groundwater erosion.

3. Specimen Preparation and Testing Program Design

3.1. Specimen Preparation

Due to the large scale of underground tunnel engineering, complex geological conditions, and the significant safety risks associated with on-site testing, conducting comprehensive and systematic studies at the prototype scale presents substantial challenges. To more effectively simulate the mechanical response of mining tunnels in real-world engineering scenarios, this study focuses on a typical straight-wall circular arch tunnel as the research subject. The prototype tunnel structure has a total height of 4 m, consisting of a straight wall section of 2 m and an arch height of 2 m. The bottom width is 4 m, and the arch employs a semicircular design with an arch radius of 2 m. The prototype structure was scaled down proportionally using the principles of geometric similarity at a ratio of 1:100 to create small-scale specimens.

The specimens used in this study were prepared from natural sandstone parent rock [27], known for its excellent integrity. Prior to specimen preparation, the parent rock was manually selected to ensure the absence of significant fractures, joints, or weathering marks, thereby enhancing the representativeness of the specimens and the stability of the experimental data. Initially, a diamond saw cutting machine was used to make rough cuts in the parent rock, shaping it into rectangular blocks. Subsequently, precision cutting and grinding equipment was employed to refine these blocks into standard rectangular specimens with dimensions of 100 mm × 30 mm × 100 mm (length × width × height). Throughout the process, strict control over processing precision was maintained, with repeated measurements to ensure dimensional deviations remained within ±0.5 mm, significantly reducing experimental errors due to size discrepancies. After processing, all specimens were rechecked for dimensions using a vernier caliper to ensure consistency.

A standardized hole was designed and fabricated in the central region of each specimen to simulate the typical tunnel structure in underground engineering. This hole features a straight-wall circular arch geometry, with a rectangular base measuring 40 mm in length and 20 mm in width. The straight wall height is 20 mm, and the top is shaped as a semicircular arch with a radius of 20 mm. The hole structure was designed according to the principles of similarity theory and mechanical model similarity, ensuring that the structural characteristics of the prototype tunnel and the stress distribution pattern in the surrounding rock are preserved. This approach guarantees that the mechanical behavior of the prototype tunnel is accurately represented in the small-scale model specimens, providing a reliable engineering analog. Through this meticulous processing procedure, we ensured the consistency, reliability, and suitability of the specimens for experimental testing.

3.2. Experimental Scheme

The experimental setup for this study consists of five main components: the loading and control system, the acoustic emission (AE) signal acquisition system, the static strain acquisition system, a high-speed camera, and vacuum saturation equipment, as illustrated in Figure 2. The loading system is a TM-B-SW-2000 microcomputer-controlled electro-hydraulic servo testing machine (Jinli Testing Technology Co., Ltd., Changchun, China) provided by Harbin University of Science and Technology, which features high-precision displacement control. The axial displacement loading rate is set at 0.002 mm/s [28]. The accuracy of the loading force measurement is ±1% of full scale, and the displacement resolution is 0.01 mm. This system satisfies the technical requirements for loading precision and control stability in rock uniaxial compression tests. Specimen saturation is achieved using the ZKSB-II vacuum saturation machine, which is equipped with a vacuum pump and a constant-pressure water tank. The vacuum level is maintained at −0.098 MPa (±0.002 MPa), and the water pressure control accuracy is better than ±0.01 MPa. Each specimen is saturated for 24 h to ensure complete saturation. Acoustic emission signals are collected using the SH-II acoustic emission system, which operates concurrently with the loading process to monitor the initiation and propagation of fractures in real time. The deformation of the metal shim is monitored by the static strain acquisition system, which has a strain gauge sensitivity of 2.0 ± 1% mV/V and a sampling frequency of 1 kHz, allowing for precise analysis of localized stress concentrations. In addition, a high-speed camera with a resolution of 1920 × 1080 pixels and a frame rate of up to 1000 frames per second is used to capture the entire loading process, enabling synchronized analysis of both the images and the acoustic emission data. Considering the inherent randomness of sandstone failure behavior during uniaxial compression tests, five specimens were subjected to non-uniformly distributed loading tests within each group to ensure the reliability and representativeness of the experimental results. The findings are detailed in Table 1.

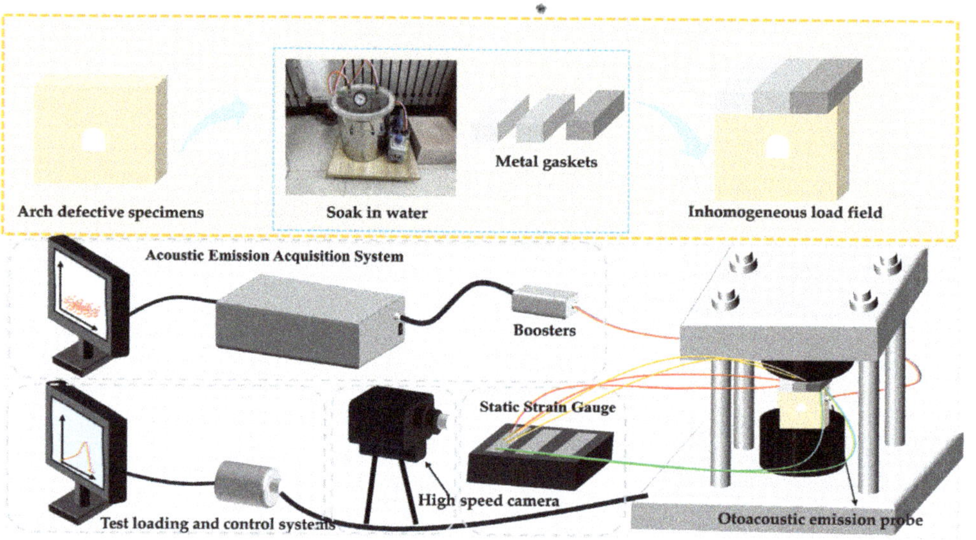

Figure 2. Diagram of the test system.

Table 1. Basic information of the test group.

Serial Number	Specimen Type	Density (kg/m³)	Elastic Modulus (GPa)	Poisson's Ratio (ν)	State of Affairs	Metal Gasket Arrangement Order	Specimen Height × Length × Width/mm
M-B-S							
M-B-S-1		2620	11.0	0.27			100.14 × 100.24 × 20.11
M-B-S-2		2630	11.2	0.26			100.02 × 100.12 × 20.05
M-B-S-3		2620	11.1	0.27		SY-2, SY-3, SY-1	100.07 × 100.13 × 20.01
M-B-S-4		2630	11.3	0.26			100.12 × 100.04 × 20.03
M-B-S-5		2620	10.9	0.27			100.06 × 100.07 × 20.06
M-S-B							
M-S-B-1		2630	11.2	0.26			100.09 × 100.13 × 20.12
M-S-B-2	Water-saturated sandstone specimens with pore defects	2620	11.0	0.28	Saturated with water		100.04 × 100.02 × 20.14
M-S-B-3		2630	11.3	0.27			100.05 × 100.02 × 20.10
M-S-B-4		2630	11.1	0.27		SY-2, SY-1, SY-3	100.07 × 100.02 × 20.10
M-S-B-5		2610	11.2	0.26			100.05 × 100.02 × 20.10
B-M-S							
B-M-S-1		2620	11.1	0.28			100.03 × 100.16 × 20.13
B-M-S-2		2630	11.2	0.27			100.13 × 100.18 × 20.06
B-M-S-3		2610	11.0	0.27			100.06 × 100.13 × 20.10
B-M-S-4		2610	11.3	0.26		SY-3, SY-2, SY-1	100.05 × 100.03 × 20.05
B-M-S-5		2620	11.1	0.28			100.02 × 100.06 × 20.02

To investigate the evolution of damage in saturated sandstone under non-uniform loading conditions, this experiment is structured into three distinct experimental groups:

M-B-S, M-S-B, and B-M-S. These groups aim to explore the fracture evolution of saturated sandstone under varying non-uniform loading distributions, as detailed in Table 1. Due to the uniform loading system employed in the experiment, direct non-uniform loading tests cannot be conducted. To overcome this limitation, three types of metal shims with different elastic moduli are utilized and arranged in three distinct configurations at the top of the specimens. In the M-B-S group, the metal shims are arranged from left to right as SY-3, SY-1, and SY-2; in the M-S-B group, the arrangement is SY-2, SY-1, and SY-3; and in the B-M-S group, the order is SY-3, SY-2, and SY-1. By applying material mechanics stress calculation formulas, the stress values in the various loading zones at the upper part of the specimen can be determined based on the relationship between elastic modulus and strain. Although the testing machine applies a uniform load to the specimen, the differing elastic moduli of the metal shims result in varying strain responses, leading to different stress transmission through the shims to the upper part of the specimen. This creates non-uniform loading regions at the top. Consequently, three distinct loading zones are formed in the upper portion of the specimen, as illustrated in Figure 3.

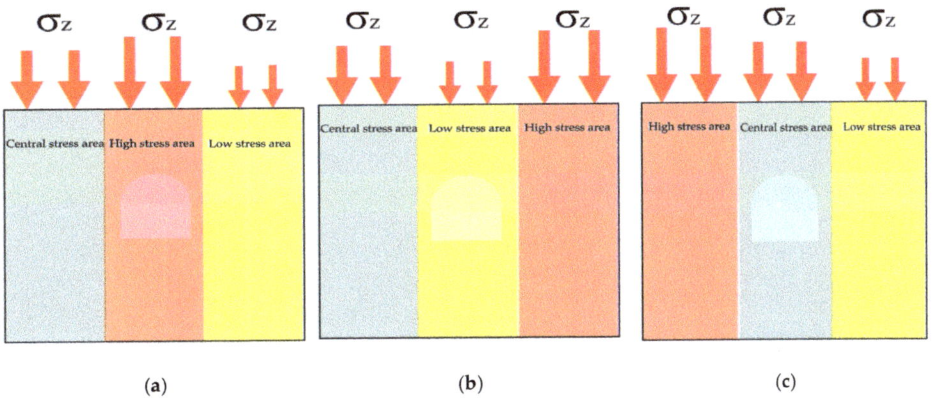

Figure 3. Schematic diagrams of heterogeneous stress distribution patterns in M-B-S, M-S-B, and B-M-S specimen groups. The red arrows indicate the magnitude and spatial distribution of the vertical stress applied to the upper surface of the specimen, simulating the non-uniform stress conditions typically experienced around underground excavations. M-B-S group: The stress distribution on the upper surface of the specimen is arranged from left to right as medium stress, high stress, and low stress. M-S-B group: The stress distribution on the upper surface of the specimen is arranged from left to right as medium stress, low stress, and high stress. B-M-S group: The stress distribution on the upper surface of the specimen is arranged from left to right as high stress, medium stress, and low stress. (**a**) Schematic diagram of group M-B-S specimens; (**b**) Schematic diagram of group M-S-B specimens; (**c**) Schematic diagram of group B-M-S specimens.

4. Experimental Results and Analysis

4.1. Non-Uniform Loading Zone

During the loading process, strain data for the metal shims were collected using a static strain acquisition system, and stress–strain curves for the metal shims were subsequently plotted. As shown in Figure 4, the stress–strain curves for the metal shims in the M-B-S, M-S-B, and B-M-S groups are compared. In the M-B-S group, the stress–strain curves of all three types of metal shims exhibit an elastic phase, with peak stress values of 34.6 MPa, 46.8 MPa, and 78 MPa, respectively, at the initiation of plastic failure. Similarly, in the M-S-B group, the stress–strain curves display an elastic phase, with the peak stress values for SY-1, SY-2, and SY-3 reaching 44.1 MPa, 64.6 MPa, and 95 MPa, respectively, at the onset of plastic failure. In the B-M-S group, the stress–strain curves for the three types of metal shims also show an elastic phase, with peak stress values of 37.5 MPa, 52.8 MPa, and 79.5 MPa, respectively, at the transition to plastic failure. Comparative analysis reveals that the stress experienced by the metal shims in the M-B-S and B-M-S groups is quite similar, whereas the metal shims in the M-S-B group experience higher stress levels than those in both the M-B-S and B-M-S groups.

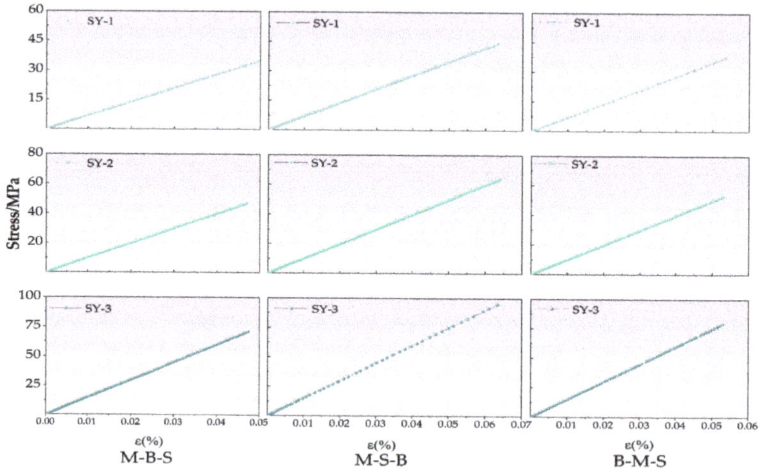

Figure 4. Vertical stress distributions in the upper loading medium of different specimen groups. This figure illustrates the stress–strain behavior of the upper loading medium under varying stress configurations across different test groups. Each column represents a specific test group (SY-1, SY-2, SY-3), characterized by different applied vertical stress levels. Each row shows the response of a particular loading medium configuration under those test conditions. The three stress distribution patterns applied to the upper boundary of the medium are: M–B–S (Medium–Low–High stress from left to right along the interface); M–S–B (Medium–High–Low); B–M–S (Low–Medium–High). Horizontal axis: axial strain (%) in the upper loading medium. Vertical axis: vertical stress (MPa). These loading schemes simulate non-uniform stress transfer scenarios representative of rock engineering environments, allowing for analysis of how different media and stress gradients influence load transmission behavior.

By monitoring the strain in the stress-transmitting medium, the distribution of the non-uniform stress field at the upper end of the specimen can be determined. As shown in Figure 5, the trends in stress distribution at the upper end of the three specimen groups are presented. Under the influence of the metal shims, which act as the stress-transmitting medium, the upper part of the specimen exhibits regions with low-, medium-, and high-stress levels. In the M-B-S group, the non-uniform stress distribution at the specimen's upper end, from left to right, is characterized by a low-stress region, a high-stress region, and a medium-stress region. The stress curve at the specimen's top shows a convex shape. In the M-S-B group, the non-uniform stress distribution at the upper end, from left to right, is marked by a medium-stress region, a low-stress region, and a high-stress region, resulting in a concave stress curve. In the B-M-S group, the non-uniform stress distribution at the specimen's upper end, from left to right, is characterized by a high-stress region, a medium-stress region, and a low-stress region, with the stress curve exhibiting a decreasing trend from left to right.

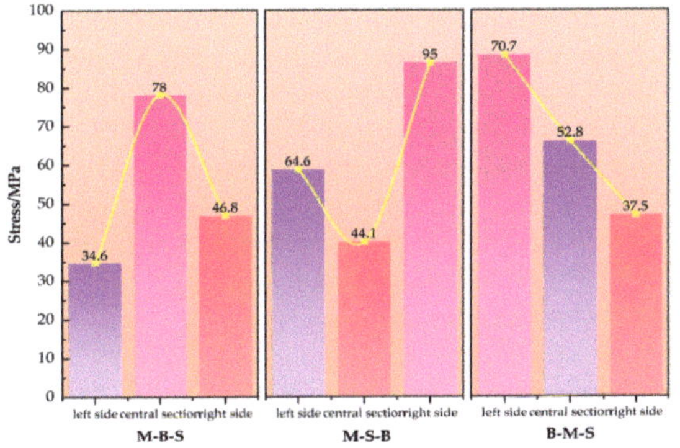

Figure 5. Distribution of stress field at the upper end of different groups of sandstone specimens.

4.2. Stress–Strain Curve of Saturated Sandstone with Pore Defects

As illustrated in Figure 6, the peak stress strengths of the specimens vary depending on the combinations of metal shims used. The relationship between the peak stress strengths for each group is as follows: M-S-B group > M-B-S group > B-M-S group. The peak stress values for the M-B-S, M-S-B, and B-M-S groups are 78 MPa, 95 MPa, and 70.7 MPa, respectively. The figure clearly demonstrates that, under the combined effects of non-uniform loading and water, the stress–strain curves of the three groups of specimens exhibit a staged phenomenon during the rock loading process. This phenomenon includes the initial compaction phase, the elastic phase, the plastic yield phase, and the plastic failure phase.

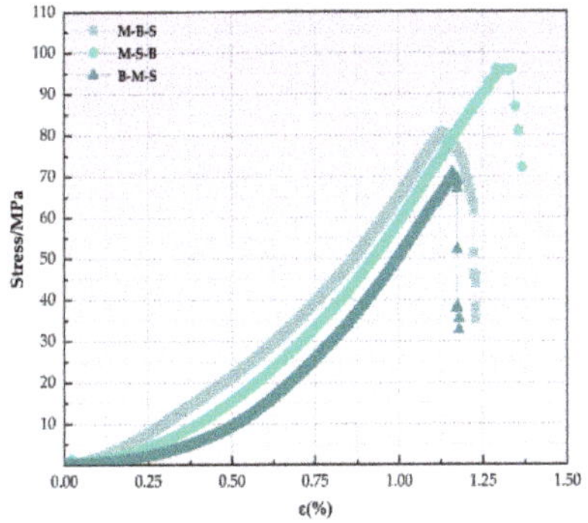

Figure 6. Stress–strain graph.

As illustrated, during the initial compaction phase, the stress–strain curves of all three groups exhibit a concave trend. At this stage, the internal particle gaps and primary fractures of the sandstone specimens are progressively compressed and compacted due to non-uniform loading, resulting in a reduction in fractures. Once the specimens enter the elastic phase, as the load gradually increases, the stress–strain curves of all three groups demonstrate a linear relationship. In this phase, the stress on the specimens remains below their elastic limit, allowing them to exhibit elastic behavior. When the load is released during this stage, the specimens return to their original state.

During the plastic yield phase, the stress–strain curves of all three groups exhibit a convex trend. At this stage, the stress on the specimens reaches their yield strength, resulting in plastic deformation. Irreversible crack propagation occurs within the specimens, and upon removal of the external force, the specimens cannot return to their original shape.

After the specimens reach their peak stress and transition into the plastic failure phase, significant damage and rupture ensue. In this phase, the specimens display pronounced crack propagation on the surface while internal cracks and fractures fully develop, interconnecting to form extensive fracture surfaces. The overall structure of the specimen is compromised, which is represented in the stress–strain curve by a rapid decline.

4.3. Analysis of Crack Propagation and Evolution in Saturated Sandstone with Void Defects

Fracture Characteristics of Saturated Sandstone with Void Defects

To investigate the crack evolution behavior of specimens with void defects under the combined effects of water saturation and non-uniform loading, the fracture evolution processes and morphologies of three specimen groups—M-B-S, M-S-B, and B-M-S—were recorded using a high-speed camera.

As shown in Figure 7, under non-uniform loading conditions, the saturated sandstone specimen in the M-B-S group is divided from left to right into medium-stress, high-stress, and low-stress zones. Figure 7a presents a physical image of crack propagation in the M-B-S specimen. To facilitate a clearer analysis of the crack propagation process and fracture morphology, a schematic diagram depicting the crack evolution and fracture pattern for the M-B-S group is shown in Figure 7b. The M-B-S specimen exhibits a tensile-shear mixed-mode failure pattern. The crack configuration is primarily characterized by a dominant through-going main crack, accompanied by secondary cracks with irregular propagation paths, as well as localized spalling zones at the crack tips. The crack propagation sequence in the M-B-S specimen begins with the initiation of tensile crack 1a in the high-stress zone. This crack propagates nearly vertically through the high-stress region, extending in the direction of maximum principal stress. It exhibits linear geometry with clean edges and a relatively stable width, indicating that tensile forces dominate the initial stage of crack growth. Subsequently, under the influence of non-uniform forces acting at the top of the specimen, tensile crack 1^a extends toward the low-stress region in the later stages of its development, triggering the formation of shear crack 2^a. Crack 2^a exhibits bending and sliding characteristics, with irregular edges and rough surfaces typical of shear slip behavior. This behavior reflects the features of stress concentration and slip, which are characteristic of shear failure modes. Subsequently, under the combined influence of tensile shear stress disturbances from cracks 1^a and 2^a, as well as stress concentration around the void defect, rock spalling occurs at the interface between the high- and low-stress zones, leading to the formation of an irregularly contoured spalling zone 3^a. Stress relief in zone 3^a causes a redistribution of the stress field around the arched void, resulting in stress concentration along the left wall of the void and the development of shear crack 4^a. During the propagation of crack 4^a, secondary microcracks with irregular morphologies form due to shear slip, acting as bypass paths that accelerate local instability and failure. Moreover, the specimen's saturated condition further weakens the rock matrix and lowers the threshold for crack propagation, increasing the susceptibility of weak planes to spalling. In the later stages of crack 4^a propagation, under the combined effects of water and shear stress, another spalling zone 5^a forms at the boundary of the low-stress region. Subsequently, a new shear crack 6^a initiates at the tip of spalling zone 3^a, propagating toward the specimen boundary and ultimately leading to the formation of a laminated spalling zone 7^a in the lower right corner.

In summary, the non-uniform loading conditions provide a dominant stress-driven pathway for specimen instability and failure, while the saturated water environment promotes crack activation. Under their combined influence, the M-B-S group demonstrates a progressive failure mode characterized by a chain-reaction process involving stress transmission, crack initiation, and rock block spalling. This process transitions from tensile fracturing to shear failure, ultimately leading to localized rock block instability and spalling.

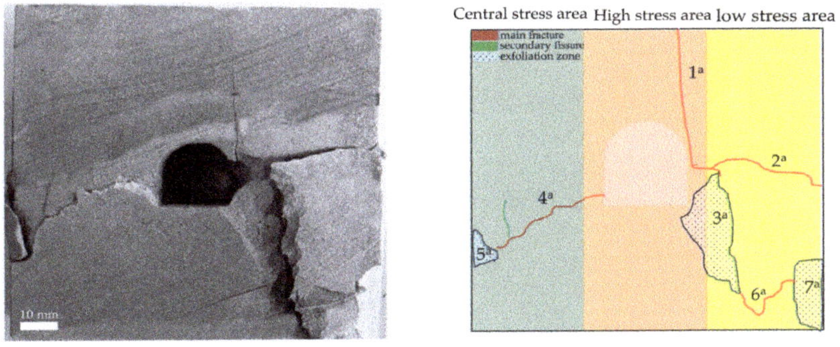

(a) Crack extension of group M-B-S specimens (b) Sketch of crack extension of group M-B-S specimens

Figure 7. Crack development and failure pattern of M–B–S specimens under non-uniform loading. (**a**) Photograph showing the final failure morphology of a specimen from the M–B–S group. (**b**) Schematic illustration of the crack propagation process: Red line: Main fracture path; Green line: Secondary crack path; Shaded areas: Exfoliation (spalling) zones formed near free surfaces. Number labels: Indicate the chronological sequence of crack propagation events. The applied stress configuration for group M–B–S involves a vertical stress distribution from medium–high–low across the top surface (from **left** to **right**).

As illustrated in Figure 8, under non-uniform loading conditions, the saturated sandstone specimen in the M-S-B group is divided from left to right into medium-stress, low-stress, and high-stress regions. Figure 8a shows the physical process of crack propagation in the saturated specimen, while Figure 8b provides a schematic diagram that illustrates the crack propagation path and fracture morphology, offering enhanced clarity for analysis.

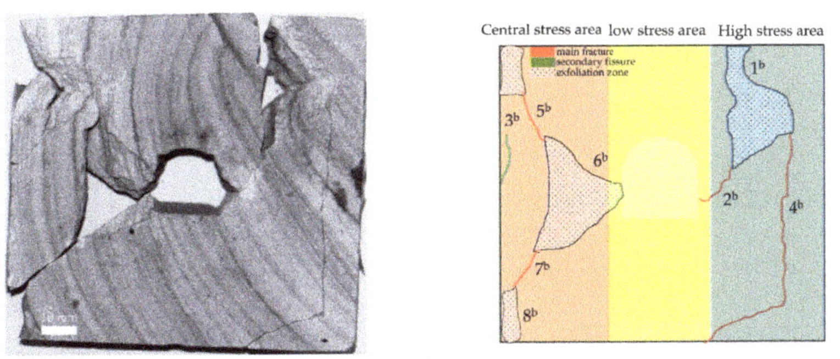

(a) Crack extension of group M-S-B specimens (b) Sketch of crack extension of Group M-S-B specimens

Figure 8. Crack development and failure mode of M–S–B specimens under non-uniform loading. (**a**) Final failure morphology of a specimen from the M–S–B group. (**b**) Schematic illustration of the crack propagation process: Red line: Main fracture path; Green line: Secondary crack path; Shaded areas: Exfoliation (spalling) zones formed near free surfaces. Number labels: Indicate the chronological sequence of crack propagation events.

At the initial stage of loading, the upper portion of the high-stress region experiences the greatest concentration of tensile stress, leading to the initiation of the primary tensile crack, denoted as crack1^b. This crack propagates in a straight trajectory, exhibiting typical tensile-dominated failure behavior. The infiltration of saturated water at the crack tip, combined with the accumulation of pore water pressure, further reduces the tensile strength of the local rock matrix, facilitating rapid crack propagation. The subsequent localized spalling phenomenon indicates that crack 1^b not only represents the point of initial tensile failure but also significantly weakens the surrounding structure, with water-induced softening contributing to more brittle spalling characteristics. The development of crack 1^b leads to the release of local stress fields, inducing stress transmission downward and resulting in the formation of a bending–shear tensile composite crack 2^b on the right wall of the void. The crack path is significantly influenced by the geometric disturbances of the void, and the presence of pore water further promotes crack extension along weak planes, thereby exhibiting a shear–tensile coupled failure mechanism. As failure in the high-stress region propagates, the medium-stress region gradually becomes the primary load-bearing zone. In the upper-left corner, compressive failure limits are reached, leading to the formation of spalling zone 3^b. This region exhibits overall block detachment at the edges, showing rapid instability under the triggering effect of pore water pressure, characteristic of brittle spalling failure controlled by compressive–shear interactions. Next, a primary tensile crack, labeled 4^b, forms along the existing weak plane in the high-stress region. The crack extends from crack 1^b and, under the weakening influence of saturated water, rapidly propagates through the material, exhibiting typical tensile–through failure. The bearing capacity of the medium-stress region gradually decreases, leading to the formation of a tortuous shear crack, designated 5^b, while a spalling zone, referred to as 6^b, emerges in the central-left region. This area represents a localized instability zone following stress redistribution, where pore water further accelerates rock deterioration, exacerbating the development of tensile–spalling coupled failure. As shear crack 5^b extends and spalling zone 6^b experiences localized instability, stress continues to concentrate in the lower-left region of the specimen. The weakening and lubricating effects of pore water contribute to the formation of shear–slip crack 7^b. The crack path is irregular, exhibiting bending and sliding features on its surface, reflecting structural slip failure under compressive–shear interactions. This process is accompanied by the activation of weak planes within the rock, with water infiltration facilitating crack propagation along these planes and reducing friction. Ultimately, based on crack 7^b, a final spalling zone, 8^b, develops in the lower-left corner. This area undergoes block-like spalling due to long-term stress accumulation and the combined weakening effects of pore water. The well-defined boundary indicates that the rock has lost its residual bearing capacity and is undergoing complete instability. The overall failure progresses into its final stage, characterized by a late-stage coupling failure mode that transitions from shear slip to localized spalling.

In summary, under non-uniform loading conditions, the fracture process in the M-S-B group saturated sandstone specimen is initiated by a tensile crack in the high-stress region. This crack progressively evolves into a composite failure involving shear, slip, and spalling. The continuous influence of pore water facilitates crack propagation, amplifies failure intensity, and accelerates rock deterioration. Ultimately, the crack system becomes fully interconnected due to the combined effects of pore water softening and stress accumulation, leading to overall rock instability and failure.

As illustrated in Figure 9, under non-uniform loading conditions, the saturated sandstone specimen in the B-M-S group is segmented from left to right into high-stress, medium-stress, and low-stress regions. The effects of non-uniform loading cause the specimen to display distinct sequential crack evolution characteristics across these varying stress regions. The cracks primarily observed include through-going tensile cracks, secondary tensile cracks, and spalling cracks, all of which are significantly influenced by the weakening effect of saturated water infiltration.

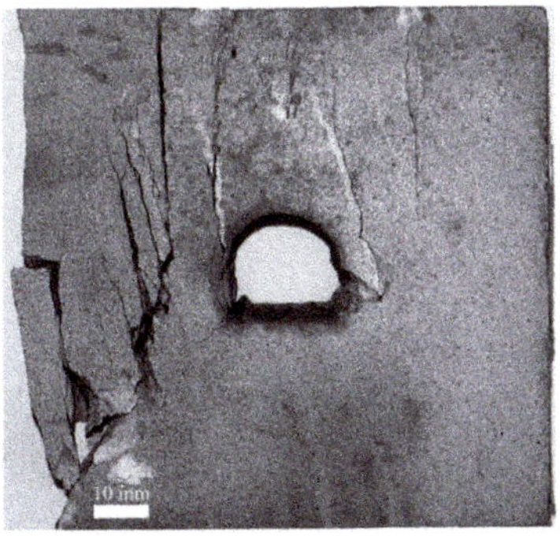

(**a**) Crack extension of group B-M-S specimens

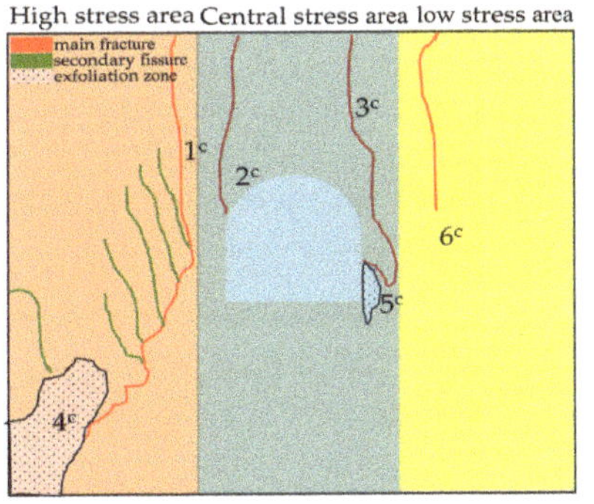

(**b**) Sketch of crack extension in specimens of group B-M-S

Figure 9. Crack development and failure mode of B–M–S specimens under non-uniform loading. (**a**) Final failure image of a B–M–S group specimen, showing typical crack propagation originating near the borehole and progressing through the specimen under heterogeneous loading. (**b**) Schematic illustration of the crack development process: Red line: Main fracture path; Green line: Secondary crack path; Shaded areas: Exfoliation (spalling) zones. Number labels: Indicate the temporal sequence of crack propagation.

Initially, during the early stages of loading, the primary crack, designated as crack 1^c, forms at the upper portion of the high-stress region. This crack is a typical through-going tensile crack characterized by a straight propagation path and rapid extension, reflecting brittle failure under localized tensile stress concentration. Under saturated conditions, pore water infiltrates the crack tip, leading to an increase in pore pressure, which further reduces the tensile strength of the local rock matrix, facilitating rapid crack propagation. This process also induces the simultaneous initiation of several secondary tensile cracks around crack 1^c. These secondary cracks exhibit a radial expansion pattern centered around crack 1^c, indicating a significant water–stress coupling acceleration effect in this region. As crack 1^c continues to propagate, it compromises the integrity of the rock mass in the high-stress region, ultimately leading to the formation of a block-like spalling zone, 4^c, which is a typical tensile–spalling composite crack. The boundaries of this spalling zone are well-defined, and the spalling failure occurs rapidly, reflecting the combined effects of water pressure accumulation at the crack tip and the interconnection of microcracks. In the medium-stress region, due to the redistribution of the stress field, tensile cracks 2^c and 3^c initiate sequentially. Crack 2^c extends toward the upper-left portion of the arched void, while crack 3^c propagates toward the right wall of the void. During the later stages of crack 3^c, it reaches the weak plane at the lower-right corner, further inducing the formation of spalling zone 5^c, which is characterized by local stress concentration and structural weakening due to the combined influence of saturated water infiltration. Finally, in the low-stress region, crack 6^c initiates during the mid-to-late loading phase. This crack has a short propagation path and extends slowly, making it a typical delayed tensile crack. The formation of crack 6^c is primarily influenced by long-term pore pressure accumulation and the degradation of rock strength, indicating that under low-stress conditions, the delayed activation of cracks due to water is more pronounced.

In summary, the B-M-S group saturated sandstone specimen subjected to non-uniform loading demonstrates a composite failure mode characterized by tensile stretching, multi-crack evolution, and localized spalling instability. The formation of cracks is influenced by variations in stress distribution and the combined effects of saturated water. Pore water significantly contributes to the instability and failure of the saturated specimen by reducing rock strength, enhancing the connectivity of microcracks, and activating weak planes.

Through the analysis of crack propagation results in saturated sandstone specimens from various groups, similarities in crack development are evident. Specifically, within the different stress regions of the specimens, the degree of crack development and propagation varies. Based on the extent of crack development, the specimens can be classified into the following stages: high-stress region—intense zone, medium-stress region—active zone, and low-stress region—quiescent phase.

4.4. Acoustic Emission Characteristics of Saturated Sandstone with Void Defects

During the instability and damage process of rocks, primary microcracks continuously expand and develop. The microscopic vibrations generated by these crack formations are transmitted as elastic waves to the surrounding environment [29–31]. Therefore, in laboratory experiments, acoustic emission (AE) monitoring equipment is used to capture the elastic wave signals emitted by rocks during the damage and instability process. By analyzing the characteristics of these signals, the failure and instability of rocks can be predicted. This method is widely used in rock mechanics.

Due to the varying stress field distributions in three groups of non-uniformly loaded sandstone specimens, this study utilizes acoustic emission signal monitoring to investigate the acoustic instability and failure characteristics of saturated sandstone under non-uniform loading conditions. The findings are expected to provide valuable insights for monitoring instability and failure in actual underground tunnels subjected to non-uniform loading.

For the uniaxial compression tests of saturated sandstone conducted under three different non-uniform loading conditions with varying stress magnitudes, the characteristics of the acoustic emission (AE) ring count signals exhibit a distinct trend. This allows the sandstone to be categorized into four stages: quiet stage (I), active stage (II), intense stage (III), and peak stage (IV). As shown in Figure 10a, this represents the stress-ring count trend curve for the M-B-S group of saturated sandstone under uniaxial compression. At the initial stage of acoustic emission (AE) monitoring, the applied load is relatively low, and the microcracks within the sandstone are in a gradual closure phase. During this period, the AE signal values remain low. To facilitate a more detailed analysis of the signal characteristics generated by the sandstone during the early loading process, the AE signal characteristics are magnified. Within the first 450 s of the uniaxial compression process, the AE ring count signals are predominantly concentrated below 500 counts. As the load gradually increases, fluctuations in the ring count signal become evident, and the overall trend of the ring count signal shows an upward trajectory, as indicated by the red arrows. When the uniaxial loading reaches 232 s, the ring count signal exhibits a local single-peak phenomenon, with the peak count reaching 500, and the stress curve shows a slight inflection. At this point, microcracks in the high-stress region of the sandstone close, and the accumulation of microcracks reaches a critical point, generating a large number of elastic waves. Due to the varying stress levels in different regions of the sandstone, the medium-stress and low-stress regions remain in the quiet stage (I). Therefore, this point is recognized as a warning characteristic, marking the sandstone's entry into the active stage. At this stage, part of the sandstone enters the active phase (II), as the overall trend of the ring count signal shows an upward trajectory, with the ring count signal being higher than in the quiet phase. However, due to the non-uniform loading in the experiment, different stress regions of the sandstone are in varying stages. As a result, compared to the quiet phase (I), the early active phase does not show a significant increase in the ring count signal. As loading progresses to the later stages of the active phase, the sandstone fully transitions into this phase, resulting in a sharp increase in the acoustic emission (AE) signal. When the loading reaches 445 s, surface spalling occurs at the arch defect in the high-stress load region of the sandstone, which corresponds to a significant inflection point in the stress curve. The propagation of cracks and surface spalling generates a substantial number of elastic waves, leading to a sharp rise in the AE ring count signal. The local peak value reaches 25,044 counts, and this peak is identified as a critical warning characteristic point. Once the surface spalling is complete, the acoustic emission (AE) ring count signal stabilizes at approximately 11,000 counts. Subsequently, surface spalling occurs at the arched defect in the medium-stress region of the sandstone, accompanied by noticeable fluctuations in the stress curve. The AE ring count signal increases again, reaching a local peak of 25,781 counts. After the spalling process concludes, the ring count signal decreases and stabilizes around 6000 counts. The sandstone then enters the peak stage (IV), during which its load-bearing capacity declines rapidly. The AE ring count signal exhibits a sharp upward trend as surface cracks extend significantly, the main crack propagates, and the sandstone approaches its ultimate compressive strength, leading to instability and failure. At this point, the AE ring count signal peaks at 28,946 counts. Observation of the M-B-S group throughout the entire process reveals that the ring count signal exhibits a stepwise pattern. During the quiet (I) and active (II) stages, as the sandstone undergoes early- and middle-stage loading, internal crack development is slow, and no significant surface cracks form, resulting in fewer acoustic emission (AE) signals. As non-uniform loading progresses, the sandstone transitions into the intense (III) and peak (IV) stages, where primary cracks initiate and propagate, leading to intensified internal crack development. The substantial number of elastic waves generated during this phase is collected, resulting in a sharp increase in the ring count signal, which peaks at the ultimate compressive failure point. The total ring count for the M-B-S group of saturated sandstone throughout the entire test is 1,321,425 counts.

As illustrated in Figure 10b, this graph depicts the stress–ring count trend for the M-S-B group of saturated sandstone during uniaxial compression. In the initial loading phase, the acoustic emission (AE) ring count signals predominantly cluster around 100 counts, exhibiting minimal fluctuations. However, when the loading duration reaches 233 s, a sharp increase in the ring count signal is observed, peaking at 869 counts. This surge indicates that the natural microcracks within the sandstone have been compacted, resulting in slight vibrations. At this juncture, the quiet phase of the sandstone concludes, marking this point as a significant warning characteristic that signifies the transition from the quiet stage (I) to the active stage (II).

As observed in the amplified figure, the characteristics of the acoustic emission (AE) signals exhibit a significant change after the warning characteristic point. Following this point, the sandstone enters the active stage (II), during which the AE ring count signals begin to display considerable fluctuations and an upward trend, as indicated by the red arrows in the figure. Similar to Figure 10a, when the loading reaches 508 s, the stress curve shows an inflection point, and microcracks start to form on the surface of the sandstone, generating a substantial number of AE signals. This results in another sharp increase in the AE ring count signal, reaching a local peak of 28,380 counts. This point is identified as the warning characteristic of the intense stage (III). Subsequently, the AE signals enter a period of stabilization. At the conclusion of the intense stage (III), approximately 575 s in, a significant increase in the AE ring count signal is observed, peaking at 29,423 counts. This indicates the end of the warning period for the intense stage (III). Subsequently, the AE signals stabilize, and a bimodal feature emerges in the AE ring count signals during this stage. Compared to Figure 10a, the acoustic emission (AE) ring count trend for the M-S-B group of saturated sandstone in the active stage (II) exhibits a higher total count than that of the M-B-S group. This indicates that crack development in the M-S-B group is more pronounced during this stage. Furthermore, the total AE ring count for the M-S-B group across all stages reaches 1,410,136 counts.

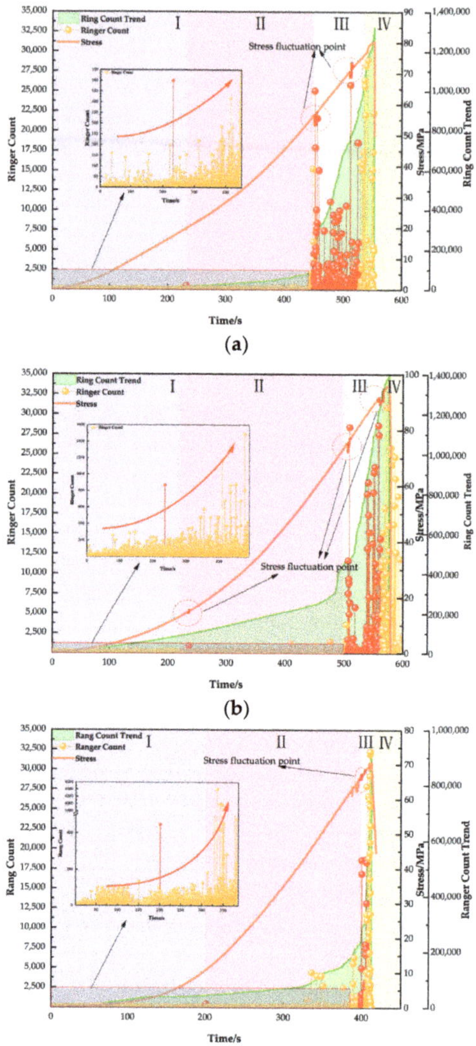

Figure 10. Time–history curves of stress and acoustic emission activity for different loading sequences of saturated sandstone. (**a**) Stress–acoustic emission data plot for group M-B-S, (**b**) Stress–acoustic emission data plot for group M-S-B, (**c**) Stress–acoustic emission data plots for groups B-M-S. Orange curve: Stress evolution curve; Yellow dots: Acoustic emission (AE) ringing count signals; Red dots: Acoustic emission (AE) warning points based on ringing count; Green-shaded region: Cumulative trend of AE ringing count signals; I: Quiet Stage—Initial stage with minimal AE activity and low-stress variation; II: Active Stage—Increased AE activity indicating microcrack initiation; III: Intense Stage—High AE activity corresponding to rapid crack propagation; IV: Peak Stage—Final stage prior to failure, marked by peak stress and AE energy release.

As shown in Figure 10c, this is the stress–ring count trend curve for the B-M-S group of saturated sandstone during uniaxial compression. During the initial and mid-loading stages, the acoustic emission (AE) ring count signals are mainly concentrated around 50 counts, with fluctuation peaks reaching up to 100 counts. This behavior is attributed to the sandstone being in a saturated state, where the water fills the microcracks within the sandstone and inhibits the crack propagation, reducing the generation of fracture-related AE signals. Particularly during the quiet phase, the microcracks expand slowly due to the hydration effect, leading to a reduction in AE signals. As compressive stress increases, the pore water pressure within the saturated sandstone gradually increases, causing microcracks to propagate along the water channels. As microcracks and localized damage accumulate, a critical point in microcrack propagation is reached at 200 s, leading to a sharp increase in AE ring count signals and the formation of a local peak at 448 counts. Due to the non-uniform loading conditions, the loads applied to different regions of the sandstone vary, causing microcracks in different areas to be at different stages of development. At this point, the sharp increase in AE signals marks the transition of cracks in some regions into the active stage (II); thus, this point is identified as a warning characteristic. When the loading reaches 400 s, the AE ring count signal exhibits a warning characteristic, with a local peak value of 18,610 counts, indicating that the sandstone has entered the intense stage (III). Due to the non-uniform stress distribution, with high stress in the left region, medium stress in the center, and low stress in the right region, significant stress concentrations occur in the left and middle regions of the saturated sandstone. Additionally, under the influence of water infiltration, cracks rapidly propagate in these regions over a short period of time, causing the sandstone to transition quickly from the intense stage (III) to the peak stage (IV). Therefore, the intense stage (III) for this group of saturated sandstone is relatively short, and the characteristic end point of the intense stage is reached rapidly, with the AE ring count reaching 18,720 counts. Similar to the M-B-S and M-S-B groups of saturated sandstone during the intense stage (III), the AE ring count signals exhibit a bimodal cyclic characteristic. Subsequently, the sandstone enters the peak stage (IV), where the AE ring count signal reaches the overall peak value of 32,294 counts. Compared to the other two groups, the total AE ring count for the B-M-S group of saturated sandstone throughout the entire process reaches 925,572 counts.

By analyzing the acoustic emission (AE) characteristics of sandstone under non-uniform loading with three different stress configurations, the AE ring count signals for the three groups of sandstone show a consistent upward trend during both the quiet stage (I) and the active stage (II), though with varying curvatures. A comparative analysis reveals the following trend: B-M-S group > M-S-B group > S-B-M group. In the intense stage (III), the AE ring count signals for all three groups of saturated sandstone exhibit a cyclic pattern. Each cycle is marked by a sharp increase in the signal, forming a single peak, followed by a decline to a stable region. During the intense stage (III), the signal characteristics cycle twice, ultimately forming a distinct bimodal pattern in the AE ring count signals throughout this phase. Upon entering the peak stage (IV), all three groups of saturated sandstone undergo instability and failure, resulting in a substantial increase in AE signals. The AE ring count signal reaches its peak during this phase. However, the total AE ring count signal varies in magnitude, following this trend: M-S-B group > M-B-S group > B-M-S group.

According to previous studies [32–34], the variation in acoustic emission (AE) peak frequency of sandstone samples under uniaxial compression with uniform loading exhibits distinct zonal characteristics. Building on this foundation, the present study investigates the AE peak frequency behavior of sandstone subjected to non-uniform loading conditions, using the uniform loading scenario as a baseline. As illustrated in the corresponding figure, the AE peak frequency signals from different sandstone groups are categorized into four distinct regions: low-frequency (I), medium-frequency (II), high-frequency (III), and ultra-high-frequency (IV). In Region I, the M-B-S group shows a lower proportion of peak frequency signals during the early loading stage, which increases significantly in the later stages. The M-S-B group demonstrates a high proportion of signals in both early and late stages, with a noticeable reduction during the intermediate phase. In contrast, the B-M-S group maintains a consistently high level of low-frequency signals throughout the entire loading process. In Region II, the M-B-S group's AE peak frequency signals are concentrated in the 125–150 KHz range, forming a narrow band that persists through all loading phases. The M-S-B group initially exhibits a broader band centered around 100–150 KHz, which progressively narrows during the mid-to-late stages and transitions into a narrower band around 125–150 KHz. The B-M-S group displays early-stage fluctuations around 125–150 KHz, followed by increased signal activity in the mid-to-late stages, dispersing across the entire medium-frequency region. In Region III, the M-B-S group's signals are predominantly distributed between 200–250 KHz and 260–290 KHz, with a signal void in the 250–260 KHz range. The M-S-B group, by contrast, exhibits no such void and presents the most active and densely distributed high-frequency signals among the three groups. The B-M-S group shows the lowest signal density in this region, with activity mainly confined to 200–225 KHz and 250–275 KHz, leaving a blank interval between 225 and 250 KHz. In Region IV, both the M-B-S and M-S-B groups exhibit similar characteristics, with high AE signal activity sustained within the 300–350 KHz range across the entire test duration. The B-M-S group presents an early-stage concentration of signals, followed by a sharp increase in mid-to-late stages, where the signals expand from 300 to 325 KHz to a broader 300 to 375 KHz range, peaking in the plastic failure phase. Figure 11b,c further reveal that the M-S-B group demonstrates extremely high signal intensities in the later loading stages, surpassing those observed in the M-B-S and B-M-S groups, whereas the B-M-S group shows a complete signal void during the early and middle phases. Compared to the uniform loading scenario, sandstone subjected to non-uniform loading exhibits more prominent AE peak frequency transition phenomena across all four frequency regions. Notably, the B-M-S group displays the most pronounced transitions, particularly between Regions I and II and between Regions II and III. This behavior can be attributed to the differential stress distributions induced by non-uniform loading, which cause coexisting states across different regions of the sample during the same time interval, thereby enhancing the signal transition effects in AE peak frequency behavior.

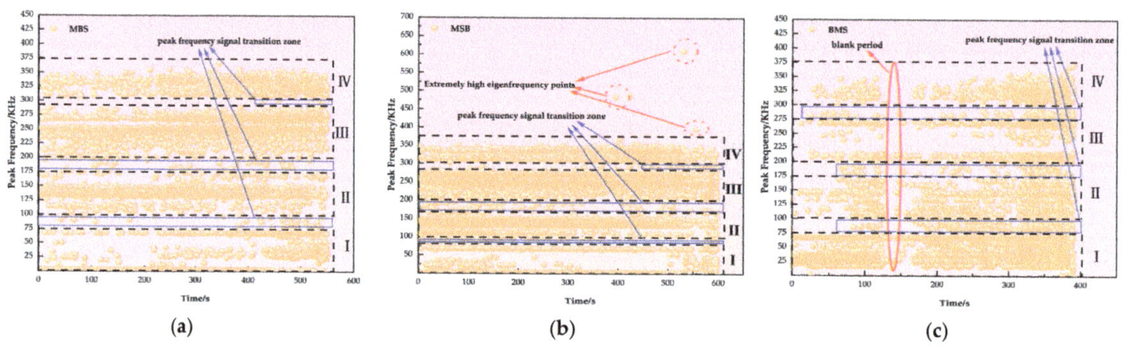

Figure 11. Acoustic emission peak frequency distributions of various saturated sandstone groups subjected to non-uniform loading. (**a**) M–B–S group, (**b**) M–S–B group, and (**c**) B–M–S group. Each scatter point represents an AE event's peak frequency over time. The horizontal axis indicates time (s), and the vertical axis shows the AE peak frequency (kHz). Four distinct frequency bands (labeled I–IV) are separated by dashed lines, indicating characteristic AE ranges. Rd dashed circles mark extremely high eigenfrequency points, possibly linked to sudden micro-cracking events. The red ellipse indicates a blank period with minimal AE activity. The blue polyline highlights the transition zone of AE signal frequencies, suggesting structural damage evolution. (**a**) Scatter plot of acoustic emission peaks of M-B-S group; (**b**) Frequency scatter plot of group M-S-B acoustic emission peaks; (**c**) Scatter plot of acoustic emission peak frequency for group B-M-S.

4.5. Numerical Simulation Results and Analysis

4.5.1. Numerical Modeling

FLAC^{3D} is a three-dimensional numerical simulation software based on the Lagrangian algorithm, extensively used for analyzing mechanical problems in geotechnical engineering. It is capable of handling complex geometries and nonlinear material behavior, enabling the simulation of rock deformation, stress distribution, and other mechanical responses under various loading conditions. FLAC^{3D} is particularly effective in identifying plastic zones, crack evolution, and failure mechanisms in rock masses subjected to loading [35,36]. In this study, FLAC^{3D} is utilized to simulate the failure process of sandstone specimens under uniaxial compression. A three-dimensional numerical model, with dimensions identical to those of the physical sandstone sample, is first constructed to generate a heterogeneous representation that closely reflects the actual material characteristics [37]. Following model construction and material property assignment, boundary conditions are applied. Given the uniaxial loading scenario, confining pressure is not considered; velocity constraints are applied solely at the base of the model, while the lateral boundaries remain free. The FISH programming language is employed to define the non-uniform loading stress distribution at the top surface of the model. A servo control mechanism is implemented to replicate the real-time loading process, as illustrated in Figure 12.

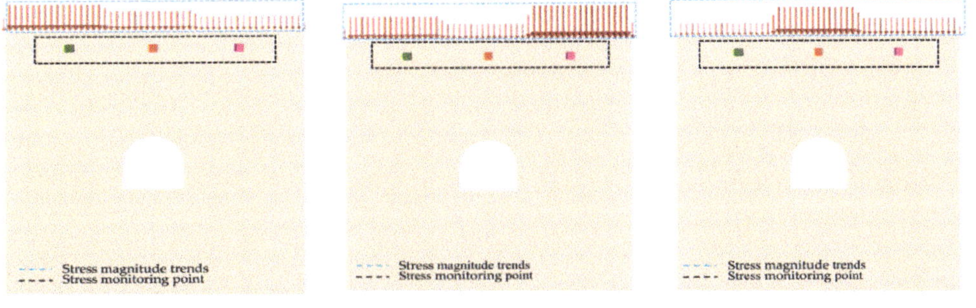

(a) (b) (c)

Figure 12. Numerical model and monitoring points. (**a**) M-B-S numerical modeling, (**b**) M-S-B numerical modeling, (**c**) B-M-S numerical modeling. The squares represent stress monitoring points set through the numerical simulation software.

Prior to performing the simulation experiments, it is essential to calibrate the parameters of the numerical model. This calibration is carried out by establishing a simulation environment that closely replicates the actual laboratory testing conditions. A trial-and-error approach [38] is employed to iteratively adjust the model parameters until the numerical results converge with the experimental data obtained from physical tests. The calibrated mechanical parameters are listed in Table 2. To monitor the evolution of the stress field across different loading zones, three sets of stress monitoring points are strategically placed within the non-uniform loading region at the top of the numerical model. To validate the mechanical parameters presented in Table 2, the simulation results of the M-S-B group specimen are compared with the corresponding laboratory measurements, as illustrated in Figure 13. The numerical simulation yields a uniaxial compressive strength of 95.1 MPa and a peak strain of 1.4%, while the laboratory test reports a uniaxial compressive strength of 95 MPa and a peak strain of 1.33%. The close agreement in both peak strength and strain confirms the reliability of the parameter calibration. Moreover, the damage distribution pattern observed in the simulation closely aligns with the macroscopic failure morphology obtained from the physical test, demonstrating that FLAC^{3D} is a valid and reliable tool for investigating the mechanical behavior of fractured rock masses.

Table 2. Simulation parameters related to sandstone specimens.

Designation	Bulk Modulus/Pa	Shear Modulus/Pa	Densities kg/m^3	Cohesion/Pa	Angle of Internal Friction/°	Tensile Strength/Pa
sandstone	12×10^9	6×10^9	2600	4×10^6	35	2×10^6

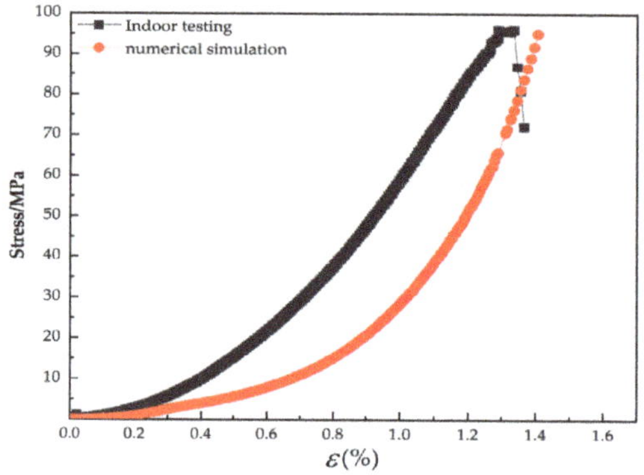

Figure 13. Numerical simulation and physical test stress–strain curves.

4.5.2. Distribution Law of Saturated Non-Uniform Stress Field of Defects Containing Holes

The application of non-uniform loading during the uniaxial compression of the sandstone model results in a distinct non-uniform stress distribution in the upper region of the specimen. As loading progresses and internal stress concentrations develop within the sandstone, various stress zones exhibit phenomena of either stress transfer or contraction. However, the overall stress distribution of the sandstone specimen follows a banded pattern that corresponds to the identified stress zones.

In the high-stress region, the stress field distribution varies depending on its location. In the BMS and MSB groups, the high-stress regions are located on the left and right sides of the sandstone model, respectively. Both groups share a common feature: stress penetration occurs within these high-stress areas, leading to stress expansion. In contrast, the high-stress region in the MBS group is located at the center of the model. Due to the arched defect in the center of the saturated sandstone model, no stress penetration occurs. Instead, stress is transferred from the high-stress region to the low- and medium-stress regions.

In the medium-stress region, the behaviors of the different groups exhibit distinct characteristics. In the MBS group, the medium-stress area is "cut off" due to stress transfer from the high-stress zone, preventing stress penetration. In the MSB group, the medium-stress region not only shows stress penetration but also exhibits an expanding trend. In the BMS group, the medium-stress area is located in the central part of the model, where stress penetration does not occur; however, a trend of stress transfer is observed.

In the low-stress region, all three saturated sandstone groups demonstrate similar behavior, regardless of the location of the stress zones. In each case, the low-stress regions show a contraction trend. Furthermore, as shown in the figure, stress concentrations are evident on both sides of the arched defect at the center of the sandstone model.

From the analysis of the figure, it is clear that the three groups of saturated sandstone do not exhibit distinct separation at the boundaries of the stress zones. Rather, a transitional pattern is observed: the high-stress regions consistently exhibit an expanding trend, while the low-stress regions show a contraction trend. The behavior of the medium-stress regions is influenced by their position, leading to either expansion or contraction. Specifically, as shown in Figure 14a, when the medium-stress region is located on the right side, it is influenced by stress transfer from the high-stress region in the center, resulting in stress contraction in the medium-stress area. In Figure 14b, when the medium-stress region is on the left side and the central region is in the low-stress zone, the medium-stress region exhibits stress transfer behavior. In Figure 14c, when the medium-stress region is centrally located, it is influenced by both stress transfer from the high-stress region on the left and stress contraction from the low-stress region on the right. Consequently, the medium-stress region displays a coexistence of stress expansion and contraction. The arched defect at the center of the saturated sandstone model shows stress concentration on both sides, unaffected by the non-uniform loading stress zones.

Plastic zones in three groups of saturated sandstone specimens (M-B-S, M-S-B, and B-M-S) under non-uniform loading conditions at ultimate bearing capacity, as illustrated in Figure 15. The results indicate that plastic failure in all three specimen groups is primarily concentrated in the high-stress regions, where the density of shear failure zones is highest, followed by the medium-stress regions, and lowest in the low-stress regions. This demonstrates pronounced stress sensitivity.

In the high-stress regions, both the M-S-B and B-M-S groups form continuous shear failure bands, exhibiting strong localization and connectivity. In contrast, the M-B-S group shows a deviation in the shear failure path due to geometric disturbances caused by the arched cavity within the high-stress zone. This prevents the formation of a fully connected shear band and leads to the expansion of plastic zones into the medium- and low-stress regions. In the medium-stress zones, the M-B-S and M-S-B groups exhibit widely distributed shear failure elements, whereas in the B-M-S group, these elements are mainly concentrated on both sides of the cavity. This reflects the guiding effect of the loading path on the reconstruction of the shear stress field. Significant differences are observed in the density of shear failure elements within the low-stress regions, following the order M-S-B group > M-B-S group > B-M-S group. This distribution pattern is closely related to the relative position of the stress regions along the loading path. In comparison, the distribution of tensile failure elements across the three groups is relatively consistent, predominantly concentrated around the arch of the central cavity. This indicates that local tensile stress concentration induced by the arched structure governs the tensile failure behavior, while the influence of the non-uniform loading path is comparatively weak.

As shown in Figure 15d–f, the spatial distribution of tensile failure plastic elements in all three groups under different non-uniform loading paths is highly consistent. These elements are primarily concentrated at the top and lateral sides of the cavity, forming a symmetric tensile stress concentration zone centered on the arched cavity. This pattern remains largely unchanged across different stress region layouts, suggesting that geometric stress concentration induced by the cavity structure dominates the tensile failure formation process rather than the loading path. Under external loading, vertical tensile stress tends to concentrate at the arch apex, thereby inducing tensile failure. Additionally, the presence of pore water in the saturated state reduces the tensile strength of the material, further enhancing failure sensitivity around the cavity. Overall, the spatial distribution of tensile failure plastic zones is primarily governed by the combined effects of geometric structure and pore water, while the influence of the non-uniform loading path remains relatively minor, indicating that the spatial evolution of tensile failure is strongly controlled by geometry.

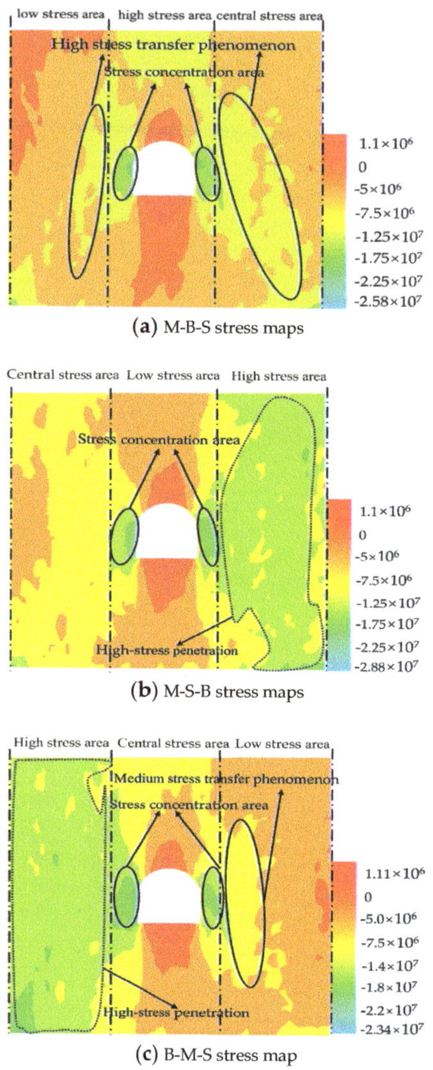

Figure 14. Stress distribution characteristics of rock specimens under different loading sequences and corresponding stress transfer mechanisms. Stress distribution maps under three different loading sequences: (**a**) M-B-S, (**b**) M-S-B, and (**c**) B-M-S. The color scale represents the magnitude of stress (in Pa), with red indicating tensile stress and green indicating compressive stress. The stress distribution maps illustrate the internal stress characteristics of the specimen. Stress concentration areas and stress transfer paths are highlighted. Attention should be focused on the stress transfer phenomenon.

In conclusion, the numerical simulation results reveal the spatial evolution patterns and controlling factors of shear and tensile failure mechanisms in saturated sandstone. The distribution of shear failure plastic zones is influenced by the combined effects of stress zone division, loading path, and geometric disturbances from cavities, showing clear stress sensitivity and path dependence. In contrast, the distribution of tensile failure elements is primarily governed by geometric stress concentrations induced by cavities, exhibiting a high degree of consistency across different loading paths and weaker dependence on loading conditions. The presence of pore water further reduces the tensile strength of the rock mass, intensifies the tensile failure response, and accelerates the development of localized failure.

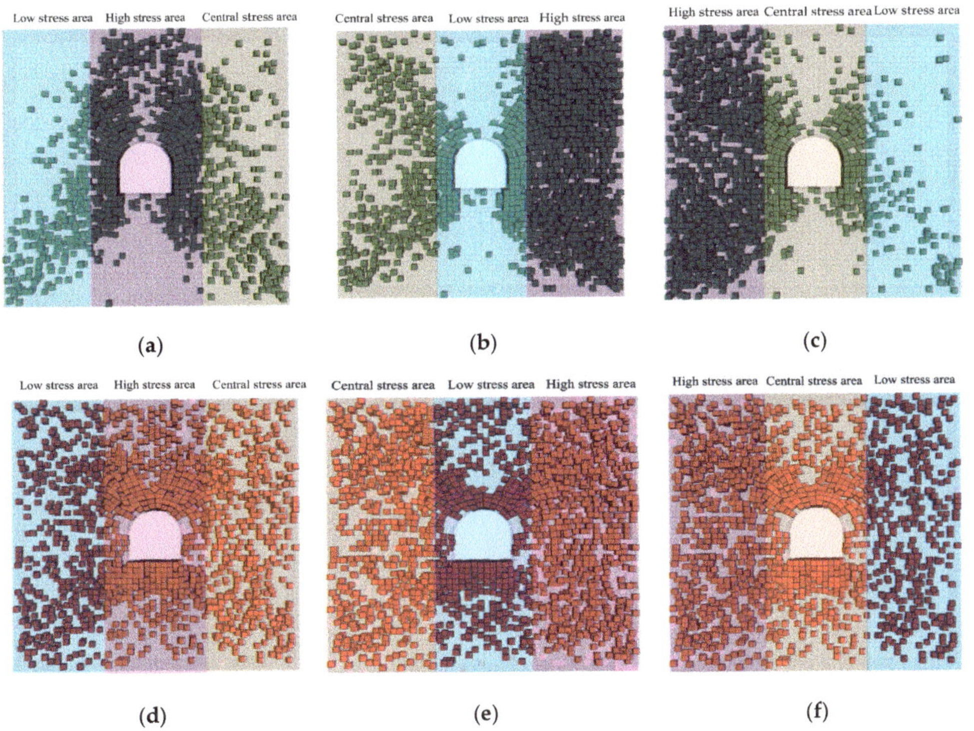

Figure 15. Spatial distribution of specimen shear–tension damage. (**a**) M-B-S shear damage zone, (**b**) M-S-B shear damage zone, (**c**) B-M-S shear damage zone, (**d**) M-B-S tensile damage zone, (**e**) M-S-B tensile damage zone, (**f**) B-M-S tensile damage zone.

4.5.3. Stress Analysis of the Upper Part of the Specimen

To analyze the stress distribution characteristics in the upper region of the specimen more clearly, stress monitoring points were set in different stress zones at the top of the numerical model. As shown in Figure 16, the stress monitoring curves for the upper regions of the M-B-S, M-S-B, and B-M-S groups of saturated sandstone specimens are presented. By observing the changes in the stress curves, the upper stress curve can be divided into three phases: the similar phase (I), the sub-discrete phase (II), and the discrete phase (III). From time steps 0 to 4000, the stress curves of the different groups are in a similar phase (I). During this period, the loads applied to the upper regions of the saturated sandstone specimens are averaged, and no non-uniform loading zones have formed. Between time steps 4000 and 6000, the stress curves enter the sub-discrete phase (II), and the effects of non-uniform loading begin to emerge. The upper regions of the saturated sandstone specimens start to exhibit stress partitioning. After time step 6000, the stress curves of the different saturated sandstone specimens show significant discrete behavior, entering the discrete phase (III). The degree of dispersion in the stress curves follows the order MSB > MBS > BMS.

Since the numerical simulation is conducted under idealized conditions, and external factors are excluded, the phenomenon of stress curve dispersion is closely related to the arrangement of non-uniform loading stress zones at the upper part of the model. As shown, in the M-B-S group, the upper stress zones from left to right are the low-stress zone, high-stress zone, and medium-stress zone, with peak strengths of 38.2 MPa, 84.7 MPa, and 60.9 MPa, respectively. In the M-S-B group, the upper stress zones from left to right are the medium-stress zone, low-stress zone, and high-stress zone, with peak strengths of 53.6 MPa, 35.7 MPa, and 96.5 MPa, respectively. In the B-M-S group, the upper stress zones from left to right are the high-stress zone, medium-stress zone, and low-stress zone, with peak strengths of 76 MPa, 56.6 MPa, and 49.5 MPa, respectively. The stress values between adjacent stress zones differ significantly, resulting in an increased degree of stress curve dispersion.

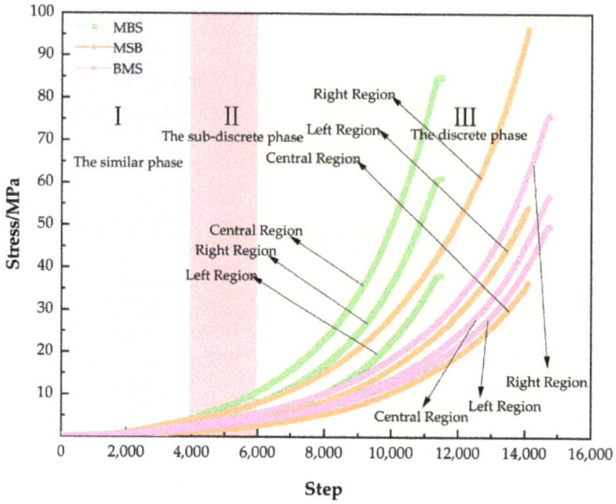

Figure 16. Upper stress field curve.

5. Discussion

Under non-uniform loading conditions, this study conducted uniaxial compression tests on saturated sandstone using three different loading sequences to simulate distinct stress zoning patterns. The experimental results revealed the complex interactions among loading sequence, stress distribution, fracture evolution, acoustic emission (AE) characteristics, and internal failure modes.

Among the tested configurations, the M-S-B group exhibited the highest compressive strength due to the central region bearing the maximum load. This suggests that concentrating stress within the structural core enhances the load-bearing capacity of the rock mass. Additionally, the spatial location of the simulated roadway within various stress zones significantly influenced the failure behavior. When the roadway was located in high- or intermediate-stress zones, intense and asymmetric fracture propagation was observed along the sidewalls. In contrast, when positioned in a low-stress zone, the fracture development was more subdued, and the overall failure behavior remained stable. These findings indicate that, in practical engineering applications, roadway placement should preferably avoid zones of high stress concentration to reduce the risk of localized failure and improve structural stability.

The AE characteristics exhibited distinct phase-based evolution throughout the loading process, which can be categorized into four stages: quiet, active, violent, and residual. Precursor signals were commonly observed during the active and violent stages, with a pronounced dual-peak AE count response emerging during the violent phase. This bimodal pattern corresponds to two critical energy release events in the fracture process and serves as a reliable indicator of impending macroscopic failure, providing a valuable reference for the development of real-time early warning systems in underground engineering.

The numerical simulation results closely matched the observed fracture propagation paths, with clear stress concentrations identified near the roadway boundaries. This further validated the effectiveness of the simulation approach in identifying structural weak zones. Shear failures were predominantly concentrated in high-stress regions, while tensile failures were more uniformly distributed, indicating that shear failure is more sensitive to localized stress environments. These observations underscore the importance of implementing targeted reinforcement strategies tailored to the dominant failure mechanisms in different zones.

Moreover, the dispersion observed in the upper stress curves revealed increased stress non-uniformity during vertical stress transmission. This phenomenon is closely associated with delayed failure onset and increased complexity in fracture propagation paths. Recognizing this behavior is critical for informing the design of staged or time-dependent support systems in deep underground projects.

This study not only provides experimental evidence for understanding the failure mechanisms of saturated rock masses under complex loading conditions but also offers theoretical insights and engineering guidance for mining hazard management and underground space design. The key implications are as follows.

(1) Optimized roadway layout:

Avoid placing roadways in high stress concentration zones to enhance surrounding rock stability.

(2) AE-based monitoring systems:

Leverage the dual-peak AE pattern as a critical precursor signal for real-time failure warning.

(3) Zonal support strategy:

Design reinforcement schemes based on the dominant local failure modes to improve efficiency and cost-effectiveness.

(4) Simulation-integrated design:

The validated stress field simulation can reliably identify structural weak zones, supporting its integration into engineering design and layout optimization.

Despite the insights gained from this study regarding the mechanical response and AE behavior of saturated sandstone under non-uniform loading, certain limitations remain. First, the loading path and stress zoning were idealized and did not fully replicate the complexity of multi-axial stress coupling in actual geological conditions, which may limit the direct applicability of the results to real-world engineering scenarios. Second, AE data analysis primarily focused on count-based indicators without incorporating multi-parameter information such as AE energy, which constrains the depth of fracture process characterization. Finally, the simulated roadway was modeled as a symmetric arched structure without considering the effects of varying geometric forms on stress perturbation and failure mechanisms. Future studies should incorporate diverse roadway cross-sectional shapes and stress disturbance scenarios to improve the generalizability and engineering relevance of the findings.

6. Conclusions

In this study, non-uniform uniaxial compression physical experiments were conducted on saturated sandstone specimens to investigate, from a macroscopic perspective, the distribution patterns of the upper stress field, the evolution of crack propagation, and the acoustic emission characteristics under non-uniform loading conditions. Additionally, by employing the FLAC^{3D} numerical simulation program, a mesoscopic analysis was carried out to examine the influence of non-uniform loading paths on the internal stress field distribution, the spatial patterns of shear and tensile failure elements, and the stress curve characteristics in the upper regions of the specimens. The main conclusions drawn from this study are as follows. According to the theory of material mechanics, using the different physical properties of the modulus of elasticity of the metal gasket, the M-B-S group, M-S-B group, B-M-S group of water-saturated sandstone specimens for the non-uniform load uniaxial compression test, the resulting stress–strain curve is still the existence of the stage characteristics, divided into the compression stage, elasticity, plastic yielding stage, the destruction of the destabilization stage, to obtain the peak uniaxial compressive strength of the peak uniaxial compressive strength, respectively, 70.7 MPa, 95 MPa, 78 MPa; non-uniform load force arrangement order on the peak uniaxial compressive strength of sandstone specimens have a significant effect.

(1) Based on the principles of material mechanics and utilizing the distinct physical properties of the elastic modulus of metal gaskets, non-uniform loading uniaxial compression tests were conducted on the saturated sandstone specimens from the M-B-S, M-S-B, and B-M-S groups. The resulting stress–strain curves exhibit clear characteristic stages, including the compaction stage, elastic stage, plastic yield stage, and failure instability stage. The peak uniaxial compressive strengths for these groups are 70.7 MPa, 95 MPa, and 78 MPa, respectively. The arrangement order of the non-uniform loading forces significantly affects the peak uniaxial compressive strength of the saturated sandstone specimens.

(2) Under the influence of non-uniform loading, crack propagation in the saturated sandstone specimens develops differently across various stress regions. Based on the extent of crack extension, the regions are classified as follows: the high-stress region exhibits intense crack propagation, the medium-stress region shows active crack propagation, and the low-stress region undergoes calm crack propagation. The arrangement of the stress regions significantly influences the failure of the central cavity defect: when the central arched cavity defect is located in the high-stress or medium-stress regions, crack propagation and rupture behavior of both walls are intense. However, when the central arched cavity defect is located in the low-stress region, crack propagation and rupture behavior of both walls are relatively calm.

(3) The acoustic emission (AE) ring counts for the different saturated sandstone groups during non-uniform loading show distinct stage characteristics. Based on the activity of the AE ring count signals, the process can be divided into the calm stage, active stage, intense stage, and residual stage. Pre-warning characteristics are observed during both the active and intense stages for all groups. In the intense stage, all three groups of saturated sandstone exhibit a bimodal trend in AE ring counts. The total AE ring count is highest for the M-S-B group, followed by the M-B-S group, and then the B-M-S group. Additionally, the AE peak frequency displays a banded zoning phenomenon, categorized into low-frequency, medium-frequency, high-frequency, and ultra-high-frequency zones. Under the influence of non-uniform loading, transitional signals in peak frequency appear between different regions, with each group of saturated sandstone showing differentiated characteristics.

(4) A comparative analysis of the internal stress field distributions among different saturated sandstone groups reveals that non-uniform loading induces complex stress transfer phenomena within the rock mass. These phenomena are strongly influenced by the spatial configuration of stress zones: high-stress regions predominantly exhibit stress expansion; medium-stress regions display diverse behaviors including expansion, contraction, or both; whereas low-stress regions consistently exhibit stress contraction. The numerically simulated stress fields are in strong agreement with the macroscopic crack propagation patterns observed in physical model experiments, thereby validating the accuracy and reliability of the simulation approach. Under all three stress distribution

patterns, stress consistently concentrates around the central arched cavity walls, identifying this area as structurally vulnerable under non-uniform stress conditions.

(5) Comparing the spatial distribution of failure modes in the different saturated sandstone groups, it is evident that the spatial distribution of shear failure elements is closely related to the different stress regions in the upper part of the sandstone specimens. The density of shear failure elements follows this order: high-stress region > medium-stress region > low-stress region. The spatial distribution of tensile failure elements is relatively uniform and unaffected by the non-uniform stress regions.

(6) A comparison of the upper stress curves of the different saturated sandstone specimens reveals a significant correlation between the degree of dispersion of the stress curves and the arrangement of the upper stress regions. The greater the difference between adjacent stress regions, the higher the degree of dispersion in the stress curves. The degree of dispersion in the upper stress curves follows the order: M-S-B > M-B-S > B-M-S.

Author Contributions: G.L. and Y.Z. (Yonglong Zan). wrote the main text of the manuscript and the experimental data processing and summaries, while D.W., S.W., Z.Y., Y.Z. (Yao Zeng), G.W., and X.S. were the co-authors responsible for collecting relevant information to organize the structure of the paper. All authors have read and agreed to the published version of the manuscript.

Funding: This research was funded by the Natural Science Foundation of Heilongjiang Province, grant number YQ2023E039; National Natural Science Foundation of China, grant number 52374199; Natural Science Foundation of Heilongjiang Province, grant number ZD2023E008. Basic Scientific Research Operating Expenses of Heilongjiang Provincial Universities and Colleges of China, grant number 2022-KYYWF-0554; Scientific and Technological Key Project of "Revealing the List and Taking Command" in Heilongjiang Province, grant number 2021ZXJ02A03, 2021ZXJ02A04.

Data Availability Statement: The datasets used and analyzed in this study can be provided by the authors Gang Liu and Yonglong Zan according to reasonable requirements.

Acknowledgments: This research was supported by the Heilongjiang Ground Pressure and Gas Control in Deep Mining Key Laboratory, Heilongjiang University of Science and Technology.

Conflicts of Interest: Authors Gang Liu and Zhitao Yang were employed by the company Baotailong NewMaterial Co., Ltd. The remaining authors declare that the research was conducted in the absence of any commercial or financial relationships that could be construed as a potential conflict of interest.

References

1. Wu, J.; Yang, X.; Song, Y.; Sun, Q.; Pei, Y. Study on a new inversion method for non-uniform distribution of rock material parameters. *J. Bull. Eng. Geol. Environ.* **2022**, *81*, 280. [CrossRef]
2. Liu, H.; Zhang, X.; Lin, H.; Chen, Y.; Yang, C.; Ou, K. Shear mechanical properties of rock joints under non-uniform load based on DEM. *J. Appl. Sci.* **2025**, *15*, 4257. [CrossRef]
3. Wang, X.; Wen, Z.; Jiang, Y.; Huang, H. Experimental study on mechanical and acoustic emission characteristics of rock-like material under non-uniformly distributed loads. *J. Rock Mech. Rock Eng.* **2018**, *51*, 729–745. [CrossRef]
4. Wang, T.; Ge, L.; Zhao, H.; Zhang, X.; Zhang, H.; Li, W. Deformation damage characteristics and constitutive model of raw coal and briquette with central hole under asymmetric loading. *J. China Coal Soc.* **2022**, *47*, 4040–4054. [CrossRef]
5. Yu, W.; Pan, B.; Wu, G.; Guo, H. Experimental study on mechanical characteristics of rock of trapezoidal cross section under the equivalent non-uniform loading conditions. *J. Min. Strat. Control Eng.* **2024**, *6*, 108–117. [CrossRef]
6. Zhao, H.B.; Hu, G.L.; Wang, F.H.; Ju, N.S. Quantitative analysis of crack expansion in specimens of coal having a single pre-existing hole. *J. China Coal Soc.* **2017**, *42*, 860–870. [CrossRef]
7. Ge, L. Study on Mechanical Properties and Damage Mechanism of Coal and Rocks with Hole Defects Under Eccentric Load. Ph.D. Thesis, Liaoning Technical University, Fuxin, China, 2022. [CrossRef]
8. Heiland, J. Laboratory testing of coupled hydro-mechanical processes during rock deformation. *Hydrogeol. J.* **2003**, *11*, 122–141. [CrossRef]
9. Li, Y.; Gao, S.; Shen, S.; Zhang, Q.; Chen, W.; Hu, L. Experimental study on influence of immersion times on rock strength and failure characteristics. *Chin. J. Rock Mech. Eng.* **2025**, *44*, 1–15. [CrossRef]

10. Duan, X.; Zhang, Q.; Cao, Y.; Zhao, D.; Wang, W.; Zhu, Q.; Liu, S. Study on water sensitivity effect and mechanical damage characteristics of Xiyu conglomerate under complex water environment and stress condition. *Chin. J. Rock Mech. Eng.* **2024**, *43*, 2992–3004. [CrossRef]
11. Zhang, T.; Zhang, L.; Li, S.; Liu, J.; Ji, X.; Pan, H. Characteristics of wave velocity and power spectral density of hole-containing specimens under water conditions. *J. China Coal Soc.* **2018**, *43*, 3387–3394. [CrossRef]
12. Zhou, Z.; Cheng, X.; Cai, X.; Zhao, Y.; Li, X.; Du, K. Mechanical and infrared radiation properties of sandstone with different water contents under uniaxial compression. *J. Cent. South Univ. Sci. Technol.* **2018**, *49*, 1189–1196.
13. Thörn, J.; Ericsson, L.O.; Fransson, Å. Hydraulic and hydromechanical laboratory testing of large crystalline rock cores. *J. Rock Mech. Rock Eng.* **2015**, *48*, 61–73. [CrossRef]
14. Cardona, A.; Finkbeiner, T.; Santamarina, J.C. Natural Rock Fractures: From Aperture to Fluid Flow. *J. Rock Mech. Rock Eng.* **2021**, *54*, 5827–5844. [CrossRef]
15. Von Planta, C.; Vogler, D.; Chen, X.; Nestola, M.G.C.; Saar, M.O.; Krause, R. Simulation of hydro-mechanically coupled processes in rough rock fractures using an immersed boundary method and variational transfer operators. *J. Comput. Geosci.* **2019**, *23*, 1125–1140. [CrossRef]
16. Hao, R.; Cao, P.; Wang, H.; Jin, J. Mechanical properties and fracture behaviors of sandstone in dry and saturated states with tip intersecting cracks. *Chin. J. Rock Mech. Eng.* **2024**, *43*, 1696–1706. [CrossRef]
17. Luo, Z.S.; Zhu, Z.X.; Su, Q.; Li, J.L.; Deng, H.F.; Yang, C. Creep simulation and deterioration mechanism of sandstone under water-rock interaction based on parallel bond model. *J. Rock Soil Mech.* **2023**, *44*, 2445–2457.
18. Chen, X.; Yu, F.; Tang, M.; Xiao, Y.; Zeng, P.; Zhong, Z. DEM simulation of progressive damage and permeability evolution of sandstone during direct-shearing. *Chin. J. Rock Mech. Eng.* **2024**, *43*, 3161–3174.
19. Luo, K.; Zhao, G.; Zeng, J.; Zhang, X.; Pu, C.Z. Fracture experiments and numerical simulation of cracked body in rock-like materials affected by loading rate. *Chin. J. Rock Mech. Eng.* **2018**, *37*, 1833–1842.
20. Tian, W.L.; Yang, S.Q.; Huang, Y.H. PFC2D simulation on crack evolution behavior of brittle sandstone containing two coplanar fissures under different confining pressures. *J. Min. Saf. Eng.* **2017**, *34*, 1207–1215. [CrossRef]
21. Zhang, G.K.; Li, H.B.; Wang, M.Y.; Li, X.F. Study on characteristics of failure strength and crack propagation of granite rocks containing a single fissure. *Chin. J. Rock Mech. Eng.* **2019**, *38*, 2760–2771.
22. Zhu, D.; Zong, Y.J.; Zhang, X.F.; Tao, X.L. Experimental study on influence of sandstone with dome shape defect on its mechanical characteristics. *J. Saf. Sci. Technol.* **2019**, *15*, 30–35.
23. Bayat, N.; Sadeghi, E.; Nassery, H.R. Evaluating the characteristics of geological structures in karst groundwater inflow, Nowsud Tunnel. *J. Mt. Sci.* **2024**, *21*, 3434–3452. [CrossRef]
24. Coli, M.; Pinzani, A. Tunnelling and hydrogeological issues: A short review of the current state of the art. *J. Rock Mech. Rock Eng.* **2014**, *47*, 839–851. [CrossRef]
25. Hack, H.R.G.K. Weathering, Erosion, and Susceptibility to Weathering. In *Soft Rock Mechanics and Engineering*; Kanji, M., He, M., Ribeiro e Sousa, L., Eds.; Springer: Cham, Switzerland, 2020; pp. 235–254. [CrossRef]
26. Xing, Y.; Kulatilake, P.H.S.W.; Sandbak, L.A. Rock mass stability investigation around tunnels in an underground mine in USA. *J. Geotech. Geol. Eng.* **2017**, *35*, 45–67. [CrossRef]
27. Pettijohn, F.J.; Potter, P.E.; Siever, R. *Petrography of Common Sands and Sandstones*; Springer: New York, NY, USA, 1987; pp. 153–217. [CrossRef]
28. Liu, G.; Wang, S.; Wang, D.; Yang, Z.; Zan, Y. Characteristics of Deformation and Damage and Acoustic Properties of Sandstone in Circular Tunnel Morphology under Varying Inundation Depths. *J. Water* **2024**, *16*, 2938. [CrossRef]
29. Song, Y.J.; Cheng, K.Y.; Meng, F.D. Research on acoustic emission characteristics of fractured rock damage under freeze-thaw action. *J. Min. Saf. Eng.* **2023**, *40*, 408–419. [CrossRef]
30. Zhang, X.; Zhang, S.; Wang, L.; Rong, T.; Zhou, R. Identification of pre-peak characteristic stress of sandstone under triaxial loading based on acoustic emission and wave velocity variation. *Chin. J. Rock Mech. Eng.* **2023**, *42*, 2711–2726. [CrossRef]
31. Liu, S.; Gengshe, Y. Study on acoustic emission signal and damage characteristics of frozen sandstone under lateral unloading. *Chin. J. Rock Mech. Eng.* **2023**, *42*, 4237–4247.
32. Garg, P.; Zafar, S.; Hedayat, A.; Moradian, O.Z.; Griffiths, D.V. A novel methodology for characterizing fracture process zone evolution in Barre granite specimens under mode I loading. *Theor. Appl. Fract. Mech.* **2023**, *123*, 103727. [CrossRef]
33. Ali, M.; Wang, E.; Li, Z.; Khan, N.M.; Sabri Sabri, M.M.; Ullah, B. Investigation of the acoustic emission and fractal characteristics of coal with varying water contents during uniaxial compression failure. *Sci. Rep.* **2023**, *13*, 2238. [CrossRef]
34. Mahanta, B.; Ranjith, P.G.; Vishal, V.; Singh, T.N. Temperature-induced deformational responses and microstructural alteration of sandstone. *J. Pet. Sci. Eng.* **2020**, *192*, 107239. [CrossRef]
35. Shang, Z.; Guo, L.; Zhu, Q. Statistical Laws for Stress and Strain Under Uniaxial Compression Test Based on FLAC^{3D}. *J. Hunan Univ. Sci. Technol. (Nat. Sci. Ed.)* **2020**, *35*, 18–26. [CrossRef]

36. Li, Y.; Liu, J.; Tang, X.; Yang, S. Simulation analysis of mechanical properties of the soft rock mass with holes and fractures. *J. Min. Saf. Eng.* **2020**, *37*, 594–603. [CrossRef]
37. Li, G.; Ma, F.; Liu, G.; Zhao, H.; Guo, A. Strain-Softening Constitutive Model of Heterogeneous Rock Mass Considering Statistical Damage and Its Application in Numerical Modeling of Deep Roadways. *Sustainability* **2019**, *11*, 2399. [CrossRef]
38. Li, X.; Li, H.; Yang, Z.; Wang, X.; He, W. Numerical simulation case study by using $FLAC^{3D}$ of the composite coal-rock on the load-free fracture. *J. Saf. Environ.* **2020**, *20*, 2187–2195.

water

MDPI

Article

A Numerical Simulation Study on the Spread of Mine Water Inrush in Complex Roadways

Donglin Fan [1], Shoubiao Li [2,*], Peidong He [1], Sushe Chen [1], Xin Zou [1,2] and Yang Wu [2,*]

1 CHN Energy Shendong Coal Group Co., Ltd., Yulin 719315, China; 20035780@ceic.com (D.F.); 10020204@ceic.com (P.H.); chensushe@sina.com (S.C.); 10019624@ceic.com (X.Z.)

2 State Key Laboratory of Water Resource Protection and Utilization in Coal Mining, National Institute of Low Carbon and Clean Energy, Beijing 102209, China

* Correspondence: 20089529@ceic.com (S.L.); 20039425@ceic.com (Y.W.)

Abstract: Emergency water release from underground reservoirs is characterized by its suddenness and significant harm. The quantitative prediction of water spreading processes in mine tunnels is crucial for enhancing underground safety. The study focuses on an underground roadway in a coal mine, constructing a three-dimensional physical model of the complex tunnel network to explore the spatiotemporal characteristics of water flow spreading after water release in coal mine tunnels. The Volume of Fluid (VOF) model of the Eulerian multiphase flow was adopted to simulate the flow state of water in the roadway. The results indicate that after water release from the reservoir, water flows along the tunnel network towards locations with relatively lower altitude terrain. During the initial stage of water release, sloping tunnels act as barriers to water spreading. The water level height at each point in the tunnel network generally experiences three developmental stages: rapid rise, slow increase, and stable equilibrium. The water level height in the tunnel area near the water release outlet rises sharply within a time range of 550 s; tunnels farther from the water release outlet experience a rapid rise in water level height only after 13,200 s. The final stable equilibrium water level in the tunnel depends on the location of the water release outlet and the relative height of the terrain, with a water level height ranging from 0.3 to 3.3 m. The maximum safe evacuation time for personnel within a radius of 300 m from the drainage outlet is only 1 h. In contrast, areas farther away from the drainage location benefit from the water storage capacity of the complex tunnel network and have significantly extended evacuation opportunities.

Keywords: underground reservoir; numerical simulation; mine water disaster; mine water inrush; water inrush time; three-dimensional model; water flow diffusion process

1. Introduction

Mine water disasters pose a significant threat to the safe production of mines. These disasters easily damage underground equipment and hinder normal mining operations [1], often requiring costly repairs and causing prolonged production delays [2]. In severe cases, uncontrolled water influx may directly lead to the abandonment of the mining area, while posing life-threatening risks to workers through sudden flooding or destabilized workings [3,4]. In recent years, numerical simulation has made remarkable progress in the field of mine safety and planning. Regarding the sudden water inrush caused by emergency water drainage in underground coal mines, researchers have deeply explored key issues such as the mechanism of sudden water inrush, the evolution process of disasters, and emergency response strategies through numerical simulation methods, which have greatly promoted the development of mine safety prevention and control technologies.

Water inrush in underground roadways is extremely hazardous, posing significant threats to miners' lives, causing substantial damage to mining equipment and disrupting production activities. Researchers have conducted extensive studies on the mechanisms of water inrush. Li et al. [5] conducted a study on the mechanism of water inrush from the floor in deep coal mine exploitation, using Fast Lagrangian Analysis of Continua (FLAC 3D) to simulate the activation and development process of floor fractures under the dynamic load of roof collapse. The large number of numerical simulations have been used to reconstruct the dynamic process of the inrush. A comprehensive investigation was performed to examine the interactive progression among stress distribution [6], deformation patterns, and fluid migration dynamics within geological formations [7,8], while critical early warning indicators were systematically identified and characterized [9,10].

In exploring the spatiotemporal evolution law of water flow spreading in roadways, researchers have carried out a great deal of work. Wu Qiang et al. [11,12] established the concept of proof and model technology of water flow spreading in roadways during mine water inrush, which can simulate and obtain the evolution of water flow in underground roadways, thereby assisting in the scientific implementation of emergency rescue for water disasters. Wang et al. [13] employed the computational fluid dynamics (CFD) software CFX to simulate the propagation dynamics of water inrush within bifurcated roadway sections. By analyzing computational data of water level, flow rate, and pressure, the study systematically examined the influence of roadway slope gradient and inflow discharge magnitude of hydrodynamic spreading patterns. Pathfinding algorithms represent a prevalent computational approach for simulating water inrush propagation, primarily incorporating mine structural configurations, graph theory principles, and the characteristic downward-to-upward propagation dynamics of water inrush events. These algorithms enable semi-quantitative computational frameworks to identify water spread pathways within specified timeframes. Li et al. [14] implemented graph-theoretic algorithms to identify water inrush propagation pathways, subsequently applying the Chézy formula to determine propagation velocities and temporal progression. Wu et al. [15] established a roadway model with adjustable gradients and carried out tests on personnel stability and the walking speed of people wading through water under various water flow conditions in roadways with different gradients.

The safe evacuation of personnel during water inrush incidents in underground mines has attracted scholarly attention [16]. The dynamic routing algorithm was developed to identify time-critical evacuation routes in temporal networks, specifically addressing emergency scenarios induced by underground mine water influx [17,18]. Ma et al. [19] established an evacuation time matrix within graph theory constructs, subsequently deriving a tactical navigation system that dynamically adjusts escape routes during mining hydraulic emergencies. Researchers have explored algorithmic approaches to optimize evacuation path planning in mining flood disasters. The researchers proposed to improve the Dijkstra algorithm, which enhances the practicality and effectiveness of the algorithm in complex scenarios. As a result, it becomes possible to calculate the optimal water-evading path when water inrush occurs underground [20,21]. There are also scholars who have combined the Dijkstra algorithm with the breadth-first search algorithm and conducted simulations in the water inrush scenarios of a three-dimensional virtual space [22,23]. Yang et al. [24] constructed a virtual simulation scene with 3D roadway and personnel models based on OpenGL, conducted a visual simulation of emergency evacuation, and evaluated and analyzed the simulation results. They subsequently developed a reference technical framework for the emergency evacuation of underground space workers. A human–environment–computer interaction system was developed, which combines the A*-3D Time algorithm with mixed reality (MR) technology [25]. Ji et al. [26]

established a visualization model for the evacuation of personnel in underground mines and simulated the evacuation process of personnel in underground mines. Through the simulation, information such as evacuation time, exit flow rate, and evacuation routes was obtained. However, the influence of dynamic fluctuations in underground tunnel water levels on personnel evacuation duration remains critically understudied.

Based on the above analysis, scholars have carried out extensive research in the field of mine water disasters, mainly focusing on aspects such as the mechanism of water inrush, the spatiotemporal evolution of water flow propagation, and the safety evacuation of personnel. Currently, the research methods mainly involve the use of numerical simulation software such as FLAC 3D and Storm Water Management Model (SWMM) [27], and the research objects are concentrated on water inrush of different scales, bifurcated roadways, and "H"-shaped roadway structures. However, due to the complexity of the underground roadway network, the research on the law of water flow propagation within the complex underground roadway network during water inrush accidents is very limited. In order to systematically understand the law of water flow spreading in the complex underground network and deepen the understanding of the fluid transport mechanism in the mining space, this study selects a roadway network with a complex underground structure as the research object. The RNG k-ε turbulence model and the Eulerian multiphase flow model (VOF) are adopted to simulate the flow state of water in the roadway. At the same time, in order to ensure the accuracy of the simulation, the volume fraction, momentum equation, and turbulence equation are all discretized using the second-order upwind scheme. This study systematically explores the spatiotemporal evolution law of water flow propagation in a complex roadway network. Taking the spatiotemporal changes in the water level elevation in the roadway as the key criterion, it provides a theoretical basis for the safe evacuation of underground personnel.

2. Numerical Method

2.1. Physical Model

The geometrically constrained physical roadway model in this study was computationally reconstructed using photogrammetric survey data from operational mining panels at Shigetai Mine. Shigetai Coal Mine is located in the northern part of the Shenmu North Mining Area of the Northern Shaanxi Jurassic Coalfield. It has experienced a marine–continental evolutionary process from the Paleozoic to the present. During the Late Paleozoic, coastal plain-type Carboniferous–Permian strata were deposited. In the Mesozoic, coal-bearing formations of the inland lake system and red rock series were deposited. In the Cenozoic, influenced by inland uplift and subsidence and arid climate, reddish-brown and yellowish-brown sub-clay, sub-sand, and sand layers were formed. The coal-bearing strata belong to the Early–Middle Jurassic Yan'an Formation, and the Triassic strata serve as the sedimentary basement of the coal-measure strata. The coal mine storage water goaf of Shigetai Mine was completed in 2016. The goaf as a whole is higher in the east and lower in the west, with a length of about 2170 m and a width of about 1865 m. The maximum water storage area of the underground reservoir is 1.36 million square meters, and the maximum water storage capacity is about 1.714 million cubic meters. The elevation of the goaf floor is 1078–1107 m. Based on the occurrence conditions and hydraulic characteristics of groundwater, the aquifers in the mining area are divided into pore phreatic water in the Cenozoic loose layers, fissure phreatic water and confined water in the Mesozoic clastic rocks, as well as fissure and karst phreatic water in the burnt-altered rocks. The process of water inrush spreading and its inundation scope directly determine the severity of the underground disaster situation. Based on the principle of hydrodynamics, the main driving forces of water flow movement are gravity (i.e., elevation difference) and pressure (i.e., water pressure difference). Among them, the mine structure determines the elevation variations at each location and is a crucial factor influencing water flow movement; the inflow conditions, including the inlet position, water inrush volume, and duration, constitute the specific driving source of water flow movement. The mine structure, the position of the water inrush inlet, the volume of water inrush and its duration, etc., jointly determine the spreading process of water inrush and the inundation degree of the mine. In this study, a physically representative model featuring a complex tunnel network was adopted. This model incorporates variations in elevation across different sections of the tunnels, reflecting the true underground environment more accurately.

The physical model can be seen from Figure 1. It adopts full-scale modeling, with a tunnel width of 5 m and a height of 3.4 m. The total length of the roadway is approximately 2170 m, and the roadway structure is relatively complex. As can be seen from Figure 2, the roadway network presents a pattern of multiple roadways in parallel. The types of each branch roadway include long roadways, bifurcated roadways, T-shaped roadways, etc., and some roadways have a circular structure. During grid division, the impact of grid size in the VOF model on the capture accuracy of gas–liquid interfaces was considered. In total, 15 and 10 grids were used to fill the tunnel width and height, respectively. The grid division adopted a poly-hexcore method, which uses polyhedral grids at the tunnel walls to ensure a good fit, and hexahedral grids inside the tunnel to ensure accurate capture of gas–liquid interfaces. The total number of grids in the tunnel is 3.17 million. The height of the underground water level is 15 m, and the size of the dam's water inlet is 1 m × 0.5 m. After the dam bursts, water flows into the tunnel in this area, flowing along the tunnel terrain at different heights. In terms of boundary condition settings, an inlet was set at the front end of the tunnel near the water-accumulated area, with an inlet water flow rate of 0.90 m^3/s. The outlet was set as a free-flow outlet to simulate water flowing out of the tunnel under natural conditions. In the simulation, kinetic energy loss of water flowing on a rough wall was considered, with a roughness coefficient of 0.5 and a roughness height of 0.01 m.

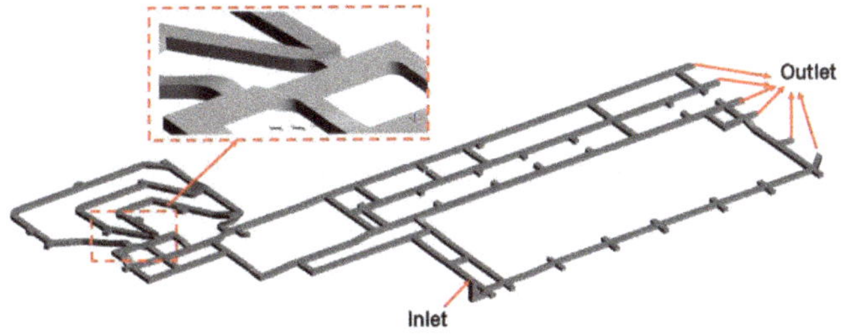

Figure 1. Schematic diagram of physical model.

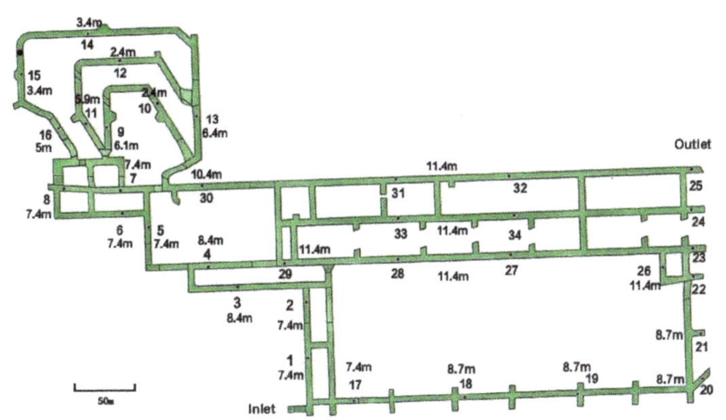

Figure 2. The layout of water level monitoring points in the tunnel.

2.2. Governing Equation

Mine water inrush exhibits flow characteristics such as high inertial force, unsteadiness, non-uniformity, transition between rapid and slow flow, and coexistence and alternation of open and full flow. Based on the flow state of water within the tunnel, the calculation model employs the Eulerian multiphase flow model (VOF) based on the liquid surface tracking method to simulate the flow state of water in the tunnel. Due to the slow exchange of velocity, temperature, and pressure between water and air, which has no significant impact on their densities, the simulation process is treated as incompressible flow. Furthermore, the flow of water and air is time-dependent, and its momentum equation is as follows:

$$\frac{\partial\left(\rho\vec{v}\right)}{\partial t}+\nabla\cdot\left(\rho\vec{v}\times\vec{v}\right)=-\nabla P+\nabla\cdot\left[\mu\left(\nabla\vec{v}+\nabla\vec{v}^{T}\right)\right]+\rho\vec{g}+\vec{F} \quad (1)$$

$\vec{v}=(x,y,z)$ represents the fluid velocity; ρ and μ represent the density and dynamic viscosity coefficient of the fluid, respectively; $\vec{g}$ is the acceleration of gravity; and $\vec{F}$ is the additional source term force at the interface. When studying the relationship between the gas–liquid interface and time, the explicit interface tracking method is commonly used in the VOF model. The volume fraction equation of the q-phase is discretized according to the following formula:

$$\frac{\alpha_q^{n+1}\rho_q^{n+1}-\alpha_q^n\rho_q^n}{\Delta t}V+\sum\nolimits_f\left(\rho_q^{n+1}U_f^{n+1}\alpha_{q,f}^{n+1}\right)=\left[S_{\alpha_q}+\sum\nolimits_{P=1}^{n}\left(\dot{m}_{pq}-\dot{m}_{qp}\right)\right]V \quad (2)$$

The volume fraction of the main phase is constrained and calculated using the following equation:

$$\sum\nolimits_{q=1}^{n}\alpha_q=1 \quad (3)$$

n+1 represents the index of the current time step; n is the index of the previous time step; α_q^{n+1} represents the volume fraction of the grid at time step n + 1; α_q^n represents the volume fraction of the grid at time step n; $\alpha_{q,f}^{n+1}$ represents the q phase volume fraction of the grid at time step n + 1; U_f^{n+1} represents the volumetric flux passing through the grid surface at time step n + 1; and V represents the grid volume. $\dot{m}_{qp}$ represents the mass transfer from phase q to phase p; $\dot{m}_{pq}$ represents the mass transfer from phase p to q; and the source term S_α equals zero.

The flow state of water in the tunnel belongs to turbulence, and the RNG k-ε turbulence model is commonly used in complex open full-flow problems. This model can handle complex water flow problems such as large curvature, strong pressure gradient, and swirling flow. The governing equation is

$$\frac{\partial}{\partial t}(\rho k)+\frac{\partial}{\partial x_i}(\rho k u_i)=\frac{\partial}{\partial x_j}\left(\alpha_k\mu_{eff}\frac{\partial k}{\partial x_j}\right)+G_k+G_b-\rho\varepsilon-Y_M+S_k \quad (4)$$

$$\frac{\partial}{\partial t}(\rho\varepsilon)+\frac{\partial}{\partial x_i}(\rho\varepsilon u_i)=\frac{\partial}{\partial x_j}\left(\alpha_\varepsilon\mu_{eff}\frac{\partial_\varepsilon}{\partial_{x_j}}\right)+C_{1\varepsilon}\frac{\varepsilon}{k}(G_k+C_{3\varepsilon}G_b)-C_{2\varepsilon}\rho\frac{\varepsilon^2}{k}-R_\varepsilon+S_\varepsilon \quad (5)$$

G_K represents the turbulent kinetic energy generation term due to the average velocity gradient; and G_b represents the turbulent kinetic energy generation term caused by buoyancy. Y_M represents the contribution of pulsating expansion in compressible turbulence to the total dissipation rate. α_k and α_ε are the reciprocal of the effective Prandtl numbers of k and ε, respectively. S_k and S_ε are user-defined source terms.

2.3. Distribution of Water Level Height Measuring Points

In order to comprehensively analyze the variation law of the height of the water level at each position of the roadway with the flow time, 34 monitoring points were set up along the distance between the roadway and taking into account the characteristics of the intersection, slope, bifurcation, and other road sections. As evident from the layout of water level monitoring points in Figure 2, the roadway exhibits characteristics such as slope, intersection, bifurcation, and corners. As can be seen from Figure 2, the lowest point of the roadway is designated as the reference point for zero height, and the corresponding relative terrain height of each water level monitoring point is indicated.

Figure 2 shows the inlet of the tunnel model, with a height of 15 m. Additionally, the zero reference plane and outlet are marked in the figure. As can be seen from Figure 2, the tunnel network has a significant slope, with an elevation difference ranging from 2.4 to 11.4 m.

2.4. Research Methodology

In this study, the flow state of the water in the roadway is turbulent, and the RNG k-ε turbulence model was adopted. This model is most commonly applied in complex free-surface and full-flow problems and can handle complex water flow issues such as large curvature, strong pressure gradients, and swirling flows. The Eulerian multiphase flow model (VOF) based on the liquid surface tracking method was selected to simulate the flow state of water in the roadway. The influence of the grid size in the VOF model on the accuracy of capturing the gas–liquid interface was considered. In total, 15 and 10 grids, respectively, are used to fill the width and height of the roadway. The grid is divided using the poly-hexcore method; that is, polyhedral grids are used at the roadway wall surface to ensure a good fit to the roadway wall, while hexahedral grids are used inside the roadway to ensure the accuracy of capturing the gas–liquid interface. The inlet condition of the model is set to 0.90 m^3/s, and the outlet is a free-flow outlet to simulate the situation where water flows out of the roadway under natural conditions. In the simulation, the kinetic energy loss of the water flow on the rough wall surface is considered, with a roughness coefficient of 0.5 and a roughness height of 0.01 m.

3. Results and Discussion

3.1. Analysis of Water Spreading Process in Tunnel

To comprehensively understand the dynamic distribution pattern of water spread in the adjacent tunnel network after underground water discharge, Figure 3 illustrates the dynamic distribution of water spread in the tunnels at different times. After underground water discharge occurs, water begins to enter the tunnel entrances near the water-accumulating areas, as shown in Figure 3a. At 400 s, water enters from the tunnel entrances. Since measuring points 1, 2, and 17 (as shown in Figure 2) are closest to the water inlet, water first accumulates in these areas and then spreads to other areas of the tunnel. As shown in Figure 3b, within 1200 s after water discharge occurs, water in the water-accumulating area still converges at the areas where measuring points 1, 2, and 17 are located, and the water level gradually rises. As shown in Figure 3c, at 3200 s, it is evident that water flows along the tunnel network towards areas with relatively lower terrain. The figure clearly demonstrates that the relative elevation of the terrain determines the direction of water flow. For example, in the areas adjacent to measuring points 18, 29, and 30 shown in Figure 2, there is no water flow, because these three measuring points have relatively high elevations of 8.7 m, 11.4 m, and 10.4 m, respectively. Areas with high relative elevation act as barriers to the spread of water. As shown in Figure 3d, at 7200 s, underground water discharge further spreads along the tunnel network, and the water level in the tunnels gradually rises. The highest water level in the flooded area reaches 1.5 m. As the time of water spread increases, the water level gradually increases in the tunnels. When the water level is higher than the slope of the tunnel, water will bypass the sloping tunnel and quickly spread towards lower water level areas. It can be seen from Figure 3e that at 11,200 s, underground water discharge bypasses the sloping tunnel where measuring point 18 is located and spreads towards areas such as measuring points 19 to 21. As illustrated in Figure 3e, the continuous propagation of water inrush demonstrates sustained fluid migration and convergence toward lower elevation zones with reduced hydraulic head. When the temporal duration reaches 19,460 s, the area spanning monitoring points 7 to 15 becomes entirely submerged by the advancing water front. Furthermore, the water depth within this inundated region exhibits a progressive increase correlated with temporal evolution. As further evidenced in the figure, despite prolonged temporal propagation of the water inrush, the roadway sector encompassing monitoring points 26 to 34 maintains persistent dryness. This phenomenon primarily arises from the region's relative elevation of 11.4 m, which exceeds the discharge outlet's elevation threshold. Consequently, hydraulic connectivity remains obstructed, preventing water ingress into this zone. However, numerical simulation indicates that through the mechanism of continuous fluid transportation over a long period of time, areas at higher altitudes than the drainage outlet will also be completely submerged. From the above analysis, it can be seen that water enters the areas where measuring points are located through the tunnel network. Due to the significant slope variations in the tunnels, the final water level in the tunnel areas is related to their relative elevation. The final equilibrium water level distribution ranges from 0.3 to 3.3 m.

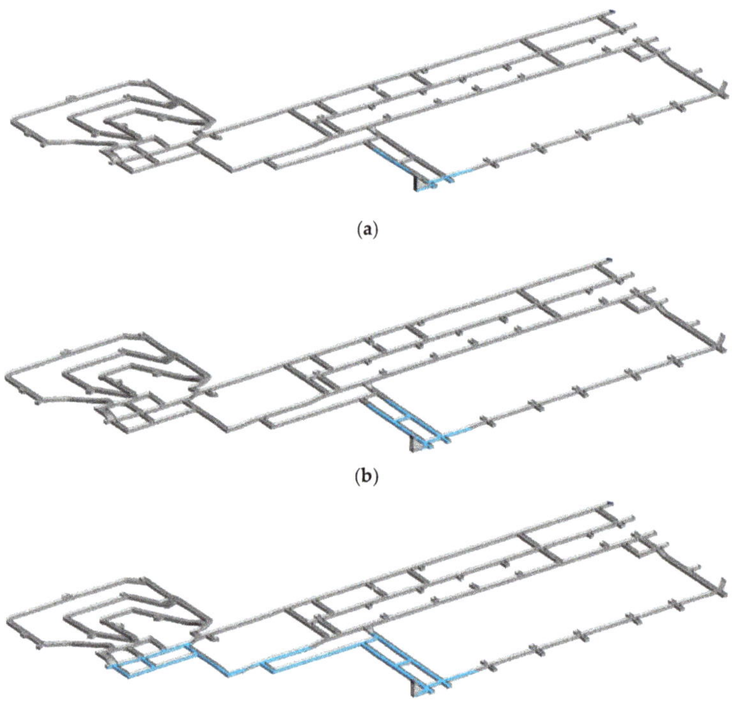

(a)

(b)

(c)

Figure 3. *Cont.*

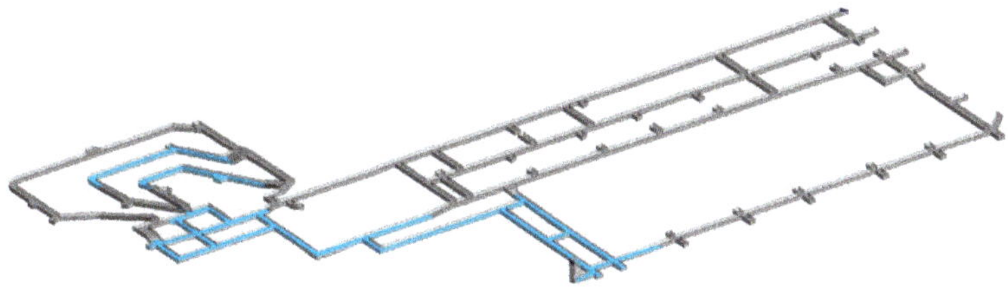

(d)

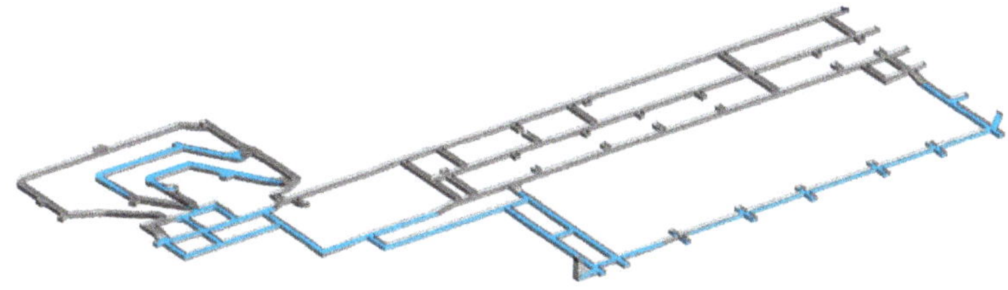

(e)

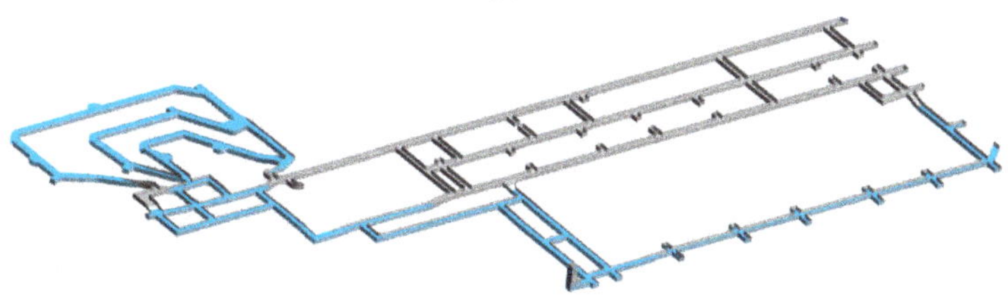

(f)

Figure 3. Dynamic development diagram of water spreading in tunnel. (**a**) Water spreads in the tunnel at 400 s. (**b**) Water spreads in the tunnel at 1200 s. (**c**) Water spreads in the tunnel at 3200 s. (**d**) Water spreads in the tunnel at 7200 s. (**e**) Water spreads in the tunnel at 11,200 s. (**f**) Water spreads in the tunnel at 19,460 s.

3.2. The Variation in Water Level Height in the Tunnel

To quantitatively analyze the pattern of water level height variation with flow time at various locations in the tunnel, monitoring points were set up at sections with multiple confluences, steep slopes, and bifurcations. Once the underground water-logged area is drained, the accumulated water will flow into nearby tunnels through the drainage outlet. The tunnel inlet mentioned in this article is shown in Figure 1. After the water flows into the nearby tunnels, it spreads rapidly through the tunnels, and the variation in water level height with time is shown in Figure 4.

As can be seen from Figure 4a, water level measurement points 1 to 7 are the areas closest to the tunnel inlet. After the start of drainage, the water level at measurement point 1 begins to rise rapidly at 550 s, reaching a height of 1.5 m around 3400 s, and then stabilizing at 1.6 m at 15,400 s. The flow velocity of water in the roadway mainly depends on the relative height of the roadway, which is consistent with the analysis results in the literature [11]. The variation pattern of water level height at measurement point 2 is similar to that at measurement point 1, i.e., the water level begins to rise rapidly at 1100 s and reaches a height of 1.1 m around 3400 s. Meanwhile, due to the influence of water level height, drainage rapidly spreads to other areas of the tunnel. The water level height at measurement points 3 to 4 begins to rise at 1600 s, reaching 0.5 m at 3400 s, and then slowly rising to 0.6 m. This is mainly because the areas where measurement points 3 and 4 are located are 1 m higher in elevation compared to measurement points 1 and 2. Subsequently, as the terrain height decreases, the water level height at measurement points 5 to 8 begins to rise rapidly at 3300 s, reaching a height of 1.3 m around 5300 s, and then stabilizing at 1.8 m at 15,400 s. The delay in the start of water level rise in this area is mainly due to the elevated positions of measurement points 3 and 4, which serve to delay the flow of water to some extent. From the above analysis, it can be seen that the water level rise pattern near the tunnel inlet is similar in the adjacent areas. The water level heights at the measurement points all undergo three states: rapid rise, slow rise, and stable equilibrium. The speed of drainage spread and the final height of water level at each measurement point are closely related to the terrain height of the area. As can be seen from Figure 2, the water level measurement points 9 and 10 of the tunnel are located in the area after the tunnel bifurcation. The so-called tunnel bifurcation means that the tunnel where measuring points 9 and 10 are located and the tunnel where measuring points 11 and 12 are located are two roadways that branch out after the bifurcation point. Compared to the height of the zero reference point, they are 6.1 m and 2.4 m, respectively. The relative height of the overall area is lower than that of the area containing measurement points 1–8. Therefore, as can be seen from Figure 4b, the water level at measurement point 10 begins to rise rapidly at 5300 s and stabilizes at 3.3 m at 13,200 s. Since measurement point 9 is located on the slope of the tunnel, under the influence of the terrain elevation difference, the water flow in the tunnel enters the area of measurement point 10 through the gap on the slope side. Therefore, when the water level at measurement point 10 is higher than that at measurement point 9, the water begins to flow into the area of measurement point 9 after 13,200 s, causing the water level at this location to rise rapidly. At around 22,000 s, the water level reaches 3 m. Figure 4c shows the variation curve of water level height over time from measuring point 11 to measuring point 16. From the change in water level height at the measuring point, it can be seen that once the water in the flooded area flows through the measuring point position, the water level height at that position will rapidly rise. Measuring point 12 is located at a relatively low height of 2.4 m, similar to the water level change pattern at measuring point 10. The water level starts to rise rapidly at 6000 s and stabilizes at 3.3 m at 16,500 s. When the water level at measuring point 12 exceeds that at measuring point 11, the water flow begins to flow into the area of measuring point 11 after 17,600 s, causing the water level at that location to rapidly rise. The water level from measuring point 13 to measuring point 16 gradually increases after 13,200 s, and finally stabilizes within the range of 2.8–3.5 m. Figure 4d shows the variation curve of water level height over time from measuring point 17 to measuring point 21. It can be seen from Figure 2 that the roadway water level measuring points 17 to 21 are located near the roadway inlet area, and the relative height of measuring point 17 is 7.4 m compared with the zero reference point, and the relative height of other measuring points is 8.4 m. Therefore, it can be seen from Figure 4d that the water level of measuring point 17 rises rapidly almost at the initial time of water discharge, reaching 1.3 m near 3300 s, and then stabilized at 1.6 m with the growth of water spreading time. Since the relative height from measuring point 18 to measuring point 21 is higher than that of measuring point 17, there is no water flow in the area for a long time after the discharge of water in the ponding area. When the height of water level measuring point 17 reaches 1.4 m, the water flow gradually flows to the area from measuring point 18 to measuring point 21. Since then, the water level in the area has not been set at 0.2 m. In addition, it can be seen from Figure 2 that the relative height of the area from measuring point 22 to measuring point 34 reaches 11.4 m. Due to its high terrain, the water flow still does not pass through the area within 22,000 s (about 6.1 h) after the emergency discharge in the ponding area. From the perspective of emergency, the area is in a relatively safe state, and subsequent analysis is not required.

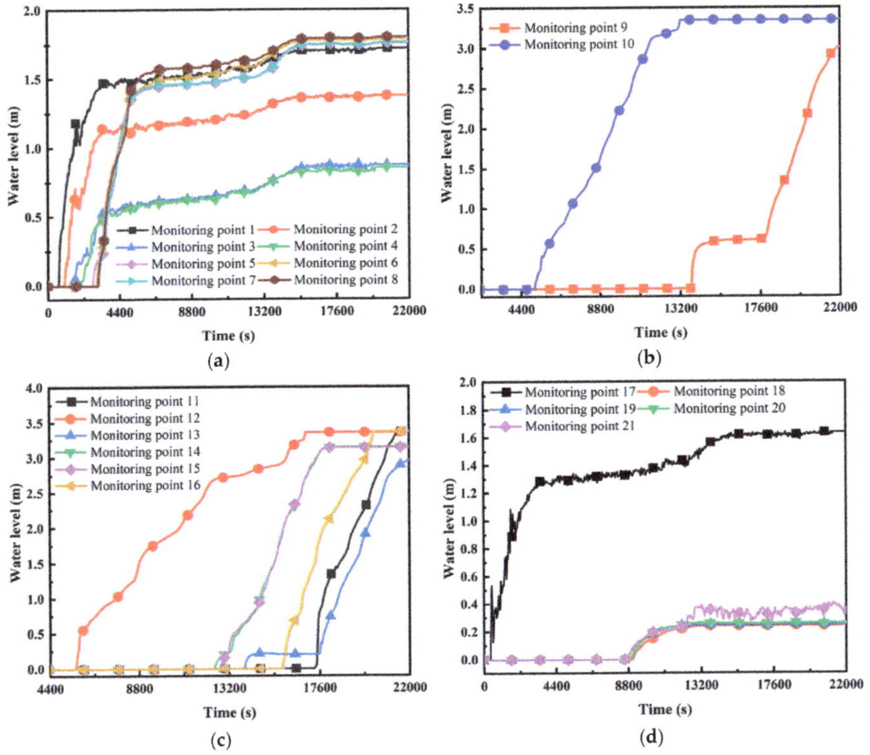

Figure 4. The variation in water level height at various monitoring points in the tunnel.(**a**) Changes in water levels over time at monitoring sites 1 to 8. (**b**) Changes in water levels over time at monitoring sites 9 to 10. (**c**) Changes in water levels over time at monitoring sites 11 to 16. (**d**) Changes in water levels over time at monitoring sites 17 to 20.

Following a water inrush event in underground mines, the tunnel network gradually becomes inundated as water propagation advances over time, with rising water levels across different zones significantly hindering safe personnel evacuation. Empirical evidence indicates that evacuation difficulty increases with proximity to the discharge outlet. Table 1 quantifies safe evacuation times for monitoring points within 1000 m of the water release location, revealing critical spatiotemporal patterns: Water depth exceeds 0.5 m within 100 m of the outlet 1555 s post-inrush, while warning-level inundation occurs at 3700 s in the 200–300 m and 200–400 m zones, 6000 s in the 400–500 m range, and 17,885 s in the 500–1000 m area. Notably, regions beyond 500 m exhibit extended evacuation windows due to the combined effects of complex tunnel geometry, elevation variations, and the formation of large water retention basins. Therefore, when setting up locations for the safe evacuation of personnel underground, factors such as accident-prone areas and evacuation routes should be comprehensively considered, and it is advisable to keep as far away from the accident area as possible [23]. An exception occurs at monitoring points 9 and 11 near the outlet, where protruding obstacles create hydrodynamic diversion pathways, allowing water to bypass these zones through adjacent tunnels and maintain lower water levels over extended durations. This spatial–temporal progression underscores the critical relationship between infrastructure topology, hydraulic behavior, and evacuation feasibility during mine flooding emergencies.

Table 1. Table of water spread influence range in roadways.

Measurement Points	Distance from Water Source (m)	Time to Reach Water Level at 0.5 m (s)
1	50	950
2	100	1555
3	175	3413
4	236	3586
5	296	3758
6	319	3672
7	330	3715
8	400	3586
9	387	14,213
10	463	5875
11	405	17,496
12	493	6005
13	567	17,885
14	750	13,651
15	844	13,478
16	913	16,200

4. Conclusions

In order to understand in detail the dynamic law of the spread of water flow in the roadway network after the water discharge in the ponding area of the coal mine, this paper selects the typical full-scale complex roadway network as the research object, selects the Euler multiphase flow model (VOF) to simulate the flow state of water in the roadway, and explores the law of roadway water spread in the process of emergency water discharge in the coal mine with the characteristics of water spread distribution and the change in water level height at each point of the roadway. The specific research results are as follows:

When the underground water discharge occurs, the water flows along the roadway network to the location with relatively low terrain. Water begins to enter the tunnel entrances near the water-accumulating areas. Due to the different relative altitudes of each roadway in the underground roadway network, the slope roadway plays a role in preventing the spread of water at the initial stage of water discharge. The equilibrium water level height of the roadway area is related to its relative height, and the stable equilibrium water level height is distributed in the range of 0.3–3.3 m.

The idea of water level rise at each point of the roadway network is similar; that is, it generally has three development states of rapid rise, slow rise, and stable balance. Once the discharge occurs, the water level in the roadway area near the discharge outlet rises sharply within 550 s; the slope roadway can slow down the speed of water spread and water level rise to a certain extent; and the water level in the roadway far from the discharge outlet begins to rise rapidly after 13,200 s. In addition, due to its high terrain, there are still some areas where the water flow still does not pass through within 22,000 s after the underground emergency discharge, and this area is an evacuable area.

Following a water inrush event, safe evacuation becomes increasingly challenging for personnel in areas closer to the discharge outlet. Those within a 300 m radius of the outlet face a maximum safe evacuation window of merely 1 h. In contrast, regions farther from the discharge location benefit from significantly extended evacuation opportunities due to the water retention capacity of complex tunnel networks, which function similarly to reservoir systems, effectively delaying water level rise and creating critical time buffers for escape. This spatial pattern highlights the inverse relationship between the proximity to the hazard source and available evacuation time during underground flooding emergencies.

Author Contributions: Data curation, P.H.; formal analysis, Y.W.; funding acquisition, P.H.; project administration, S.C.; software, Y.W. and X.Z.; validation, S.L.; writing—original draft, D.F.; writing—review and editing, Y.W. and S.L. All authors have read and agreed to the published version of the manuscript.

Funding: This work was supported by the National Energy Group's Science and Technology Innovation Project, Grant number: GJNY-21-26-04.

Data Availability Statement: The original contributions presented in this study are included in the article, and further inquiries can be directed to the corresponding author.

Acknowledgments: A special acknowledgment is made to the anonymous reviewers for their constructive and valuable comments. We thank them for their guidance despite their busy schedules.

Conflicts of Interest: Authors Donglin Fan, Peidong He, Sushe Chen and Xin Zou were employed by the company CHN Energy Shendong Coal Group Co., Ltd., Yulin 719315, China. The remaining authors declare that the research was conducted in the absence of any commercial or financial relationships that could be construed as a potential conflict of interest.

References

1. Niu, C.; Tian, Q.; Xiao, L.; Xue, X.; Zhang, R.; Xu, D.; Luo, S. Principal causes of water damage in mining roofs under giant thick topsoil–lilou coal mine. *Appl. Water Sci.* **2024**, *14*, 146. [CrossRef]
2. Qiang, W.; Bo, L.; Yulong, C. Vulnerability assessment of groundwater inrush from underlying aquifers based on variable weight model and its application. *Water Resour. Manag.* **2016**, *30*, 3331–3345. [CrossRef]
3. Xu, Z.; Sun, Y. Mine water safety and environment: Chinese experience. *Water* **2024**, *16*, 2833. [CrossRef]
4. Gu, H.; Shang, X.; Zhao, H. Theory and technology for the prevention of mine water disasters. *Water* **2024**, *16*, 2952. [CrossRef]
5. Li, H.; Bai, H. Simulation research on the mechanism of water inrush from fractured floor under the dynamic load induced by roof caving: Taking the Xinji Second Coal Mine as an example. *Arab. J. Geosci.* **2019**, *12*, 466. [CrossRef]
6. Cui, F.; Wu, Q.; Lin, Y.; Zeng, Y.; Zhang, K. Damage Features and Formation Mechanism of the Strong Water Inrush Disaster at the Daxing Co Mine, Guangdong Province, China. *Mine Water Environ.* **2018**, *37*, 346–350. [CrossRef]
7. LaMoreaux, J.W.; Wu, Q.; Wanfang, Z. New development in theory and practice in mine water control in China. *Carbonates Evaporites* **2014**, *29*, 141–145. [CrossRef]
8. Li, B.; Wu, Q. Catastrophic Evolution of Water Inrush from A Water-Rich Fault in Front of Roadway Development: A Case Study of The Hongcai Coal Mine. *Mine Water Environ.* **2019**, *38*, 421–430. [CrossRef]
9. Zhang, X.; Lei, M.; Wang, S.; Zhang, X.; Chen, H.; Jiang, X. Risk Assessment and Mechanism of Water Inrush in Water-rich Deep-buried Karst Tunnel. *Nat. Hazards Earth Syst. Sci. Discuss.* 2024; *preprint*.
10. Guo, J.; Zhao, H.; Ma, F.; Li, K.; Zhao, C. Investigating the permeability of fractured rock masses and the origin of water in a mine tunnel in Shandong Province, China. *Water Sci. Technol.* **2015**, *72*, 2006–2017. [CrossRef]
11. Wu, Q.; Zhang, X.; Zhao, Y.; Li, P.; Yao, Y.; Du, Z.; Xu, H.; Wang, X. Numerical Simulation of Inrush Water Spreading Through a Mine: A Case Study of the Beixinyao Mine, Shanxi Province, China. *Mine Water Environ.* **2022**, *41*, 487–503. [CrossRef]
12. Zhang, X.; Wu, Q.; Zhao, Y.; Liu, S.; Xu, H. Spatiotemporal modeling of water inrush spreading in mine roadway networks. *Water Sci. Technol.* **2022**, *85*, 872–886. [CrossRef] [PubMed]
13. Qin, W.; Yicheng, Y.; Ying, J. Simulation on Mine Water-inrush Characteristics of Cross Roadway Based on CFX. *Procedia Eng.* **2011**, *26*, 759–764. [CrossRef]
14. Li, C.; Li, J.; Li, Z.; Hou, D. Establishment of spatiotemporal dynamic model for water inrush spreading processes in underground mining operations. *Saf. Sci.* **2013**, *55*, 45–52. [CrossRef]
15. Wu, Q.; Yao, Y.; Du, Y.; Li, H.; Zhao, Y.; Xu, H.; Liu, S.; Wang, X. Physical simulation test study on the stability of wading personnel inmine water inrush roadway. *J. China Coal Soc.* **2025**, *50*, 13–22.
16. Hamilton, K.; Demant, D.; Peden, A.E.; Hagger, M.S. A systematic review of human behaviour in and around floodwater. *Int. J. Disaster Risk Reduct.* **2020**, *47*, 101561. [CrossRef]
17. Wu, Q.; Du, Y.; Xu, H.; Zhao, Y.; Zhang, X.; Yao, Y. Finding the earliest arrival path through a time-varying network for evacuation planning of mine water inrush. *Saf. Sci.* **2020**, *130*, 104836. [CrossRef]
18. Du, Y.; Wu, Q.; Zhao, Y.; Zhang, X.; Yao, Y.; Xu, H. A parallel time-varying earliest arrival path algorithm for evacuation planning of underground mine water inrush accidents. *Concurr. Comput. Pr. Exp.* **2020**, *32*, e5644. [CrossRef]
19. Ma, X.W.; Xu, H.L. Dynamic planning of mine water avoidance path based on strategy. *Min. Metall.* **2023**, *32*, 12.

20. Huang, Y. Research on the Improvement of Dijkstra Algorithm in the Shortest Path Calculation. In Proceedings of the 2017 4th International Conference on Machinery, Materials and Computer (MACMC 2017), Xi'an, China, 27–29 November 2017; Volume 141.
21. Chen, J.; Li, F.; Lian, Z. The hydrogeology of mine water inrush period using Dijkstra's algorithm. *Arab. J. Geosci.* **2020**, *13*, 825. [CrossRef]
22. Zhao, X.; Wang, Q.; Xie, B.; Pan, S.; Ji, M. A dynamic rescue route planning method based on 3D network in mine water inrush hazard. *Geomat. Nat. Hazards Risk* **2019**, *10*, 2387–2407. [CrossRef]
23. Hong, Y.; Li, D.; Wu, Q.; Xu, H. Priority-Oriented Route Network Planning for Evacuation in Constrained Space Scenarios. *J. Optim. Theory Appl.* **2019**, *181*, 279–297. [CrossRef]
24. Yang, H.S.; Xu, H.; Ju, Z.W. Simulation and research on 3D tunnel pedestrian evacuation based on agent. *Comput. Technol. Dev.* **2019**, *29*, 132–137.
25. Li, W.; Wang, Y.; Ye, Z.; Liu, Y.A.; Wang, L. Development of a mixed reality assisted escape system for underground mine- based on the mine water-inrush accident background. *Tunn. Undergr. Space Technol.* **2024**, *143*, 105471. [CrossRef]
26. Ji, J.; Zhang, J.; Chen, J.; Wu, S. Computer simulation of evacuation in underground coal mines. *Min. Sci. Technol.* **2010**, *20*, 677–681. [CrossRef]
27. Khatooni, K.; Hooshyaripor, F.; MalekMohammadi, B.; Noori, R. A new approach for urban flood risk assessment using coupled SWMM–HEC-RAS-2D model. *J. Environ. Manag.* **2025**, *374*, 123849. [CrossRef]

water

MDPI

Article

A Study on the Deformation Mechanism of a Landslide Reinforced with an Anti-Slip Pile Under the Effect of Reservoir Water Level Decline

Gang Yang [1], Zhuolin Wu [1], Lin Zhang [1], Jingfeng Hou [1], Shen Tong [2], Fei Liu [2,*] and Yong Zheng [3]

[1] Chongqing Survey and Design Institute of Water Resources, Electric Power and Architecture Co., Ltd., Chongqing 401120, China; hjf0802@163.com (J.H.)
[2] School of Civil and Transportation Engineering, Henan University of Urban Construction, Pingdingshan 467000, China; 20201001@huuc.edu.cn
[3] College of River and Ocean Engineering, Chongqing Jiaotong University, Chongqing 400074, China
* Correspondence: liufei1991c@163.com

Abstract: The fluctuation of reservoir water levels is a critical factor influencing the evolution of reservoir landslide–anti-slide pile systems. To investigate the reinforcement mechanism of anti-slide piles in reservoir landslides under the effect of reservoir water level fluctuations, this study employs numerical simulation methods to establish a three-dimensional slope model, simulating the drawdown process of the reservoir water level from 175 m to 145 m. The displacement and strain fields of the reservoir landslide during the water level drawdown are analyzed. Furthermore, the strain characteristics of the anti-slide pile-reinforced reservoir landslide under stress–seepage coupling are studied, and the prevention effectiveness of the landslide–anti-slide pile interaction system is explored. The results indicate that the drawdown of the reservoir water level can lead to the gradual expansion of the strain and displacement zones in the landslide, as well as a reduction in the safety factor. Under the effect of anti-slide piles, the maximum deformation of the reservoir landslide is significantly reduced. The optimal reinforcement effect is achieved when the anti-slide piles are arranged in the middle of the reservoir landslide, with a pile spacing of four times the pile diameter and an embedded depth reaching the critical depth. The findings of this study can provide a scientific basis for analyzing the instability mechanisms and mitigation of reservoir landslides.

Keywords: anti-slip piles; reservoir water level; reservoir landslide; safety factor

1. Introduction

The long-term cyclical fluctuations in reservoir water levels induce corresponding periodic variations in groundwater levels. This hydrological dynamic leads to intermittent saturation conditions, where portions or even the entire reservoir landslide experience alternating periods of saturation and unsaturation [1,2]. This results in cyclical fluctuations of pore water pressure and shear strength in the soil and rock mass of the landslide, thereby impacting the landslide's stability. According to the research results, even a 5% cyclical alteration in shear strength parameters can cause a remarkable reduction in the safety factor as compared to the limiting safety factor, thus highlighting the significant effect of reservoir water level changes on the long-term stability of reservoir landslides [3,4]. For example, since the Three Gorges Reservoir impounded water to 175 m in 2008, more than 400 new and old landslide deformations have occurred throughout the reservoir area, with more

 https://doi.org/10.3390/w17091390

than 100 collapses and 50 km of unstable reservoir banks, closely related to the cyclical fluctuations of the reservoir water level.

Given the significant impact of reservoir water level fluctuations on the stability of reservoir landslides, scholars both domestically and internationally have conducted relevant research on this issue. Jia et al. [5] conducted model tests to obtain the deformation and failure characteristics and mechanisms of reservoir landslides caused by water level fluctuations. Huang et al. [6] systematically studied monitoring data from a buoyancy-reduced landslide in the Three Gorges Reservoir area, finding that landslide deformations were larger during high water levels and smaller during low water levels. Tang et al. [7] conducted statistical analyses of numerous landslides along the Three Gorges Reservoir, identifying the deformation patterns and mechanisms of landslides with different sliding surfaces and permeabilities under the influence of water level fluctuations. Zhao et al. [8] established a seepage–stress coupling model for the Huangnidaba landslide using the finite element method, indicating that the landslide was significantly affected by the drop in reservoir water levels, exhibiting creep deformation. During the water level decline, the pore water pressure within the landslide decreased, while the effective creep stress increased, accelerating the creep deformation of the landslide. Zangerl et al. [9] investigated the effects of reservoir water level fluctuations on deep-seated landslides at the Gepatsch dam and revealed that the seasonal variation in landslide deformation rates was closely related to the rise in reservoir water levels.

Anti-sliding piles, as an effective landslide support structure, have been widely adopted for landslide stabilization and large landslide remediation in reservoir areas due to their strong deep anti-sliding capability, minimal disturbance to the sliding mass, and ease of construction [10–13]. Many researchers have conducted extensive studies on the mechanisms of anti-sliding pile reinforcement and design calculation methods. Song et al. [14] revealed the response characteristics of anti-sliding piles and the landslide deformation field under strong rainfall based on field tests and numerical simulations. Through centrifuge model tests, Wang et al. [15] investigated the influence of anti-sliding piles on the deformation field of landslides. Their research demonstrated that anti-sliding piles are capable of efficiently inhibiting the development of sliding surfaces. Zhang et al. [16] highlighted the formation and progressive development of sliding surfaces, noting that the influence of anti-sliding piles decreases as the distance to the failure point increases. Lirer et al. [17] examined the impact of drainage piles on the displacement field of landslides through field monitoring and numerical simulations. Won et al. [18] analyzed the distribution characteristics of earth pressure acting on anti-sliding piles under different parameters using numerical simulations.

As mentioned above, numerous scholars have conducted extensive research on the stability of reservoir landslides under varying water levels, as well as the interaction between anti-slide piles and soil masses. However, there are relatively few studies on the effectiveness of anti-slide piles in reinforcing reservoir landslides under the condition of declining reservoir water levels. This study investigates the interaction mechanism between anti-slide piles and surrounding soil under rapid reservoir drawdown conditions, using a representative landslide in the Three Gorges Reservoir area as a case study through finite element numerical analysis. First, the simulation software ABAQUS [19] was used to explore the variations in the shear strain field, displacement field, and landslide stability of the reservoir landslide in response to water level changes. Subsequently, the effectiveness of anti-sliding piles in reinforcing the reservoir landslide under low reservoir water levels is analyzed. Finally, a sensitivity analysis of the design parameters of the anti-sliding piles is conducted, and an optimized design scheme is proposed.

2. Numerical Methodology

2.1. Numerical Model

(1) Mathematical formulation

For reservoir landslides, periodic fluctuations in the reservoir water level create a hydraulic head difference between the interior and exterior of the landslide. Periodic water level fluctuations in reservoirs induce hydraulic head differentials between the interior and exterior of landslides. This hydraulic gradient drives seepage flow through the slope body, modifying the groundwater flow regime. Consequently, variations in infiltration pressure and pore water pressure occur, leading to changes in stress distribution and subsequent deformation of the landslide mass. Variations in the stress field exert a profound influence on the soil's stress–strain behavior, thereby triggering alterations in the soil's permeability coefficient and porosity. These changes, in turn, further transform the seepage field within the landslide mass. Subsequently, the evolving seepage field prompts additional adjustments to the stress field, giving rise to a fluid–solid coupling effect. This effect represents a dynamic interplay between hydraulic and mechanical processes, playing a pivotal role in governing landslide stability. The equations for stress balance and seepage continuity are presented as follows:

$$I^N - P^N = 0 \tag{1}$$

$$\int_V \delta u_w \frac{1}{J}\frac{d}{dt}(Jp_w n_w)dV + \int_V \delta u_w \frac{\partial}{\partial x}[p_w\left(-k\frac{\partial \psi}{\partial x}\right)]dV \tag{2}$$

where I^N and P^N are the internal and external force matrices, respectively; δu_w represents variable fractions of pore water pressure; V represents the change in volume of the soil; p_w is the density of water; k denotes the permeability coefficient of porous media; ψ is the gauge head, which is the sum of the position head and the pressure head; and $\frac{\partial \psi}{\partial x}$ represents the permeability gradient.

(2) Boundary conditions

Figure 1 presents the built numerical model established in this paper. For the landslide model, the static boundary conditions impose constraints on the displacements of nodes in the X, Y, and Z directions at the bottom. The displacements of nodes in the X direction are restricted at the front and rear boundaries. The displacement of nodes in the Y direction is constrained on both sides. Given that the front edge of the landslide is submerged by the water level of the Yangtze River, the top surface and the front surface of the model are designated as permeable boundaries, while the remaining surfaces are defined as impermeable boundaries.

The entire landslide model is 105 m long, 55 m high, and 15 m wide. The anti-sliding piles are circular piles with a diameter D = 2.0 m. The anti-sliding piles are anchored in the bedrock, with a length L_p = 24 m and an anchoring depth L_y = 9 m in the bedrock layer. The horizontal distance from the anti-sliding piles to the toe of the landslide L_x = 30 m, the horizontal distance from the top of the landslide to the toe of the landslide L = 60 m, and the pile spacing S = 3D. By virtue of symmetry, only half of the model needs to be analyzed.

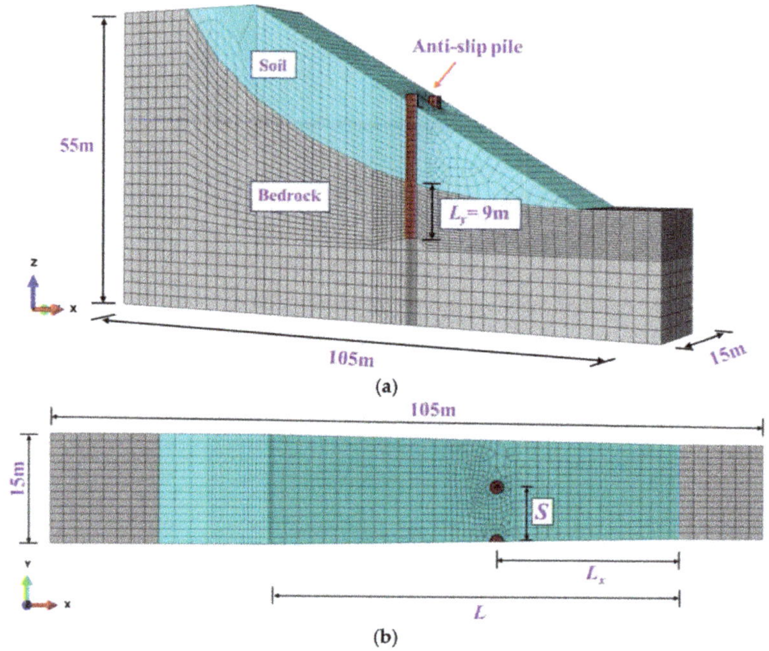

Figure 1. Numerical model: (**a**) front view, (**b**) top plan view.

(3) Element types and sizes

The landslide model (i.e., the slide mass and bedrock model) was discretized into 8-noded Lagrangian brick elements with both displacement and pore pressure degrees of freedom (C3D8P) for coupled flow analysis. The anti-sliding piles were discretized using Lagrangian C3D8R elements. The maximum and minimum mesh sizes for the landslide model were set at 2.2 m and 0.36 m, respectively.

(4) The landslide parameters

The relevant parameters of the soil and bedrock slopes are shown in Table 1, while the parameters of the anti-sliding piles are presented in Table 2. It should also be noted that, to date, there has been no experimental research on the unsaturated soil properties of landslides in the Three Gorges Reservoir area. Therefore, the unsaturated permeability coefficient of the landslide can only be derived using an engineering analogy method, which is based on the saturated permeability coefficient and the grain size distribution of the soil to formulate its permeability function.

Table 1. Properties of the soil and bedrock.

Material	Density ρ (kN/m³)		Friction Angle φ (°)		Cohesion c (kPa)		Permeability Coefficient k (m/d)	Elasticity Modulus G (MPa)	Poisson's Ratio υ
	Natural	Saturated	Natural	Saturated	Natural	Saturated			
Soil	16.8	22.0	20	18	15	3	7.5	22.0	0.33
Bedrock	26.0	27.0	26	24	40	36	0.01	28,000	0.25

Table 2. Properties of the anti-sliding piles.

Parameter	Symbol	Unit	Value
Diameter	D	m	2.0
Density	ρ	kN/m^3	24.0
Elasticity modulus	G	MPa	26,000
Effective Poisson's ratio	v		0.13

(5) Contact surface parameters

The general contact functionality is employed to account for significant sliding and separation between the anti-slip piles and the soil surfaces in this study. The friction coefficient between the anti-slip piles and the soil is set at 0.5.

2.2. Simulation Scheme

Based on the specific geological conditions of the Three Gorges reservoir area and relevant research findings, numerical models were used to study the variations in the shear strain field, displacement field, and landslide stability of the reservoir landslide under declining water levels. The following scenario was simulated using the numerical model: the reservoir water level was initially maintained at a constant height of 175 m until the landslide reached its natural state, after which the water level dropped from 175 m to 145 m. During the numerical calculations, the boundary conditions were as follows: (i) impermeable with no vertical displacement at the bottom; (ii) no horizontal displacement on the left side; and (iii) free drainage and displacement at the slope surface.

3. Results and Discussion

3.1. Analysis of Landslide Deformation Mechanism

During the decline of the reservoir water level, six calculation conditions were established, with the water level decreasing from 175 m to 170 m, 165 m, 160 m, 155 m, 150 m, and 145 m. The calculation results for the landslide displacement field and equivalent plastic strain field are presented for reservoir water levels of 175 m, 160 m, and 145 m, respectively.

Figure 2a illustrates the accumulated equivalent plastic strain (PEEQ) of the landslide when the reservoir level is maintained at 175 m. From this figure, it is evident that numerous shear plastic zones are generated in the middle of the landslide. However, no continuous plastic zone is formed. This is due to the high hydrostatic pressure exerted on the leading edge of the landslide from the 175 m reservoir water level, which prevents plastic deformation in that area. Conversely, the rise in groundwater level, combined with the softening effect on the landslide from prolonged submersion, reduces both the slip resistance and shear strength of the landslide. As a result, plastic damage primarily occurs in the middle section and gradually extends toward the rear edge of the landslide.

Figure 2b,c show the accumulated equivalent plastic strain (PEEQ) of the landslide when the reservoir water level decreases to 160 m and 145 m. As the water level falls, the position of the landslide's plastic zone progressively shifts toward the base, and its distribution range widens. This is particularly evident in the middle and lower sections of the landslide, where sliding occurs along the slip surface due to changes in effective stress resulting from fluctuations in groundwater level and the effects of infiltration forces. During the decline of the reservoir level, groundwater infiltration into the slope exhibits a hysteresis effect, resulting in the development of larger dynamic water pressure directed toward the critical surface, which increases the instability of the slope. As the hydrostatic pressure acting on the slope surface decreases, the downward forces on the slope increase. Consequently, within the range of water level changes, the slope experiences larger displacements toward the critical surface. Above the water level, the enhancement effect of geotechnical properties exhibits a lagging behavior, resulting in larger horizontal displacements of the slope body during water level fluctuations. As the water level decreases, the pore water pressure within the geotechnical body is reduced, leading to an increase in the effective stress. This results in the compression of the geotechnical body above the water level, contributing to the development of a section of the landslide area that approaches the formation of a continuous plastic zone.

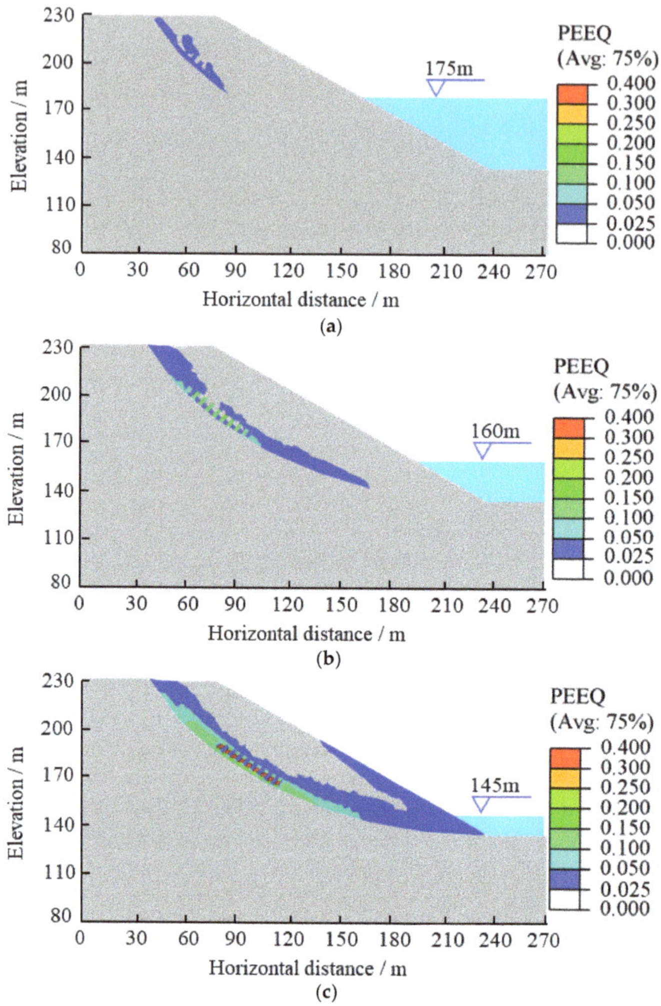

Figure 2. Plastic zone distribution of landslide during reservoir water level: (**a**) 175 m; (**b**) 160 m; (**c**) 145 m.

Figure 3 shows the displacement maps of the landslide as the reservoir water level decreases to three different elevations. The maximum total displacements of the landslide are 23 mm, 55 mm, and 85 mm when the water level is at 175 m, 160 m, and 145 m, respectively. As the water level decreases, the distribution of the maximum total displacement of the landslide shifts upward, and the range of displacement distribution continuously expands.

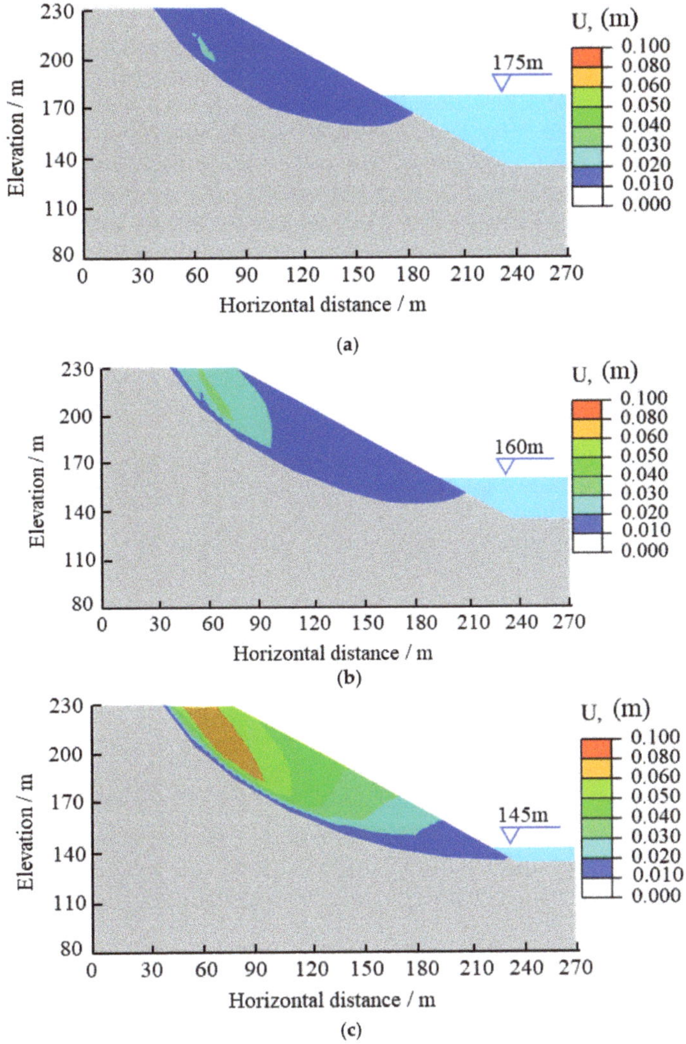

Figure 3. Displacement field distribution of landslide during reservoir water level: (**a**) 175 m; (**b**) 160 m; (**c**) 145 m.

The shear strength reduction method proposed by Zienkiewicz et al. [20] and Dawson et al. [21] is implemented in ABAQUS software (v.6.14) to calculate the slope safety factor. In ABAQUS, the process for obtaining the slope safety factor can be divided into the following steps:

(1) Define the field variable, usually the strength reduction factor.
(2) Define the material properties that vary with the field variables.
(3) Set the boundary conditions and obtain the numerical equilibrium.
(4) Change and increase the field variable (i.e., the reduction factor) until the numerical calculation does not converge.

The method for calculating the slope safety factor in ABAQUS has been utilized by many scholars and experts to elucidate the effectiveness of landslide reinforcement measures [22–27].

The variation curve of the maximum total displacement of the landslide during the decline in water level is shown in Figure 4. It is evident that the lowering of the reservoir water level significantly affects landslide displacement, which increases from 20 mm to 80 mm during the period of water level decline. In contrast, the safety factor of the landslide exhibits an opposite trend. As the water level decreases from 175 m to 145 m, the safety factor shows a downward trend, as illustrated in Figure 5. When the reservoir water level drops to 135 m, the safety factor of the landslide reaches a minimum value of 1.11, representing a reduction of approximately 15% compared to the safety factor at the static water level of 175 m. This indicates an increased risk of instability and failure in the reservoir landslide.

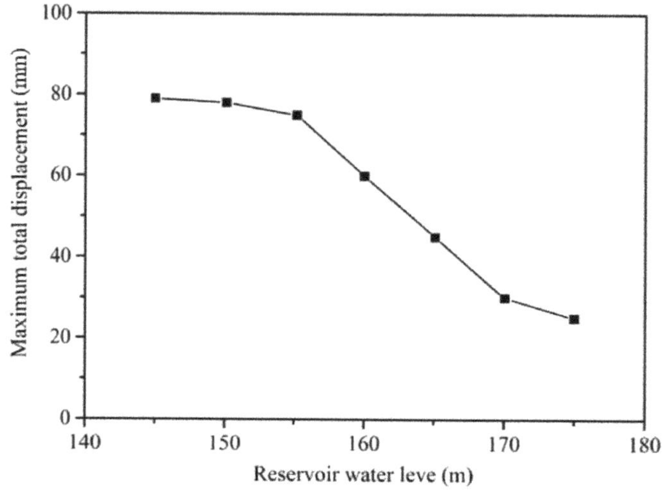

Figure 4. Maximum total displacement of the landslide with different reservoir water levels.

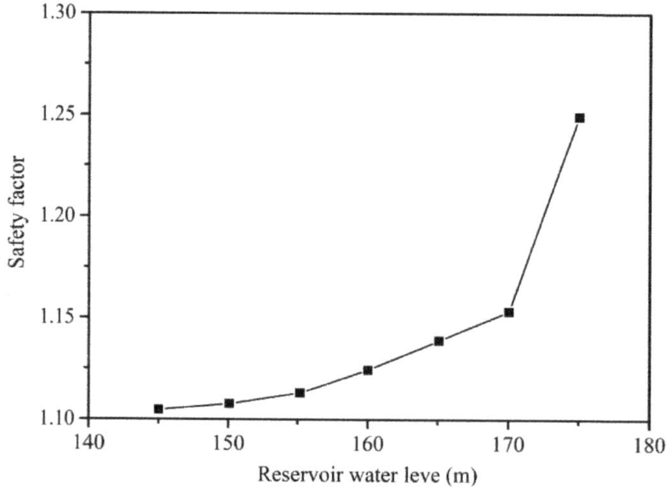

Figure 5. Safety factor of the landslide during the reservoir water level drop.

3.2. Analysis of Anti-Slide Pile Reinforcement Effects

From the analysis above, it is evident that a decline in the reservoir water level can lead to significant plastic damage to the landslide. When the water level drops to 145 m, the safety factor of the landslide reaches its minimum. Therefore, in landslide stabilization projects within the reservoir area, it is crucial to consider the impact of lowered water levels on the landslide stability. As an effective landslide retaining structure, anti-slide piles are widely used in landslide disaster mitigation due to their simple design, high stiffness, substantial resistance, and ease of construction. The following section will analyze the effectiveness of anti-slide pile reinforcement on the stability of the reservoir landslide during sudden changes in water level.

Figures 6 and 7 show the distribution contours of the plastic zone and total displacement after reinforcing the reservoir landslide with anti-slide piles at a water level of 145 m, respectively. From Figure 6, it is evident that, after reinforcing the reservoir landslide with anti-slide piles, the maximum shear strain increment is limited near the location of the piles. Additionally, the total displacement of the reservoir landslide has been reduced to 14 mm (see Figure 7), demonstrating a significant decrease compared to the situation without the use of anti-slide piles (see Figure 3c). This is because when anti-slide piles are used for landslide reinforcement, the soil body slides downward relative to the piles. On one hand, the anti-slide piles are embedded in a hard and relatively stable bedrock layer; on the other hand, the physical properties of the piles are stronger than those of the landslide soil. When a portion of the landslide soil experiences uneven deformation or displacement, the cohesion and friction within the sliding mass create frictional resistance against the uneven movement of the soil. This interaction causes soil particles to exert a mutual "wedge effect", resulting in stress deflection near the anti-slide piles. Consequently, a more pronounced soil arch forms around the base of the neighboring piles, as illustrated in Figure 8. The presence of the soil arch effect causes the anti-slide piles to be embedded in a hard and relatively stable bedrock layer. Additionally, the physical properties of the anti-slide piles are stronger than those of the landslide soil. This soil arch effect makes it difficult for the soil to flow between the piles, thereby inhibiting the downward sliding of the landslide.

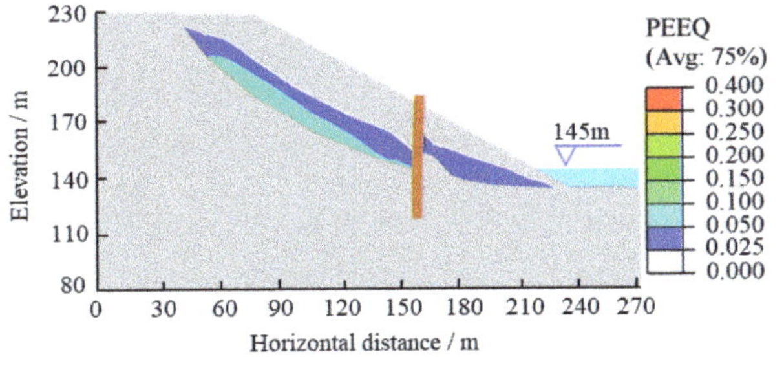

Figure 6. Plastic zone distribution of landslide.

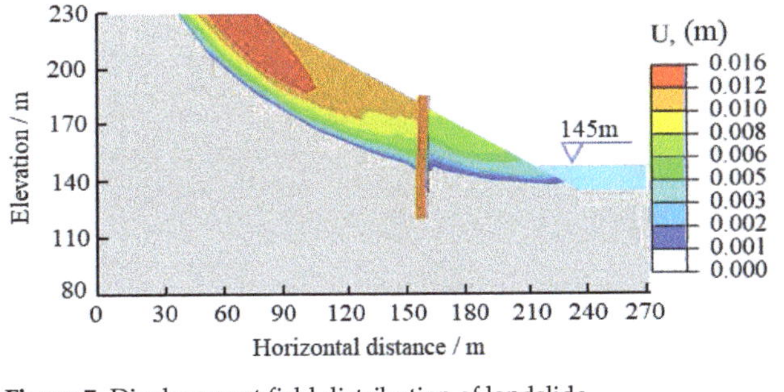

Figure 7. Displacement field distribution of landslide.

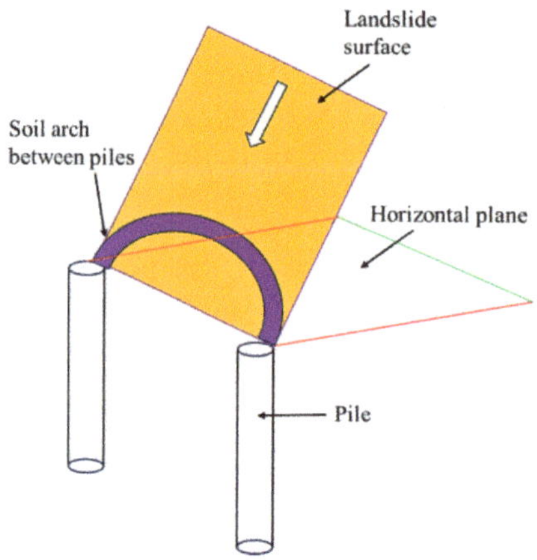

Figure 8. Soil arch effect of anti-slide pile.

3.3. Parameter Analysis

The design parameters of anti-slide piles are directly related to the effectiveness and cost of landslide remediation. A significant challenge is how to optimize and determine these parameters reasonably, minimizing economic costs while ensuring that the remediation effects meet standards and are technically feasible. This is particularly complex for reservoir landslides that involve seepage–stress coupling. Therefore, in this section, numerical simulation methods will be employed to account for the effects of seepage-stress coupling, optimizing the design parameters of anti-slide piles, and proposing an improved design scheme for landslide remediation.

(a) Pile spacing

The spacing between anti-slide piles is a critical design parameter. If the spacing is set too wide, the soil can easily slide out between the piles, causing them to behave similarly to single piles. Conversely, if the spacing is set too narrow, the economic costs will increase without a significant improvement in stabilization effectiveness. Therefore, studying the impact of pile spacing on landslide stability and designing an appropriate spacing is essential for enhancing the effectiveness of landslide remediation and reducing engineering costs.

Figure 9 shows the equivalent plastic strain and displacement field contour of the landslide for different S, respectively. It is evident from Figure 9a that when $S = 2D$, the sliding surface of the landslide is effectively separated into upper and lower sections at the position of the anti-sliding piles. This indicates that the anti-sliding piles, due to the smaller spacing, act like a retaining wall, isolating the upper and lower soil masses and preventing the formation of a continuous sliding surface from top to bottom, resulting in a lower equivalent plastic strain for the entire landslide. When $S = 5D$, as shown in Figure 9b, the sliding surface develops into a continuous curve extending from bottom to top, indicating that the piles have not mobilized the surrounding soil to resist sliding together, and each pile in the group piles plays the role of resisting sliding individually, and the effect of the anti-sliding pile group is not activated, resulting in a high equivalent plastic strain of the landslide.

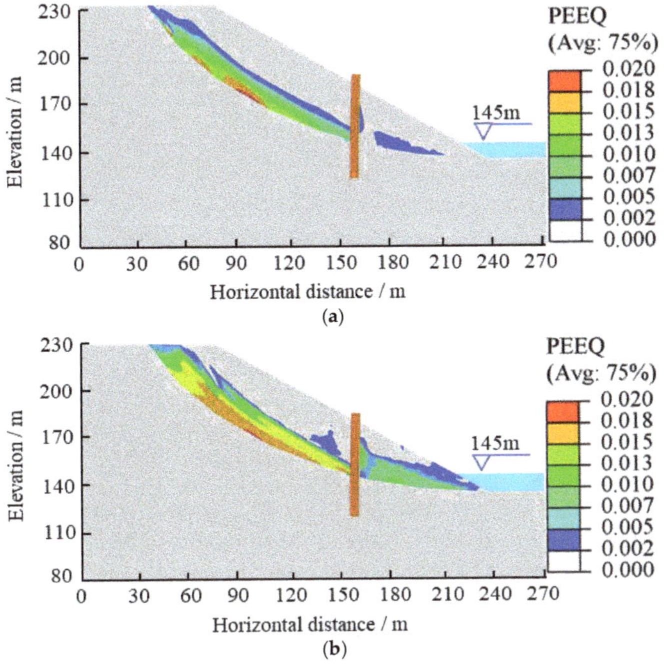

Figure 9. Plastic zone distribution of landslide: (**a**) $S = 2D$, (**b**) $S = 5D$.

Figures 10 and 11, respectively, show the deformation and bending moment results of anti-slide piles corresponding to different pile spacings when the reservoir water level is 145 m. It can be seen from Figure 10 that the pile deflection first increases and then decreases as the pile spacing increases. When the $S \leq$ 4D, the pile spacing increases, and the pile deflection increases slightly. When the $S \geq$ 4D, the pile deflection decreases with the increase in pile spacing. The deflection of anti-slide piles is closely related to the effect of soil arch between anti-slide piles: when no soil arch can be formed between the anti-slide piles, the deflection of the anti-slide piles decreases. According to Figure 11, the larger the S, the larger the bending moment of the anti-slide piles. This is because, with the increase in the pile spacing, the landslide thrust borne by each anti-slip pile also increases gradually.

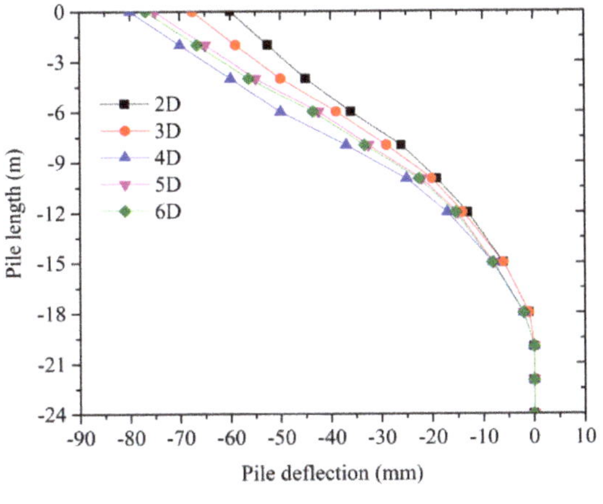

Figure 10. Deflection of the anti-slide pile for different pile spacings.

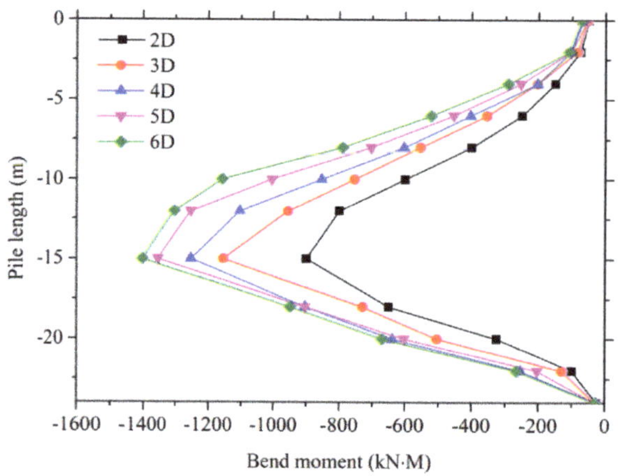

Figure 11. Bending moment of the anti-slide pile for different pile spacings.

Figure 12 shows the calculation results of the maximum displacement of the landslide for different S during the reservoir water level drops. It can be seen from Figure 12 that the maximum displacement increases from 15 mm to 26 mm as the pile spacing increases from $2D$ to $4D$ when the reservoir level drops to 145 m; the maximum displacement of the landslide is better controlled, and the change amplitude is also smaller. When the $S > 4D$, the maximum displacement of the landslide increases continuously, which indicates that the anti-slip piles have a poor anti-slip effect and the overall stability of the landslide is not well controlled.

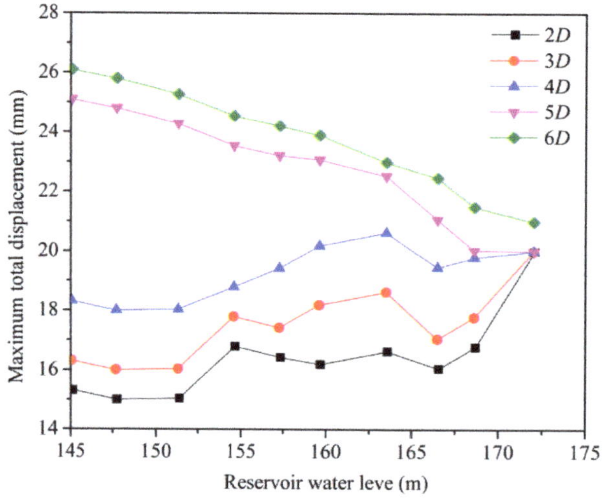

Figure 12. Maximum total displacement of landslide for different pile spacings during the reservoir water level drop.

Figure 13 presents the variation curve of the safety factor of the anti-sliding piles for different values of S. It can be seen that, as the S increases, the safety factor of the landslide gradually decreases. When the S is $5D$ and $6D$, the safety factor of the landslide is relatively low. When S is $3D$ and $4D$, the difference in safety factors between the two is minimal. Therefore, considering economic factors and the ease of construction, an S of $4D$ is sufficient to achieve effective landslide remediation. Based on the comprehensive analysis, a pile spacing of $4D$ is deemed reasonable for stabilizing the landslide in this paper.

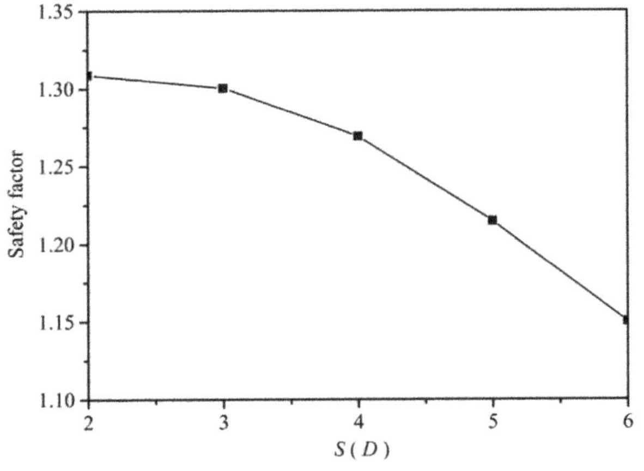

Figure 13. Safety factor of the landslide at different pile spacings.

(b) Embedment depth

The embedment depth L_y of the anti-slip piles also determines the effectiveness of landslide management and the economic cost, which is an important parameter for the design of anti-slip piles. In this section, the influence of the L_y of anti-slip piles on the effect of landslide stabilization when the reservoir water level is reduced to 145 m elevation will be discussed, to comprehensively determine the more reasonable embedment depth of anti-slip piles.

Figure 14 shows the equivalent plastic strain contour map of the landslide when L_y/L_p is 0.17 and 0.55, respectively. During the service life of the anti-slide piles, the portion of the piles above the embedded section transfers the load to the stable rock mass within the embedded section under the action of landslide thrust. This utilizes the embedded effect and passive resistance of the stable stratum to counterbalance the landslide thrust resulting from landslide deformation. When the embedment depth is shallow, as shown in Figure 14a, the portion of the anti-slide piles above the embedded section carries most of the load, resulting in the increased displacement of the piles. This makes it challenging to effectively counteract the landslide thrust, leading to the overall instability and deformation of the landslide. Conversely, when the embedment depth is deep, as depicted in Figure 14b, the anti-slide pile can transfer the landslide thrust to the stable rock layer beneath the sliding surface. By leveraging the embedded effect and passive resistance of the stable stratum, the landslide thrust resulting from slope deformation is counterbalanced, thereby improving the overall stability of the landslide.

Figures 15 and 16 show the results of the deformation and bending moment of the anti-slide pile for different embedment depths, respectively. It can be seen from Figure 15 that the pile deflection shows an obvious decreasing trend with the growth of the embedment depth. When the $L_y/L_p = 0.17$, the embedment depth of the anti-slide pile is shallow, and the displacement of the pile end also begins to appear, and the anti-slide pile slides with the landslide soil. This shows that the anti-slip pile reinforcement effect is poor when the embedding depth is insufficient. In addition, it can be seen from Figure 15 that when $L_y/L_p > 0.38$, the pile displacements are nearly identical. This indicates that once the embedment depth reaches a certain threshold, further increases in depth do not significantly enhance the reinforcement effect of the anti-slide piles. From Figure 16, it is evident that as the embedment depth increases, the bending moment of the anti-slide pile exhibits an upward trend. However, this growth rate gradually slows with further increases in embedment depth. When the anti-slide pile crosses the slip surface, the maximum bending moment is generally reached at the same position along the pile's cross-section.

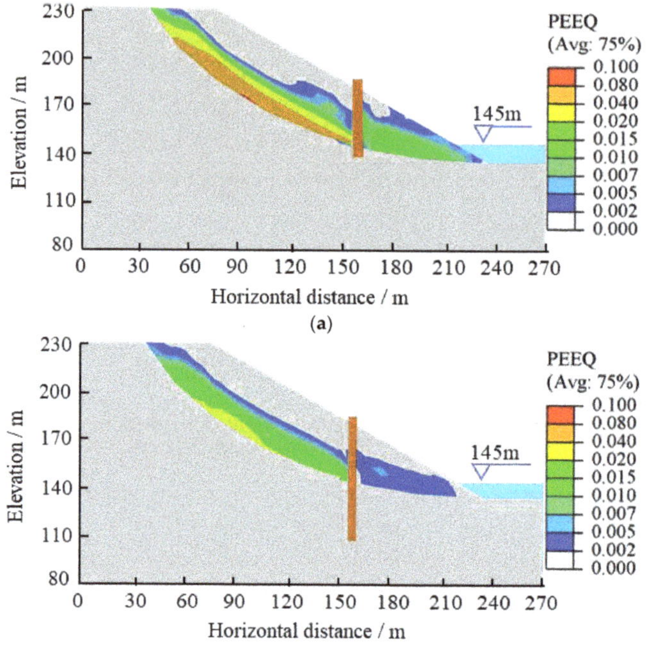

Figure 14. Plastic zone distribution of landslide for (a) L_y/L_p = 0.17, (b) L_y/L_p = 0.55.

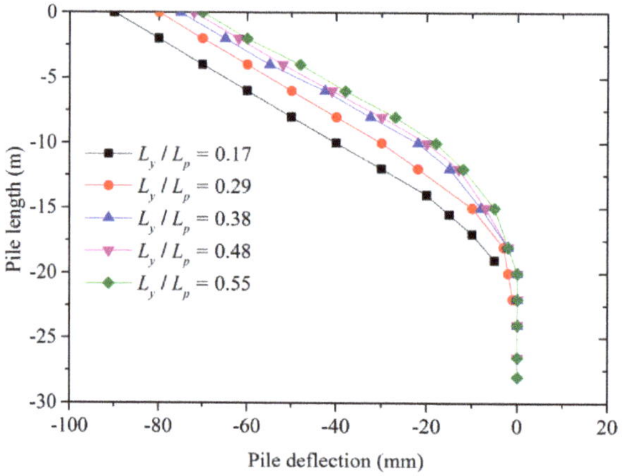

Figure 15. Deflection of the anti-slide pile for different embedment depths.

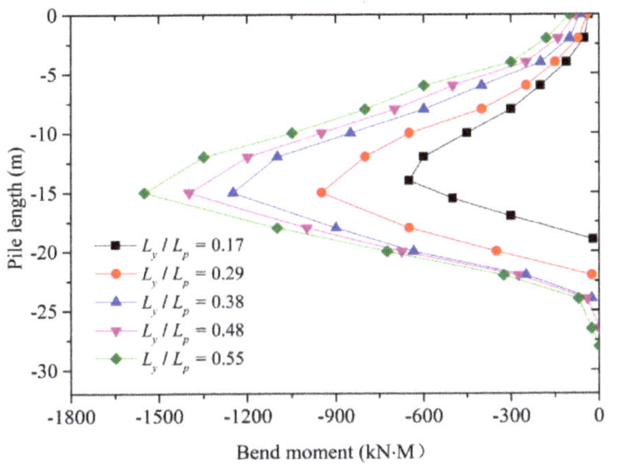

Figure 16. Bending moment of the anti-slide pile for different embedment depths.

Figure 17 illustrates the maximum landslide displacement for various embedment depths of the anti-slide piles during the drop in reservoir water level. The embedment depth significantly impacts landslide displacement. The maximum displacement of the landslide decreases from 18 mm to 29 mm when the L_y/L_p is increased from 0.17 to 0.55. When the $L_y/L_p \geq 0.38$, the maximum displacement shows little variation during the decline of the reservoir water level, indicating that the anti-slide piles can effectively maintain the stability of the landslide in this range.

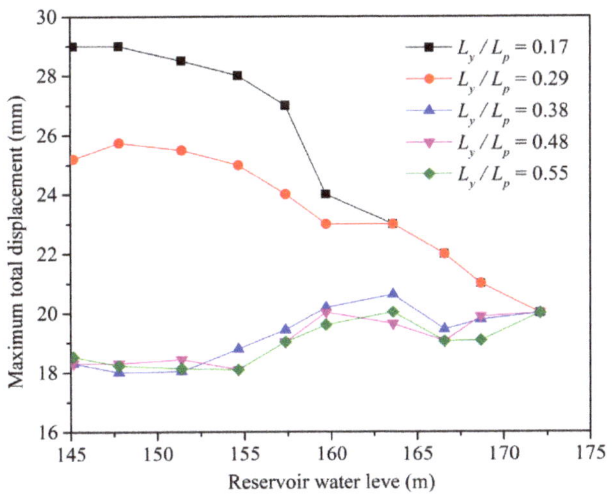

Figure 17. Maximum total displacement of landslide for different embedment depths of the pile during the reservoir water level drops.

Figure 18 shows the variation curve of the safety factor of the anti-sliding piles for embedment depth ratio L_y/L_p. Clearly, as the L_y/L_p increases, the safety factor of the landslide also increases. When the $L_y/L_p \geq 0.48$, the increase in the safety factor tends to level off. This is attributed to the presence of an effective embedment depth for the anti-slide piles. Once this effective depth is exceeded, the anti-slide effect does not continue to increase, and the overall remediation impact on the landslide is not significantly enhanced. Therefore, while ensuring safety reserves, it is advisable to minimize the materials used for the anti-slide piles to reduce remediation costs. Consequently, L_y/L_p =0.48 can be used as the reasonable anchoring depth ratio of anti-slide piles in landslide restoration in this study.

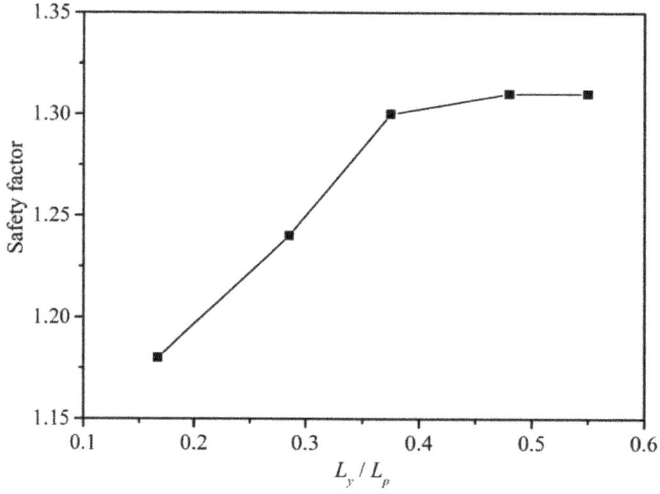

Figure 18. Safety factor of the landslide at different embedment depths.

(c) Pile position

The selection of locations for anti-slide piles is crucial for the effectiveness of landslide remediation. Many researchers believe that different displacements of anti-slide piles result in varying failure modes of landslides, which subsequently impact the stability of the slope in different ways. Therefore, optimizing the placement of anti-slide piles is essential.

In this section, the L_x was varied to investigate the effect of the choice of anti-slip pile location on the landslide stability at a reservoir level of 145 m, while maintaining the pile spacing S = 4D and the embedment ratio L_x/L = 0.38. Figure 19 shows the equivalent plastic strain contour map of the landslide when L_x/L = 0.20 and 0.80, respectively. The sliding body slides along the bedrock interface regardless of where in the landslide the anti-slip piles are arranged. When the anti-sliding piles are positioned at the toe of the landslide (i.e., L_x/L = 0.2), the equivalent plastic strain in the soil near the piles increases and extends toward the top of the landslide. The toe cannot withstand the weight of the upper soil, leading to the gradual failure of the entire landslide. This results in the formation of a continuous zone of soil movement from around the piles to the top of the landslide, causing sliding. At this point, the anti-sliding piles experience almost no horizontal displacement and cannot effectively perform their sliding resistance function. When the anti-sliding piles are positioned at the top of the landslide (i.e., L_x/L = 0.8), the sliding zone area is significantly reduced, ensuring the stability of the lower part of the landslide. In this situation, the sliding resistance of the soil below the piles helps maintain the stability of the entire anti-sliding pile landslide.

Figures 20 and 21 show the results of the displacement and bending moment of the anti-slip pile at different pile positions, respectively. From Figures 20 and 21, it can be seen that the pile displacement and bending moment show a trend of increasing and then decreasing with the increase in L_x/L. When the anti-slip piles are arranged in the middle of the landslide (i.e., L_x/L = 0.5), the pile displacement and bending moment reach the maximum value. This is because, on one hand, when the anti-slide pile is positioned near the foot of the landslide, its embedment depth is shallow, and the corresponding critical slip surface does not intersect with the anti-slide pile. On the other hand, when the anti-slide pile is located near the top of the landslide, the volume of the landslide body decreases, leading to a reduction in the landslide thrust acting on the anti-slide pile. Therefore, when $L_x/L = 0.5$, the pile bending moment and displacement of the skid-resistant pile are both maximized. In addition, it is worth noting that when $L_x/L = 0.2$, the displacement curve of the pile is approximately linear, which indicates that the anti-slip pile produces rotation and overturning damage may occur. This is because when the anti-slip pile is arranged at the foot of the landslide, the thickness of the landslide slip body is larger, and thus the embedment depth of the anti-slip pile is smaller and the anchorage is insufficient.

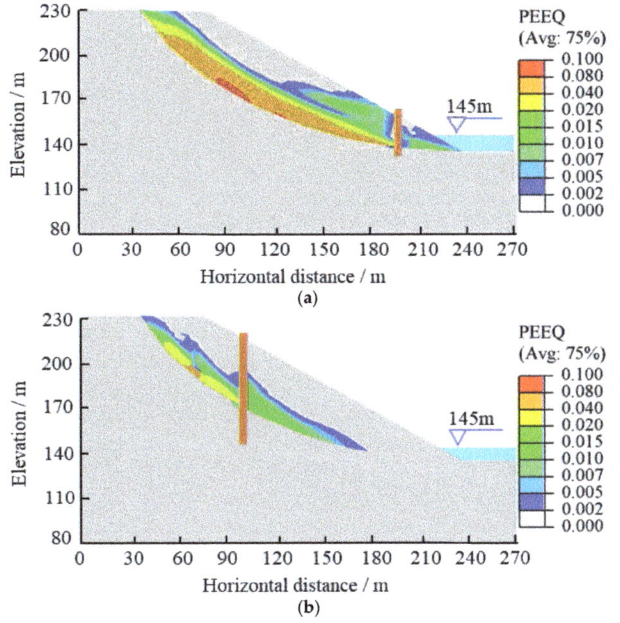

Figure 19. Equivalent plastic strain contour at different pile positions: (**a**) $L_x/L = 0.20$, (**b**) $L_x/L = 0.80$.

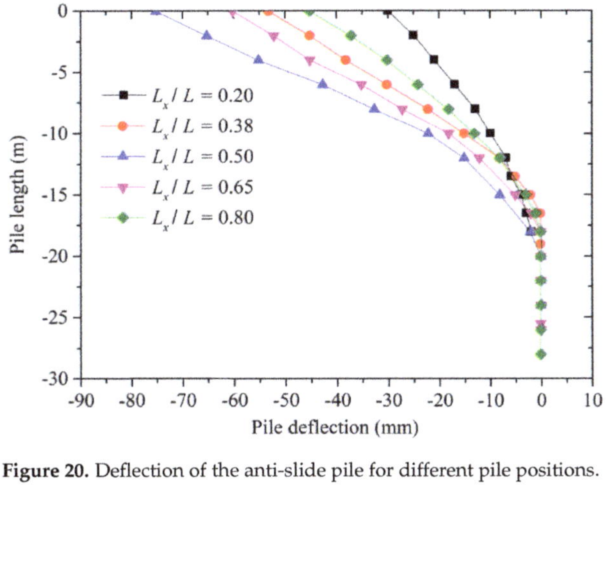

Figure 20. Deflection of the anti-slide pile for different pile positions.

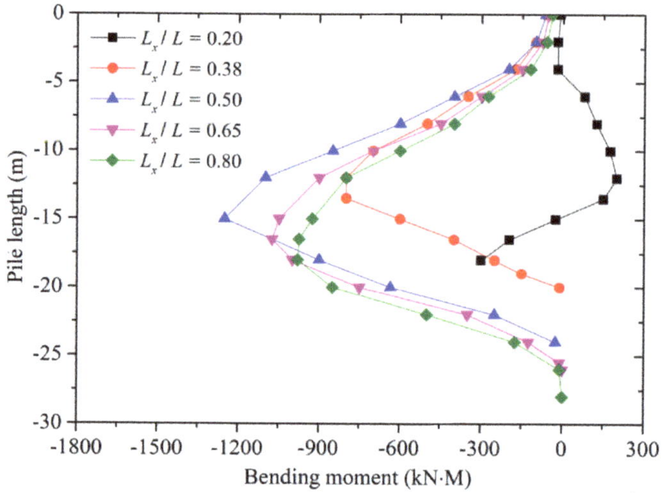

Figure 21. Bending moment of the anti-slide pile for different pile positions.

Figure 22 shows the calculation results of the maximum displacement of the landslide for different pile locations during the reservoir water level drops. When the anti-slip piles are arranged in the middle of the landslide (i.e., $L_X / L = 0.5$), the maximum displacement of the landslide during the water level decline is small, and the landslide stability is effectively controlled. With the upward and downward movement of the pile position, the landslide displacement of the landslide gradually increases.

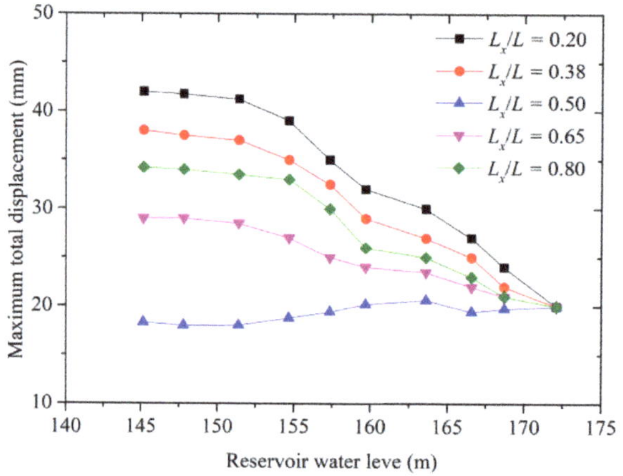

Figure 22. Maximum total displacement of the landslide for different pile locations during the reservoir water level drops.

Figure 23 shows the variation curve of the safety factor of the anti-sliding piles for different pile locations. When the anti-slide pile is located in the middle of the landslide (i.e., $L_X / L = 0.5$), the safety factor of the landslide is the largest. Therefore, from the point of view of the maximum displacement and safety factor, it is appropriate to set the anti-slide pile in the middle of the landslide in this paper.

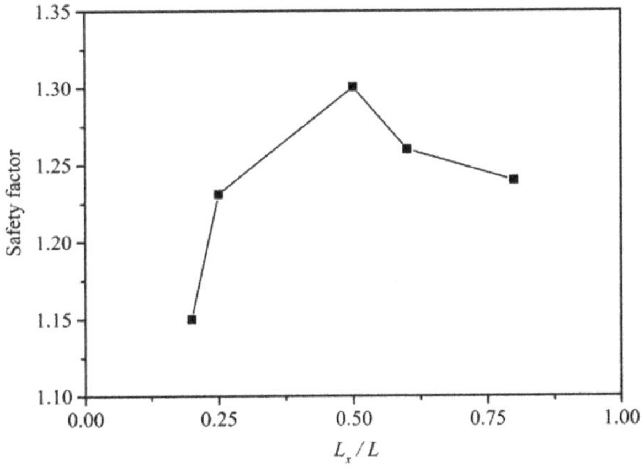

Figure 23. Safety factor of the landslide at different pile locations.

4. Conclusions

In this study, numerical simulations were conducted using the finite element software Abaqus to investigate the deformation and failure mechanisms of reservoir bank landslides during water level drawdown. Additionally, the stabilizing effect of anti-slide piles on landslides was analyzed, with a particular focus on the influence of pile spacing, embedment depth, and arrangement on the maximum landslide displacement and safety factor. The key findings are summarized as follows:

1. The decline in the reservoir water level reduces the landslide safety coefficient and affects the distribution of the reservoir landslide displacement field and equivalent plastic strain field. In the process of reservoir landslide management and design, the influence of the decline in the reservoir water level should be considered.

2. The interactions between anti-slip piles and soil to form the soil arch effect, with anti-slip piles blocking the slide effect of reservoir landslides, leading to prevention and control, is obvious, and can effectively reduce the reservoir landslide deformation and improve the reservoir landslide safety coefficient.

3. The pile spacing, embedding depth, and location of anti-slip piles have a significant effect on the stability of stabilized reservoir landslides. The determination of a reasonable pile spacing requires a combination of factors such as reinforcement effectiveness and economics. The embedment depth of anti-slip piles should ensure that they can penetrate below the slip surface, but after reaching the critical embedment depth, the reinforcement effect of the anti-slip piles will not be enhanced with the increase in embedment depth. Anti-slip piles should be installed in the middle of the landslide to maximize their reinforcement effect.

It has to be recognized that this paper still has some limitations. For example, the anti-slip piles in this paper are conventional circular piles, and other shapes of anti-slip piles, such as H-shaped and rectangular, still need to be investigated for their effectiveness in stabilizing landslides during the period of declining reservoir water level. In addition, this paper only considers the case of reservoir water level decline; when considering the coupled effect of rainfall and reservoir water level changes, the landslide stability has to face a more complicated situation. These problems are to be studied in the future.

Author Contributions: G.Y.: methodology, software, investigation, validation, data curation, funding acquisition. Z.W.: methodology, supervision, resources, writing—review and editing. L.Z.: writing—review and editing, software, funding acquisition. J.H.: writing—review and editing. S.T.: software, funding acquisition. F.L.: writing—review and editing, funding acquisition. Y.Z.: writing—review and editing. All authors have read and agreed to the published version of the manuscript.

Funding: This work was financially supported by the National Natural Science Foundation of China (No. 52108304), the Natural Science Foundation of Chongqing (Grant No. 2023NSCQ-MSX3411), the Scientific and Technological Research Program of Chongqing Municipal Education Commission (Grant No. KJQN202301307), the Henan Province Science and Technology Research Project (Grant No. 242102320010, 252102320028), and the Henan Province housing and urban and rural construction science and technology plan project (Grant No. HNJS-2022-K3, HNJS-2024-R2).

Data Availability Statement: The data presented in this study are available from the corresponding author upon request.

Conflicts of Interest: Author Gang Yang, Zhuolin Wu, Lin Zhang and Jingfeng Hou were employed by the Chongqing Survey and Design Institute of Water Resources, Electric Power and Architecture Co., Ltd. The funders had no role in the design of the study; in the collection, analyses, or interpretation of data; in the writing of the manuscript; or in the decision to publish the results. The remaining authors declare that the research was conducted in the absence of any commercial or financial relationships that could be construed as a potential conflict of interest.

References

1. Zong, X.Z.; Tao, L.; Shu, Z.; Duan, H.; Li, S.; Wang, J.; Deng, Y. A novel method to evaluate the time-dependent stability of reservoir landslides: Exemplified by Outang landslide in the Three Gorges Reservoir. *Landslides* **2023**, *20*, 1731–1746.
2. Li, Y.; Xie, X.; Jin, B.; Chen, L.; Liang, X.; Yin, K. Comprehensive risk management of reservoir landslide-tsunami hazard chains: A case study of the Liangshuijing landslide in the Three Gorges Reservoir area. *Landslides* **2024**, *22*, 671–691. [CrossRef]
3. Farooq, M.A.; Falak, S.; Ehtisham, M. Evaluation of landslide hazards potential at Dasu dam site and its reservoir area. *Environ. Earth Sci.* **2023**, *82*, 183.
4. Muhammad, K.; Xiewen, H.; Awais, M.H.; Sanaullah, M.; Ali, R. Dynamic Response and Deformation Behavior of Kadui-2 Landslide Influenced by Reservoir Impoundment and Rainfall, Baoxing, China. *J. Earth Sci.* **2023**, *34*, 911–923.
5. Song, K.; Wang, F.; Yi, Q.; Lu, S. Landslide deformation behavior influenced by water level fluctuations of the Three Gorges Reservoir (China). *Eng. Geol.* **2018**, *247*, 58–68. [CrossRef]
6. Xia, M.; Ren, G.M.; Ma, X.L. Deformation and mechanism of landslide influenced by the effects of reservoir water and rainfall, Three Gorges, China. *Nat. Hazards* **2013**, *68*, 467–482. [CrossRef]
7. Jia, G.; Zhan, T.L.; Chen, Y.; Fredlund, D. Performance of a large-scale landslide model subjected to rising and lowering water levels. *Eng. Geol.* **2009**, *106*, 92–103. [CrossRef]
8. Huang, X.; Guo, F.; Deng, M.; Yi, W.; Huang, H. Understanding the deformation mechanism and threshold reservoir level of the floating weight-reducing landslide in the Three Gorges Reservoir Area, China. *Landslides* **2020**, *17*, 2879–2894. [CrossRef]
9. Tang, M.; Xu, Q.; Yang, H.; Li, S.; Iqbal, J.; Fu, X.; Huang, X.; Cheng, W. Activity law and hydraulics mechanism of landslides with different sliding surface and permeability in the Three Gorges Reservoir Area, China. *Eng. Geol.* **2019**, *260*, 105212. [CrossRef]
10. Zhao, N.; Hu, B.; Yan, E.; Xu, X.; Yi, Q. Research on the creep mechanism of Huangniba landslide in the Three Gorges Reservoir Area of China considering the seepage–stress coupling effect. *Bull. Eng. Geol. Environ.* **2019**, *78*, 4107–4121. [CrossRef]
11. Zangerl, C.; Eberhardt, E.; Perzlmaier, S. Kinematic behaviour and velocity characteristics of a complex deep-seated crystal-line rockslide system in relation to its interaction with a dam reservoir. *Eng. Geol.* **2010**, *112*, 53–67. [CrossRef]
12. Lin, J.; Yu, S.; Luo, Y.; Xu, T.; Lin, Y.; Zheng, W.; Li, W.; Wang, Y. Research on Stability of Dam Substation on Inclined Soft Soil Foundation Reinforced by Pile Foundation. *Water* **2023**, *15*, 3527. [CrossRef]
13. Wen, Z.; Xu, W.; Yuan, B.; Zhang, L.; Liang, Z. Numerical Analysis of Anti-Slide Pile Reinforcement for Landslide Stability Under Rainfall Conditions. *Buildings* **2025**, *15*, 638. [CrossRef]
14. Zhang, Q.; Hu, J.; Du, Y.; Gao, Y.; Li, J. A laboratory and field-monitoring experiment on the ability of anti-slide piles to prevent buckling failures in bedding landslides. *Environ. Earth Sci.* **2021**, *80*, 44. [CrossRef]
15. Zhang, H.; Xing, H.-F.; Xue, D.-R.; Tannant, D. Centrifuge and numerical modeling of h-type anti-slide pile reinforced soil-rock mixture landslide. *J. Mt. Sci.* **2023**, *20*, 1441–1457. [CrossRef]
16. Song, Y.; Hong, W.; Woo, K. Behavior and analysis of stabilizing piles installed in a cut landslide during heavy rainfall. *Eng. Geol.* **2012**, *129–130*, 56–67. [CrossRef]
17. Wang, L.; Zhang, G. Centrifuge model test study on pile reinforcement behavior of cohesive soil landslides under earthquake conditions. *Landslides* **2014**, *11*, 213–223. [CrossRef]

18. Zhang, G.; Wang, L. Integrated analysis of a coupled mechanism for the failure processes of pile-reinforced landslides. *Acta. Geotech.* **2016**, *11*, 941–952. [CrossRef]
19. Lirer, S. Landslide stabilizing piles: Experimental evidences and numerical interpretation. *Eng. Geol.* **2012**, *149–150*, 70–77. [CrossRef]
20. Won, J.; You, K.; Jeong, S.; Kim, S. Coupled effects in stability analysis of pile–landslide systems. *Comput. Geotech.* **2005**, *32*, 304–315. [CrossRef]
21. ABAQUS. *Abaqus 6.14 Standard;* Dassault Systemes Simulia Corp: Providence, RI, USA, 2014.
22. Zienkiewicz, O.C.; Humpheson, C.; Lewis, R.W. Associated and non associated viscoplasticity in soil mechanics. *Geotechnique* **1975**, *25*, 671–689. [CrossRef]
23. Dawson, E.M.; Roth, W.H.; Drescher, A. Slope stability analysis by strength reduction. *Geotechnique* **1999**, *49*, 835–840. [CrossRef]
24. Yin, Y.; Huang, B.; Wang, W.; Wei, Y.; Ma, X.; Ma, F.; Zhao, C. Reservoir-induced landslides and risk control in Three Gorges Project on Yangtze River. China. *J. Rock Mech. Geotech.* **2016**, *8*, 577–595. [CrossRef]
25. Luo, S.-L.; Huang, D.; Peng, J.-B.; Aierken, A.; Li, Z.; Kuang, X.-B.; Tomás, R. Performance and application of a novel drainage anti-slide pile on accumulation landslide with a chair-like deposit-bedrock interface in the Three Gorges Reservoir area, China. *Comput. Geotech.* **2023**, *155*, 105199. [CrossRef]
26. Pulko, B.; Majes, B.; Mikoš, M. Reinforced concrete shafts for the structural mitigation of large deep-seated landslides: An experience from the Macesnik and the Slano blato landslides (Slovenia). *Landslides* **2014**, *11*, 81–91. [CrossRef]
27. Chen, L.; Guo, H.; Gong, P.; Yang, Y.; Zuo, Z.; Gu, M. Landslide susceptibility assessment using weights-of-evidence model and cluster analysis along the highways in the Hubei section of the Three Gorges Reservoir Area. *Comput. Geosci.* **2021**, *156*, 104899. [CrossRef]

water

MDPI

Article

Monitoring and Analysis of Waterproof Coal Pillars Under the Influence of Goaf Water

Xiaoqian Yuchi [1,*], Helong Gu [2], Xuanhong Du [1] and Pan Shu [1]

[1] College of New Energy, Longdong University, Qingyang 745000, China; 15349419043@163.com (X.D.); shu18392581239@163.com (P.S.)
[2] College of Energy Science and Engineering, Xi'an University of Science and Technology, Xi'an 710054, China; helonggu@xust.edu.cn
* Correspondence: 13484519960@163.com

Abstract: Performing stability studies of waterproof coal pillars is one of the key measures for preventing mine water disasters. As some areas of the coal pillar were affected by goaf water in the Nanhu Second Mine, the coal pillar and surrounding roadway were somewhat deformed. To investigate whether the pillar can ensure safe production in the mine, the source of goaf water and the direction of water infiltration were analyzed using exploration holes, and it was concluded that the goaf water originated from the V_3 aquifer and was static. Thus, a theoretical analysis was carried out to determine the relationship between the mechanical parameters of the coal and rock structures affected by water. On this basis, a numerical simulation was employed to examine the key changes in the coal pillar and roadway affected by goaf water. The simulation results showed that the plastic area was 6–11 m and the elastic area in the middle was 6–8 m after excavating the working faces on both sides of the coal pillar, and the water flow vector of the aquifer could not pass through the pillar. Finally, in situ monitoring using ground-penetrating radar, deformation measurement, and loosening circle detection revealed that the development degree of internal cracks in the coal pillar was relatively light; thus, the pillar could effectively prevent water damage. These monitoring and analysis methods comprehensively evaluate the stability of the coal pillar and provide a guarantee for the safe mining of the working face.

Keywords: waterproof coal pillar; numerical simulation; in situ monitoring; ground-penetrating radar; loosening circle detector

1. Introduction

As a common disaster in coal mine production, water damage occurs frequently in major mining areas [1–3]. Due to the extremely complex hydrogeological conditions in China, water damage accidents have inevitably become a serious problem for many mines [4–7]. Waterproof coal pillars are crucial for maintaining the safe operation of stopes and preventing goaf water [8–10]. However, due to the intricate nature of the coal–rock body formation process, hydrogeology, and occurrence conditions, waterproof coal pillars, as underground load-bearing barriers, will inevitably have a large number of primary cracks [11,12]. When influenced by mining, the cracks will continue to expand, reducing water prevention and load-bearing capacity. Therefore, guaranteeing the safety and stability of pillars is meaningful for reducing the loss of coal resources and controlling water damage [13,14].

In the current study, mechanical analysis is the first part to appear in the entire pillar study. Most scholars tend to develop corresponding theoretical models based only

 https://doi.org/10.3390/w17010065

on the bearing structural characteristics of traditional section pillars [15], such as the pillar stress distribution model proposed by the British scholar Wilson, the pressure arch model suggested based on the properties of rock cover, the limit balance theory used when the pillar is gradually destroyed, etc. [16,17]. On this basis, the weakening and infiltration of water into coal pillars are gradually mentioned [18–20]. In this regard, some scholars have conducted rock mechanics experiments by collecting coal samples affected by water to research the permeability and weakening effects of water on a microscopic level [21–23]. Therefore, studies on coal pillars through the three aspects of goaf water infiltration weakening, coal pillar fracture development, and mining stress transfer can provide a better theoretical reference basis for stability analysis [24–27].

With the advancements of various numerical simulation methods, including finite elements and discrete elements, numerical simulation has emerged as an essential method for scholars to analyze the reliability of waterproof coal pillars under different geological conditions [28–30]. Such research can effectively solve the defects that theoretical analysis struggles to structurally model and avoid the disadvantages of empirical formulas, which are difficult to apply to various field conditions. To some extent, it stresses different migration processes of the surrounding rock, which are very important for analyzing the loading of coal pillars [31–33]. On this basis, geophysical technology has made significant progress, and it is no longer difficult to detect the characteristics of fracture development in coal pillars and the impact of goaf water on coal pillars [34]. Thus, utilizing advanced geophysical methods such as ground-penetrating radar (GPR), peephole monitoring, and loosening circle detection also has become a research trend in analyzing the stabilization of waterproof coal pillars and the degradation process based on disturbance intensity [35].

Therefore, this study detects goaf water through exploration holes and analyzes the source of goaf water and the direction of infiltrating water flow. On this basis, a theoretical analysis is conducted to determine the connection between the mechanical properties of coal and rock when affected by water. The fluid–solid coupling theory and integral criterion of Mohr–Coulomb are used in FLAC^{3D} to assess the stress variation characteristics and elastic–plastic zone distribution of a waterproof coal pillar in the Nanhu Second Mine under mining disturbance, and the evolution process of the water flow vector is determined. Finally, based on the combination of geological radar, deformation monitoring, and loosening circle detection, the development degree and stability of coal pillar fractures are evaluated, providing strong technical support for subsequent safe production in the mine.

2. Research Background and Disaster Source

2.1. Working Face and Coal Pillar Characteristics

The average buried depth of the 1303 working face is 255.5 m in the Nanhu Second Mine. The overall shape of the coal seam is a syncline structure. The working face is mainly a mining 3# coal seam. The coal seam is relatively stable. Its thickness is between 6 and 9 m; the average value is 6.2 m. The inclination angle of the 3# coal seam is 8°.

The roof and bottom plate hardness of the 3# coal seam is small, as the roof is mainly siltstone, and some areas are mixed with mudstone. There are many internal joints and structural surfaces of the roof, with poor stabilization. The bottom plate is mainly mudstone, with obvious softening characteristics in water.

An 18 m coal pillar is present between the previous working face (1301) and the 1303 working face. However, the 1301 working face mining expands more to the crack, and part of the fissure is connected to the upper aquifer, so that there is a large amount of water in some areas of the 1301 goaf. Thus, the pillar must serve as a barrier against goaf water. The working face mining layout and the location of the goaf water can be seen in Figure 1.

Due to the influence of goaf water and mining disturbance, the waterproof coal pillar and 1303 auxiliary transportation roadway exhibit some deformation within the goaf water area. Through on-site monitoring, the average amount of top sink can reach 21 cm, and the average amount of bottom bulge can reach 18 cm. With the stope (1303) approaching the goaf water, it is urgent to study and evaluate the stabilization of the waterproof coal pillar and offer assistance in guaranteeing safe mining.

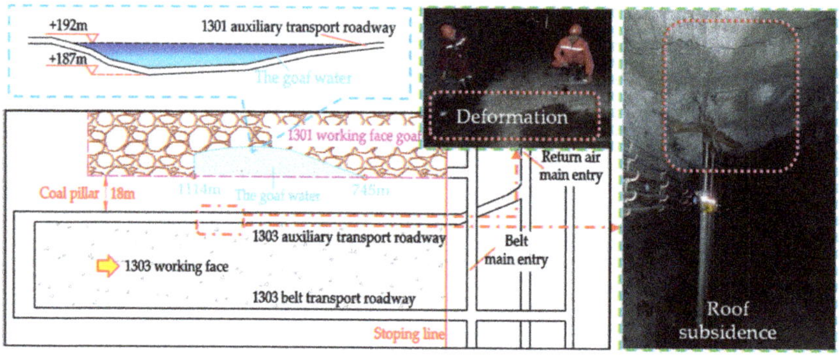

Figure 1. The 1303 working face mining layout and the location of the goaf water.

2.2. Hydrogeological Characteristics of the Study Area

According to on-site investigation, the water filling source affecting the mining of the 1303 working face mainly comes from the aquifer above the 3# coal seam. The extraction of the 3# coal seam creates a fracture zone that allows water to flow into the goaf. Because water has an obvious softening effect, the roof–floor strata characteristics of the 1303 working face become weak. Through the analysis of mine hydrology and upper-section mining data, the roof aquifer of the 3# coal seam is recognized as a weak aquifer group located in the middle part of the Xishanyao Formation from the Jurassic period that is divided into four sandstone aquifers, as shown in Table 1.

Table 1. Geological characteristics of each aquifer.

Name of Aquifer	Thickness (m)	Position	Lithology	Vertical Distance from 3# Seam
V_1	35	Upper 1# coal seam	Crude-grain sandstone is mainly used	Upper coal seam 95 m
V_2	4	Between the 1# and 2# coal seams	Medium-grain sandstone and conglomerate intersection are mixed	Top 55 m
V_3	6	Between the 2# and 3# coal seams	Siltstone is mainly mixed with mudstone	Top 20–30 m
V_4	11	Between the 3# and 4# coal seams	Grainy sandstone and conglomerate intersection are mixed	Lower roof plate is 9 m

2.3. Analysis of the Characteristics of Goaf Water

To find the specific morphological characteristics of the goaf water, workers constructed 4 exploration holes along the 1303 auxiliary transport roadway. Goaf water was found between 745 m and 1114 m along the roadway, while the horizontal elevation of the roadway in that region varied from +183 m to +205 m. The 4 holes were located in the goaf water area, and their horizontal elevations were +201 m, +192 m, +187 m, and +199 m, as shown in Table 2.

Table 2. Characteristics of exploration holes in the goaf water zone.

Exploration Hole Number	Horizontal Elevation	Location from the 1303 Roadway	Characteristic of the Hole
EH-1	+201 m	818 m	Drilling depth 19 m Drilling diameter 52 mm
EH-2	+192 m	891 m	
EH-3	+187 m	964 m	
EH-4	+199 m	1037 m	

The average hydraulic pressure of each exploration hole can be observed in Figure 2, which shows that the the goaf water pressure in the detection area is relatively stable, and the fluctuation range is very small. It can be concluded that the goaf water in this area is static. According to the elevation of each water detection hole (in Figure 3), a linear relationship exists between average hydraulic pressure and exploration hole elevation. Combined with the calculation formula of static hydraulic pressure, $P = \rho g(H - h_0)$, information can be obtained by analysis and calculation (H is the elevation of the horizontal plane, and h_0 is the horizontal elevation of the exploration hole). The goaf water of the Nanhu Second Mine is static, and the water accumulation elevation is +209 m. The maximum static hydraulic pressure is 0.26 MPa.

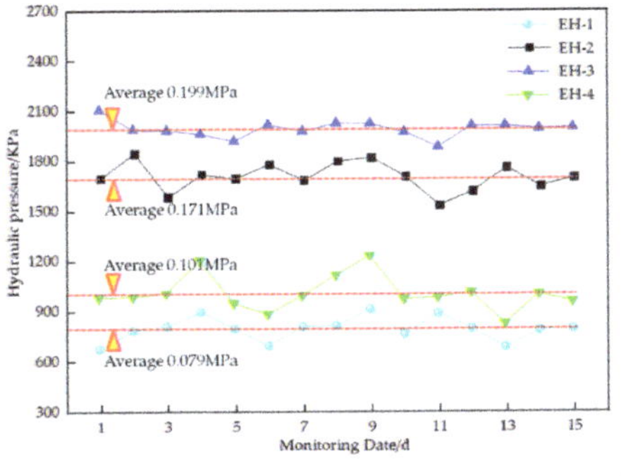

Figure 2. Hydraulic pressure measurement curve of exploration holes.

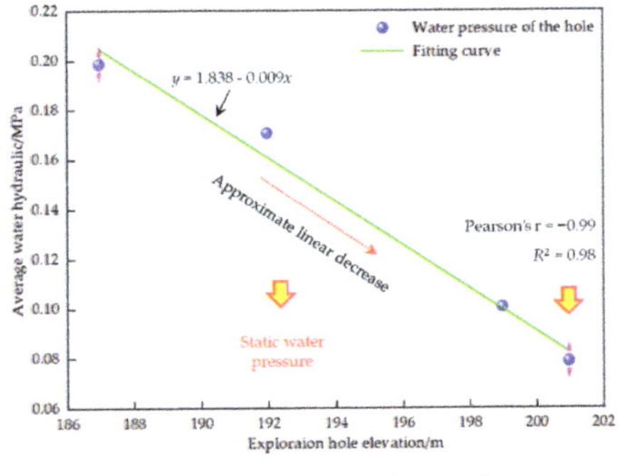

Figure 3. Relationship between hydraulic pressure and hole elevation.

Therefore, according to the +183 m level of the lowest point of the roadway in the area affected by water, the water level is 26 m above the 3# coal seam, which is the location of the V_3 aquifer. Combined with the lithologic characteristics of the 3# coal seam, it is known that the water resistance of the 3# coal seam direct roof, which is mainly formed of siltstone, is poor. Thus, the separation area formed after excavating the 1301 working face easily becomes the water accumulation space. With the connection of the fracture zone and the V_3 aquifer, goaf water is finally produced. Moreover, the bottom plate of the 3# coal seam is mudstone, which has fewer pores. When the mudstone swells, it will greatly hinder the goaf water from penetrating further. Finally, a water accumulation area similar to a reservoir forms (in Figure 4).

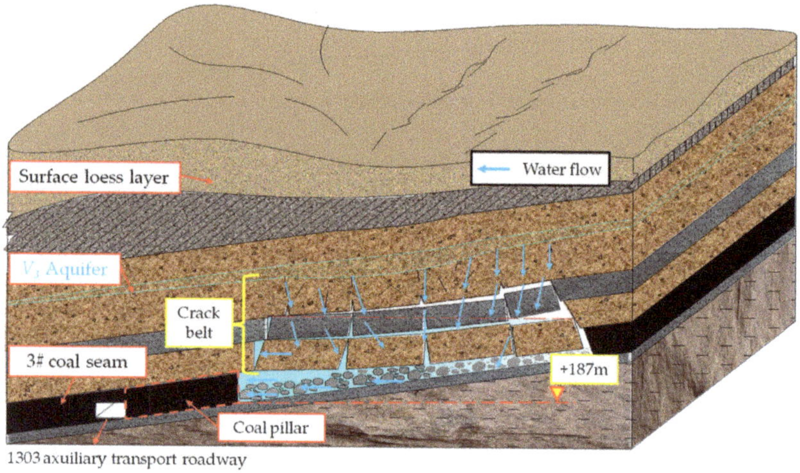

Figure 4. Schematic diagram of overburden migration and goaf water formation.

3. Analysis of Coal and Rock Damage Parameters Affected by Goaf Water

By comparing the literature [36,37], the uniaxial compressive rock strength (σ_c) and elastic modulus (E) demonstrate a linear decrease as water content increases:

$$E = a + bw \tag{1}$$

$$\sigma_c = e + fw \tag{2}$$

In Equations (1) and (2), a, b, e, and f are assumed constants.

The moisture content of coal and rock is related to their properties and the magnitude of stress. The bulk density of the rock under the influence of a stress state can be expressed as follows:

$$r_{d0} = r_d - r_d\varepsilon_v = r_0(1 + 0.01w_0) \tag{3}$$

In Equation (3), ε_v is the volumetric strain caused by stress; r_{d0} is the natural bulk density; r_d is the water-saturated bulk density; r_0 is the drying density; and w_0 is the natural moisture content of the rock.

According to rock mechanics and related studies [36], when the stress experienced by a rock is less than its critical stress value, the rock undergoes elastic deformation. When the stress on the rock exceeds the critical value, the rock undergoes dilatancy. Therefore, the volumetric strain is the sum of elastic strain and dilatancy, indicating a relationship between the stress on the rock and parameters such as volumetric strain and elastic strain as follows:

$$\sigma_x + \sigma_y + \sigma_z = \left[\varepsilon_v + 3T_0\left(\frac{\sigma_\tau}{f_3}\right)^n\right]\frac{E_0}{1-2\mu} \tag{4}$$

In Equation (4), μ is Poisson's ratio; σ_x, σ_y, and σ_z are stress components; σ_τ is the generalized shear stress; T_0, n, and f_3 are rock dilatancy parameters; and E_0 is the elastic modulus.

When $\sigma_\tau > f_3$, the rock undergoes dilatancy, and the water content under stress will exhibit the following relationship with the water-saturated bulk density and water-saturation coefficient of the rock:

$$w_0(r_s - r_d) = k_s r_w r_d r_s \tag{5}$$

In Equation (5), r_w is the bulk density of water; r_s is the bulk density of rock; and k_s is the saturation coefficient.

By substituting Equations (3) and (4) into Equation (5), the connection between stress and moisture content can be determined as follows:

$$w(\sigma) = \frac{k_s r_w (f_3^n + 3T_0\sigma_\tau^n)}{f_3^n r_0(1+0.01w_0)} - \frac{(1-2\mu)k_s r_w \sigma}{r_0(1+0.01w_0)E_0} - \frac{k_s r_w}{r_s} \tag{6}$$

In Equation (6), σ is the combined stress exerted on the rock.

By combining Equations (1) and (2), the equations for the rock elastic modulus and uniaxial compressive strength affected by water can be obtained:

$$E = a + b\left[\frac{k_s r_w (f_3^n + 3T_0\sigma_\tau^n)}{f_3^n r_0(1+0.01w_0)} - \frac{(1-2\mu)k_s r_w \sigma}{r_0(1+0.01w_0)E_0} - \frac{k_s r_w}{r_s}\right] \tag{7}$$

$$\sigma_c = e + f\left[\frac{k_s r_w (f_3^n + 3T_0\sigma_\tau^n)}{f_3^n r_0(1+0.01w_0)} - \frac{(1-2\mu)k_s r_w \sigma}{r_0(1+0.01w_0)E_0} - \frac{k_s r_w}{r_s}\right] \tag{8}$$

From this, given the known stress conditions and moisture content of the rock, its mechanical parameters under the influence of water can be determined.

4. Numerical Analysis of Coal Pillar Stability

The software FLAC^{3D} 3.0, which was developed by the ITASCA company in Minneapolis, MN, USA, can accurately simulate the elastoplastic change law of the coal pillar from the load to the final yield limit, and according to the grid characteristics and the constraints imposed by the model, the stress–strain value of a unit in different stages can be effectively determined. Thus, through a field investigation, we obtained the relevant parameters required for numerical simulation and substituted the coal and rock parameters affected by goaf water into the numerical model. Finally, using the solid-flow coupling criteria built into FLAC^{3D}, we can observe how seepage water affects the rock and coal and find the direction of the water flow vector.

4.1. Construction Model

According to the theoretical analysis presented above, rock mechanical parameters would weaken under the influence of goaf water. Therefore, given the initial parameters of coal and rock affected by water, the impact of water on the structure of the coal pillar and roadways can be modeled to judge the stabilization characteristics of the pillar. Using the geological bar chart, mining scheme, and parameters of the Nanhu Second Mine, each rock layer is optimized and merged, and the mining numerical calculation model is established (in Figure 5).

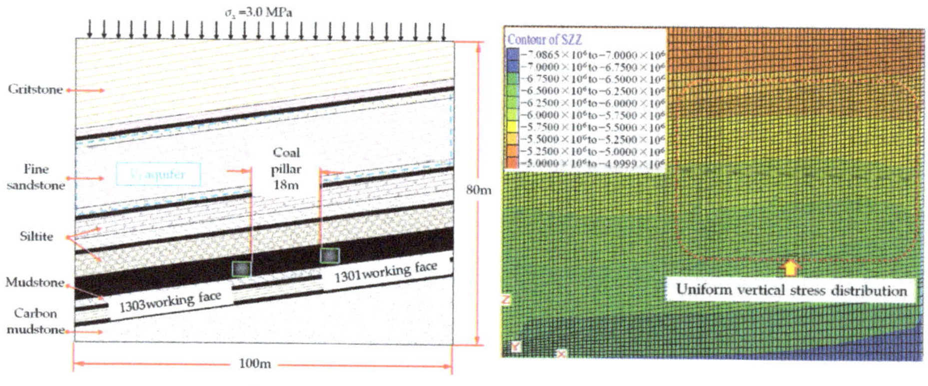

(a) (b)

Figure 5. Schematic diagram of the numerical calculation model. (**a**) Model diagram; (**b**) initial equilibrium state.

The dimensions of the model are 100 × 80 m, and the coal pillar width is 18 m. According to the mine data, the parameters are refined. To simplify the model, the typical depth at which the coal working face is buried is 220 m, and the rock covering density is 2500 kg/m^3; the upper boundary loads 3.0 MPa of force with horizontal displacement limitations on the model's left and right edges. The mining scheme is as follows: step 1, excavation of the 1301 lower roadway and 1303 upper roadway; step 2, excavation of the 1301 working face; step 3, excavation of the 1303 working face.

4.2. Results and Analysis

Mining disturbance on the working faces causes stress redistribution, resulting in a local stress concentration that causes the coal and rock to approach their strength limit; this will induce plastic deformation. The mine pressure appearance and roadway deformation in mining engineering are the result of the plastic deformation of coal and rock.

(a) Features of the coal pillar after roadway excavation

The change in the stress field is the force source for coal and rock deformation. The displacement amount and plastic area of the roadways are essentially caused by stress concentration, so it is particularly important to determine the influence of the coal pillar on the stress distribution characteristics of the surrounding rock during the mining process (as shown in Figure 6).

Thus, under the influence of roadway excavation, the coal–rock stress state is broken, the original rock stress field changes, and the vertical stress is mainly concentrated on the roadways. The pressure area is about 8.57 MPa, and the stress superposition state of the coal vertical stress state presents an arch shape; thus, the stress on both sides of the roadway roof is reduced. The horizontal stress is distributed in the coal pillar, which produces a small stress concentration zone, and the average stress value is nearly 5.76 MPa. The displacement changes in overburden and the coal pillar are relatively small. Overall, the entire horizontal stress of the model does not change much, and the stress is distributed relatively uniformly. The change mainly occurs locally in the model, indicating that the roadways and coal pillar are currently less affected by horizontal stress, and the pillar remains stable. The distribution rule of the plastic zone indicates that the primary plastic zone extends 1 m around roadways. Additionally, the plastic zone of the 1301 roadway is slightly larger than that of 1303 roadway. This indicates that the 1301 roadway experiences greater stress due to factors such as the inclination angle of the 3# coal seam.

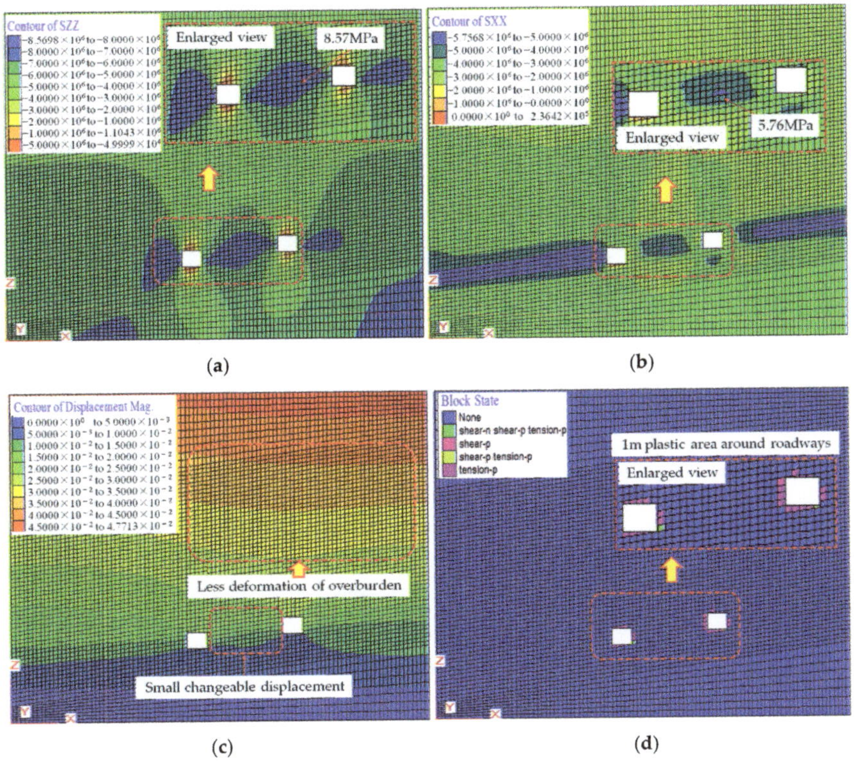

Figure 6. Numerical simulation characteristics after roadway excavation. (**a**) Vertical stress characteristics; (**b**) horizontal stress characteristics; (**c**) displacement distribution characteristics; (**d**) characteristics of the elastoplastic area.

(b) Stability features of the coal pillar after excavation on one side

With the excavation of the 1301 working face (in Figure 7), the overburden is destroyed, as clearly observed on the side of the working face. The situation results in a stress concentration, but the maximum support pressure position does not completely coincide with the coal wall. The stress on the right side of the coal wall, approximately 2 m deep, is about 16 MPa. Figure 7 illustrates that the vertical stress distribution is gradually transmitted to the depth of the coal floor.

In the horizontal direction, the stress concentration in the coal pillar deviates from that of the excavated roadway, reaching nearly 10 MPa. From the displacement of the overburden after the 1301 area excavation, the movement of the overburden on the mining face is clear, but the overall displacement change of the pillar approaches 0 compared with that seen before. Finally, through the characteristics of changes in the plastic area, following the excavation of the 1301 mining face, a triangular plastic area appears on the roof of the 1301 goaf, and the plastic zone of the pillar near the working face significantly increases. The width of the plastic area expands from 1 m to 3 m during excavation, while the plastic area on the roadway side increases only slightly, reaching approximately 1–2 m.

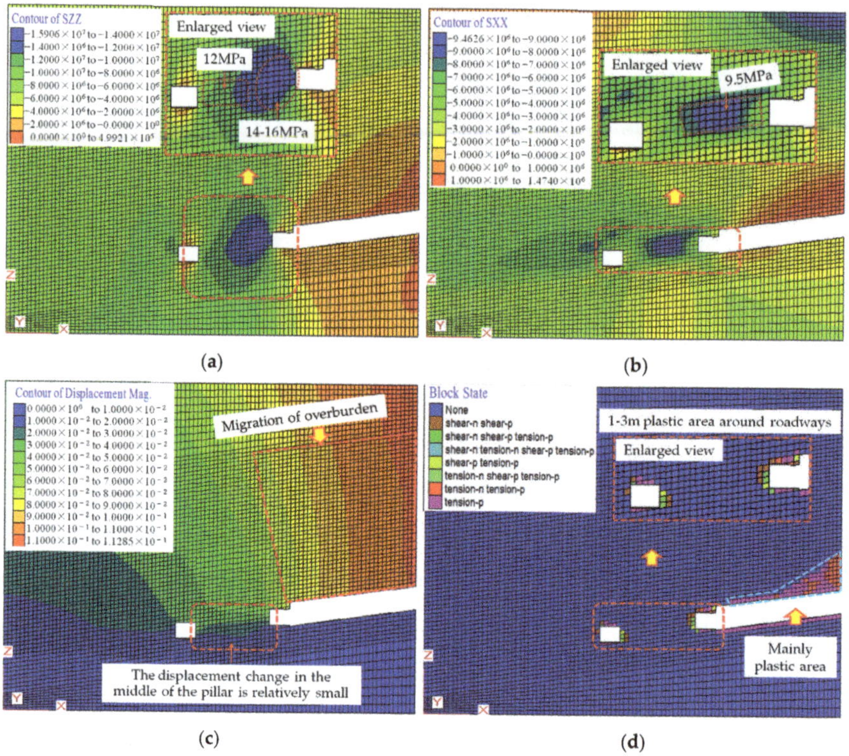

Figure 7. Numerical simulation characteristics after the excavation of the 1301 working face. (**a**) Vertical stress characteristics; (**b**) horizontal stress characteristics; (**c**) displacement distribution characteristics; (**d**) characteristics of the elastoplastic area.

(c) Stability features of the coal pillar after the excavation of both sides

After the excavation of both sides of the working face, the overburden weight on the goaf is transferred to the coal pillar (in Figure 8). The vertical stress concentration appears in the middle of the pillar, and it is gradually transmitted to the bottom plate. And the maximum value of vertical stress reaches 35.3 MPa, which is close to that of the 1301 roadway. Horizontal stress follows a butterfly-shaped distribution, and the maximum value is 23 MPa. Moreover, the displacement of the roof is relatively large. But the overburden deformation on the coal pillar is significantly smaller, indicating that its structure is relatively intact.

According to the plastic area analysis after 1303 working face mining, the pillar exhibits shear damage at both the top and bottom. Furthermore, a combination of shear and tensile damage is observed on either side of the pillar. The width of plastic area near the 1303 working face is 3–5 m, the width near the 1301 face is 3–6 m, and the elastic zone width of the coal pillar is about 6–8 m.

(d) Analysis of the fissure water flow vector model

Due to the on-site mining situation, the 1301 working face has been completed, and the 1303 working face is currently being mined. Figure 9 accurately illustrates the impact of goaf water on surrounding rocks, roadways, and the coal pillar in this situation. The figure shows a flow vector diagram of the V_3 aquifer with the working face excavation. According to the figure, the tops of the 1301 and 1303 working faces are the main flow directions of crack water, where the water located in the 1301 goaf extends laterally along the 3# coal seam bottom plate, and the final part of the water flow may enter from the 1303 auxiliary transport bottom plate. Thus, during the whole process, no goaf water crosses the coal pillar into the roadway, indicating that the coal pillar plays an effective barrier role in the numerical simulation.

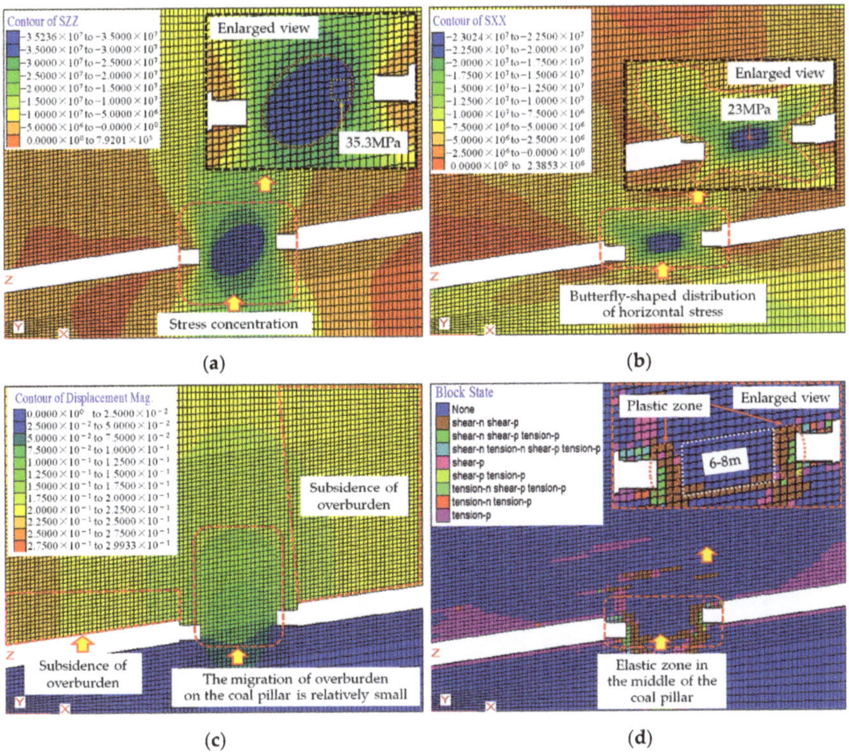

Figure 8. Numerical simulation characteristics after excavation of all working faces. (**a**) Vertical stress characteristics; (**b**) horizontal stress characteristics; (**c**) displacement distribution characteristics; (**d**) characteristics of the elastoplastic area.

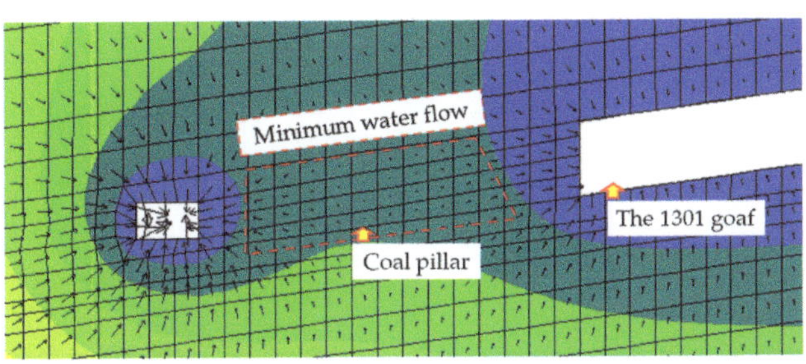

Figure 9. Flow vector diagram of the V_3 aquifer after the excavation of the 1301 working face.

According to Wilson's theory, the coal pillar must be more than two times the mining height (the mining height is 2.7 m). Combined with the numerical calculation results, it can be seen that the width of the pillar's central elastic area is more than 5.4 m. Therefore, the waterproof coal pillar can guarantee the stabilization of self-structure. Due to significant differences in on-site engineering, targeted monitoring methods should be further applied to assess the stability of the pillar.

5. Field Monitoring and Evaluation

5.1. Monitoring Principle and Scheme

In the geological exploration of mines, acoustic and electromagnetic wave detection techniques have shown good results in analyzing the degree of rock fracture development, especially in determining the loose circle of the rock around roadways, which has attracted the attention of many scholars [38]. Therefore, to further assess the structural features of the coal pillar affected by the goaf water, ground-penetrating radar (GPR) and loosening circle detection were utilized to evaluate the stability of the pillar in the study zone. The detection distribution is illustrated in Figure 10. Among these techniques, GPR is a commonly used means of scanning surrounding rock structures on a large scale, and its working principle is mainly based on the reflection of electromagnetic waves in different media for geological evaluation.

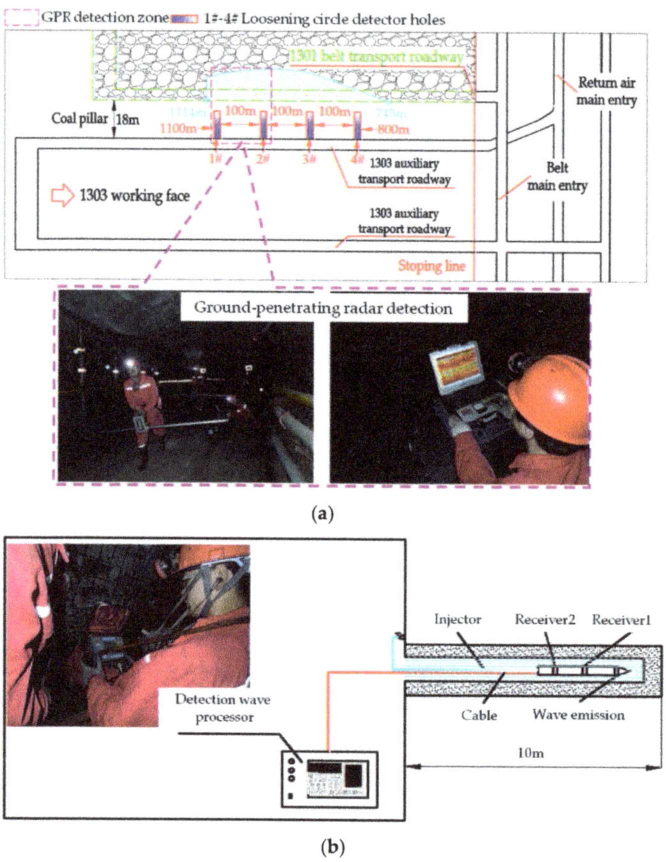

Figure 10. Schematic diagram of coal and rock exploration in the area affected by goaf water. (**a**) Overall detection map; (**b**) schematic diagram of the loosening circle of the probe hole.

Furthermore, loosening circle detection mainly detects the broken surrounding rock range based on the different velocities of sound waves in different media [39], and the wave velocity increases as the number of cracks in the surrounding rock decreases (in Figure 11a). Before detection, the borehole should be filled with water to ensure the stable propagation of sound waves (in Figure 10b). Subsequently, the transmitter of the loosening circle detector is excited by electrical pulse signals to generate transient sound wave signals.

These signals propagate in the rock and are received by two receivers. Based on the time difference between the arrival of sound waves at different receivers, the wave velocity of the coal pillar at different depths can be obtained (in Figure 11b).

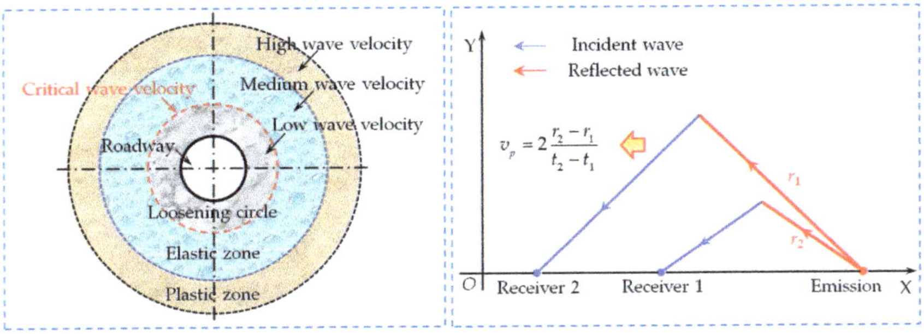

(a) (b)

Figure 11. Loosening circle theoretical analysis and detection for surrounding rocks. (**a**) Distribution characteristics of surrounding rock loosening circle; (**b**) working principle of loosening circle detection.

5.2. Analysis of Ground-Penetrating Radar Results

Figure 12 reflects the GPR detection results of the roadway within the range of 990–1110 m, where Figure 12a,b reflects the structural characteristics of the coal body in the southern portion of the roadway (unaffected by goaf water), and Figure 12c,d reflects the structural characteristics of the coal body in the northern section of the roadway (the goaf and coal pillar).

When examining the coal pillar and goaf, obvious waveform changes can be observed, and one can see that the reflected wave shows negative polarity. From 5 m to about 12.5 m of monitoring depth, the reflected wave begins to shift from high speed to low speed and tends to stabilize, as indicated by the surrounding rock elastic zone. In the 12.5–19 m range, the waveform is messier, as the figure shows with red and black horizontal bars. This indicates that media with obviously poor polarity are mixed here with each other. Judging from this, we can find the goaf water entering the coal pillar crack growth zone. Within the range of 19–26 m, the waveform changes significantly, and the graphic color also changes sharply to white, which shows that the relative dielectric constant increases in this zone, as a more uniform medium. Therefore, it can be concluded that this is the goaf water zone. Since the width of the central elastic area is above 5 m, there is only a small chance that goaf water would penetrate the coal pillar and cause water damage accidents.

As a reference, in the coal area unaffected by goaf water, the waveform change in the figure is relatively small, the waveform distribution characteristics are relatively stable, and only some zones affected by roadway excavation have a certain degree of in-phase axis deflection. Moreover, it can be seen from the scanning feature diagram that there are relatively few regions with sharp changes in the dielectric constant, showing that there is only a small zone of water in the coal body. The overall figure is mainly black and red, indicating that the goaf water does not affect the coal body around the coal pillar; thus, the coal pillar has a good waterproofing effect.

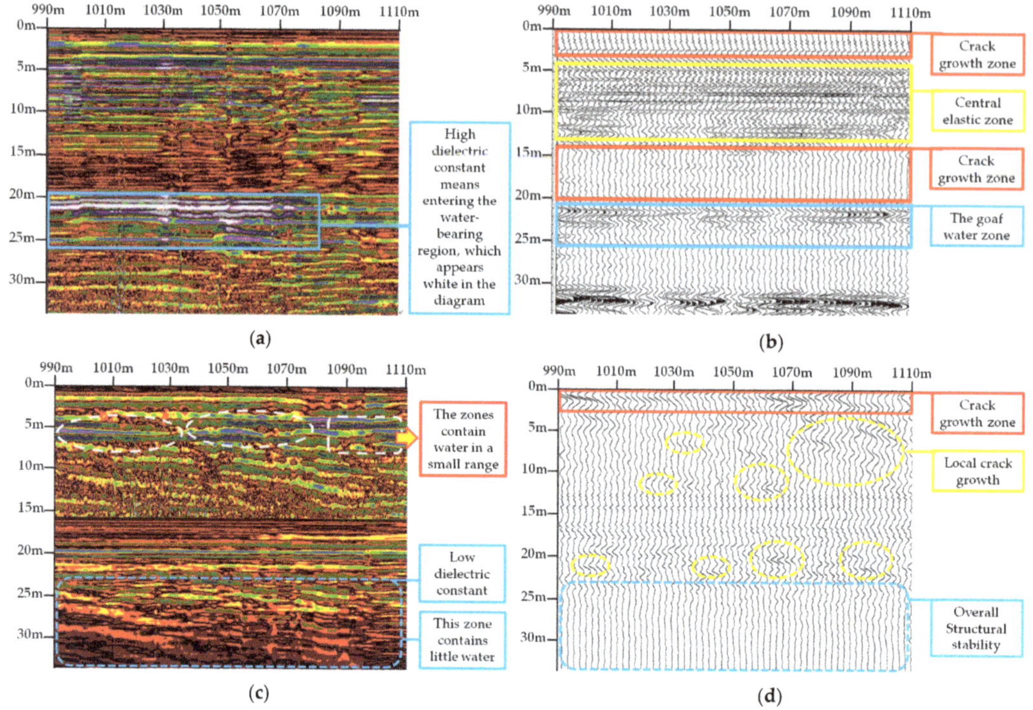

Figure 12. Ground-penetrating radar detection result diagram. (**a**) Scan feature map of the zone affected by goaf water; (**b**) oscillogram of the zone affected by goaf water; (**c**) scan feature map of the zone unaffected by goaf water; (**d**) oscillogram of the zone unaffected by goaf water.

5.3. Analysis of the Results of Loosening Circle Detection

Figure 13 and Table 3 show the changes in wave velocity of the pillar at different zones recorded using the loosening circle detector. From on-site monitoring, it can be concluded that when the wave velocity is less than 3.2 km/s (critical wave velocity), the coal–rock structure will experience some level of degradation, where the lower the wave velocity, the higher the degree of degradation. The 1# and 2# probe holes are located in the zone heavily affected by goaf water, and the 3# and 4# probe holes are relatively less affected by goaf water. The monitoring results indicate that the loosening range of the zone of 1# hole is relatively high, with an average wave velocity of 3.074 km/s, slightly lower than the critical value of wave velocity at the time of failure. The wave velocities at depths of 0–3 m and 9–10 m in the 1# hole are both around 2.6 km/s, indicating that this is the loosening range of the surrounding rock. The wave velocities at other positions in the 1# hole fluctuate around 3.2 km/s, meaning that the structure in the middle of the coal pillar is still intact. The loosening range of the 2# hole is slightly smaller than that of the 1# hole, but there are still some plastic failures in the local zone. The average wave velocity is 3.683 km/s in the 2# hole, which is higher than the critical wave velocity. At a depth of 6–10 m in the 2# hole, the wave velocity is greater than 3.774 km/s, showing that it is a completely surrounding rock zone.

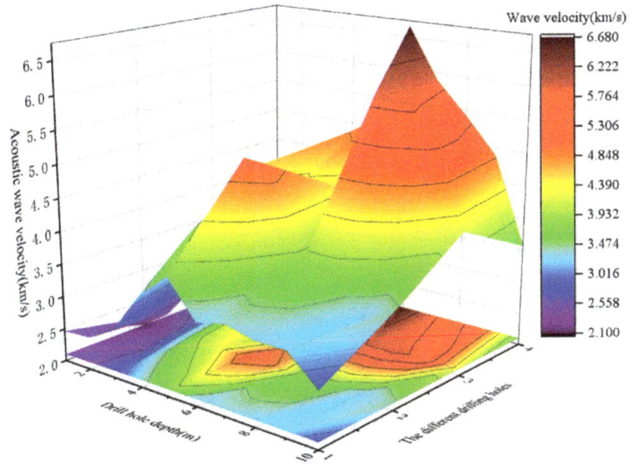

Figure 13. Wave velocity characteristics in the loosening range of the coal pillar affected by goaf water.

Table 3. Results of loosening circle wave velocity detection of 1#–4# probe holes.

Serial Number	Drilling Depth (m)	1# Probe Hole (km/s)	2# Probe Hole (km/s)	3# Probe Hole (km/s)	4# Probe Hole (km/s)
0	0–1	2.469	2.105	2.128	2.740
1	1–2	2.500	2.410	2.740	3.175
2	2–3	2.778	3.846	3.704	4.545
3	3–4	3.390	4.255	4.878	4.167
4	4–5	3.774	5.128	5.163	3.571
5	5–6	3.226	4.878	6.667	4.444
6	6–7	3.333	3.774	5.882	5.882
7	7–8	3.226	3.774	5.405	5.882
8	8–9	3.226	3.226	4.878	5.556
9	9–10	2.817	3.431	3.571	4.167
Average wave velocity		3.074	3.683	4.502	4.413

Probe holes 3# and 4# are located in a zone where the effect of goaf water is relatively light. The analysis results indicate that the average wave velocity of the zone of the 3# hole reaches 4.502 km/s, an increase of 40.7% compared to the critical wave velocity. The average wave velocity of the 4# hole reaches 4.413 km/s, an increase of 37.9% compared to the critical wave velocity. Therefore, the average wave velocity measured in the zones of the 3# and 4# holes is significantly higher than the critical wave velocity, with an average wave velocity of 4.458 km/s, which is 1.39 times the critical wave velocity, indicating effective stabilization of the pillar in this zone. Therefore, the elastic zone of the pillar is detected by the loose circle as 2–9 m, averaging over 6 m, which is in line with the theoretical study.

Finally, according to Figure 13, the evolution of wave velocity at different probe holes of the coal pillar can be observed. The blue low-wave-velocity area in the figure mainly appears in the zone significantly affected by goaf water and mining disturbances. As the area gradually moves away from the zone affected by goaf water, the blue low-wave-velocity area gradually transitions to the green–yellow medium wave velocity level, while the red high-wave-velocity area is the original rock stress zone that has not been disturbed by various factors. As the area gradually moves away from the middle zone of the goaf water, there are hardly any blue areas in the figure. Thus, the degree to which the coal pillar is affected by goaf water is limited, and the elastic zone of the middle of the pillar is relatively large, fully capable of blocking water damage and supporting the upper strata.

5.4. Deformation Analysis of the Roadway near the Coal Pillar

On the basis of geophysical monitoring, to more intuitively observe the structural deformation of the coal pillar affected by water, two cross-section measuring zones are set in the roadway near the coal pillar (in Figure 14). The 1# cross-section is unaffected by goaf water, and the 2# cross-section is located in the zone affected by water. After 16 days of monitoring, the data can be summarized.

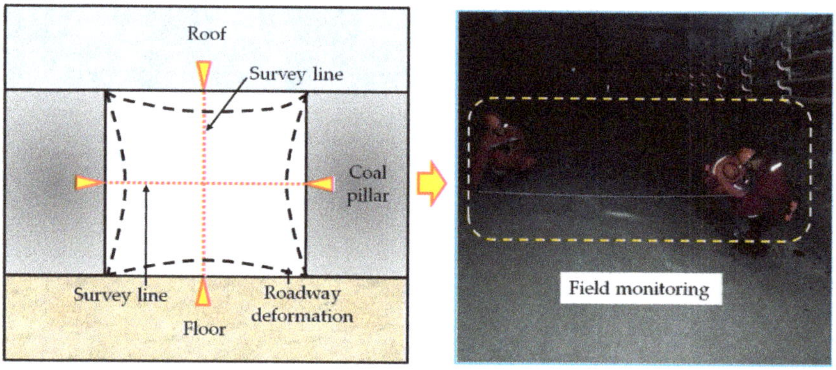

Figure 14. Deformation monitoring of the roadway near the coal pillar.

It can be seen from Figure 15 that the deformation processes of the roadway and coal pillar in different zones exhibits three stages: rapid deformation, slow deformation, and a gradual stop in deformation. The monitoring results of the 1# cross-section show that the average deformation rate of the coal pillar survey line is 4.4 mm/d, and the deformation rate of the roof survey line is 6.2 mm/d; the deformation monitored by the two lines gradually reaches 0 in 9–10 days. On the other hand, the monitoring results of the 2# cross-section show that the average deformation rate of the coal pillar line is 5.4 mm/d, which is 18.5% higher than that of the 1# cross-section. The deformation rate of the roof survey line is 7.6 mm/d, which is 18.4% higher than that of the 1# cross-section. The deformation process of the two survey lines gradually stops by the 13th day. It can be seen that the coal and rock affected by goaf water have more obvious creep characteristics. After engineering disturbance, the deformation time of the coal and rock structures will be longer. Due to the strong water isolation effect of the coal pillar, the change degree of coal and rock in different areas has little difference.

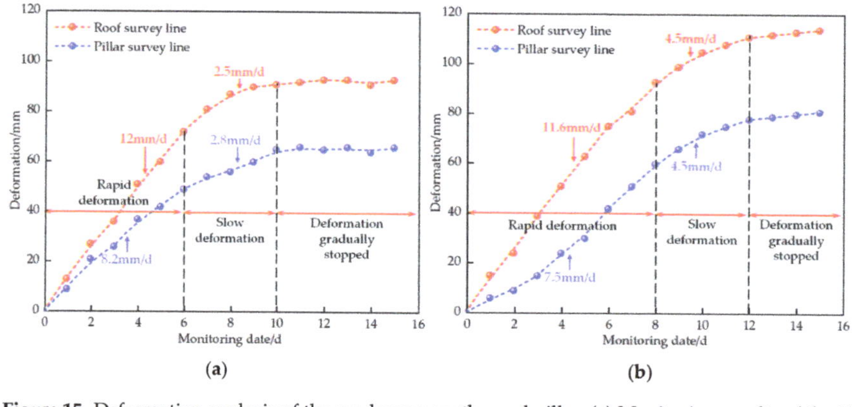

Figure 15. Deformation analysis of the roadway near the coal pillar. (**a**) Monitoring results of the 1# cross-section; (**b**) monitoring results of the 2# cross-section.

Based on the above studies, it can be found that due to the complexity and diversity of geological conditions in different mining areas, especially with the gradual involvement of coal mining in the deep region, coal mining safety has been seriously challenged, and the problems of high stress and high water pressure faced by coal pillar become increasingly prominent. Thus, it is difficult to accurately solve the instability problems of coal pillars under various complex conditions through theoretical analysis or computer simulation technology. Therefore, the targeted application of multi-dimensional monitoring technology in the field can not only provide real data on the study area dynamically and in real time but also determine the geological parameters of the coal pillar more accurately to make up for the limitations of theoretical research. In summary, the application of multivariate monitoring methods to coal pillar stability analysis can not only monitor the supporting structural characteristics of coal pillars for a long time but also provide a useful reference for various mines to achieve the accurate acquisition of underground geological information.

6. Conclusions

(1) Through the combination of exploration holes and geological investigation, it is determined that the goaf water comes from the V_3 aquifer, and the water-flowing fracture zone formed by mining disturbance is the main water flow channel. It is clear that the goaf water is static, and the maximum water pressure is 0.26 MPa. Based on this, theoretical analysis is used to determine the relationship between the mechanical parameters of the coal and rock affected by water, and the damaging effect of water on rock is further clarified.

(2) By establishing a FLAC^{3D} numerical simulation model, the model simulates the stress distribution and elastic–plastic zone distribution of the waterproof coal pillar during the excavation process of working faces and roadways. The results indicate that the width of the elastic zone in the pillar is 6–8 m, and the width of the plastic zone is 6–11 m after all excavations are completed. Moreover, through the flow vector migration process in the coal and rock, it is shown that there is no water-flowing fracture in the coal pillar, which enables the evaluation and analysis of the stabilization of the waterproof coal pillar and supports preliminary conclusions on the stability of the coal pillar.

(3) Using ground-penetrating radar, loosening circle detection, and deformation measurement to monitor the structure characteristics of the waterproof coal pillar and the distribution zone of the goaf water shows that the interior of the pillar can be categorized into a short plasticity zone, central elastic zone, and long plasticity zone; outside of this is the goaf water zone. This indicates that the internal fissure in the 2.6–8.6 m section inside the pillar is lower and is basically in the state of the in situ stress zone. Therefore, under the combined effects of the goaf water and the disruptions caused by mining, the pillar still has water resistance and loading ability.

Author Contributions: Data curation, P.S.; Formal analysis, H.G.; Funding acquisition, X.D.; Project administration, X.D.; Software, X.Y. and H.G.; Validation, P.S.; Writing—original draft, X.Y.; Writing—review & editing, X.Y. and H.G. All authors have read and agreed to the published version of the manuscript.

Funding: This work was supported by the school–enterprise cooperation project of Longdong University in Gansu Province, Grant number: HXZK2424 and HXZK2451.

Data Availability Statement: The original contributions presented in this study are included in the article, and further inquiries can be directed to the corresponding author.

Acknowledgments: A special acknowledgment is made to the anonymous reviewers for their constructive and valuable comments. We thank them for their guidance despite their busy schedules.

Conflicts of Interest: The authors declare no conflicts of interest.

References

1. Liu, Q.; Xue, Y.; Ma, D.; Li, Q. Failure characteristics of the water-resisting coal pillar under stress-seepage coupling and determination of reasonable coal pillar width. *Water* **2023**, *15*, 1002. [CrossRef]
2. Niu, C.; Tian, Q.; Xiao, L.; Xue, X.; Zhang, R.; Xu, D. Principal causes of water damage in mining roofs under giant thick topsoil–lilou coal mine. *Appl. Water Sci.* **2024**, *14*, 146. [CrossRef]
3. Gu, H.L.; Lai, X.P.; Tao, M.; Momeni, A.; Zhang, Q.L. Dynamic mechanical mechanism and optimization approach of roadway surrounding coal water infusion for dynamic disaster prevention. *Measurement* **2023**, *223*, 113639. [CrossRef]
4. Jiang, C.; Gao, X.; Hou, B.; Zhang, S.; Zhang, J.; Li, C.; Wang, W. Occurrence and environmental impact of coal mine goaf water in karst areas in China. *J. Clean. Prod.* **2020**, *275*, 123813. [CrossRef]
5. Xu, Z.; Sun, Y. Mine water safety and environment: Chinese experience. *Water* **2024**, *16*, 2833. [CrossRef]
6. Michieka, N.M. Energy and the environment: The relationship between coal production and the environment in China. *Nat. Resour. Res.* **2014**, *23*, 85–98. [CrossRef]
7. Wu, W.D.; Bai, J.B.; Wang, X.Y.; Yan, S.; Wu, S.X. Numerical study of failure mechanisms and control techniques for a gob-side yield pillar in the Sijiazhuang coal mine, China. *Rock Mech. Rock Eng.* **2019**, *52*, 1231–1245. [CrossRef]
8. Lai, X.P.; Yuchi, X.Q.; Gu, H.L.; Yang, W.H. Determination of critical damage size of inclined waterproof coal pillar under asymmetric load. *Water* **2024**, *16*, 1233. [CrossRef]
9. Xie, H.P. Research review of the state key research development program of China: Deep rock mechanics and mining theory. *J. China Coal Soc.* **2019**, *44*, 1283–1305.
10. Dong, L.; Tong, X.; Ma, J. Quantitative investigation of tomographic effects in abnormal regions of complex structures. *Engineering* **2021**, *7*, 1011–1022. [CrossRef]
11. Chen, D.; Wang, X.; Bai, J.; Li, M. Characteristics of waterproof failure and optimal width of narrow coal pillars under the coupled effects of mining, excavation and seepage. *Geomech. Geophys. Geo-Energy Geo-Resour.* **2024**, *10*, 100. [CrossRef]
12. Li, A.; Ji, B.N.; Ma, Q.; Liu, C.; Wang, F.; Ma, L. Design of longwall coal pillar for the prevention of water inrush from the seam floor with through fault. *Geofluids* **2021**, *2021*, 5536235. [CrossRef]
13. Wang, Y.Q.; Wang, X.; Zhang, J.S.; Yang, B.S.; Zhu, W.J.; Wang, Z.P. Similar experimental study on retaining waterproof coal pillar in composite strata mining. *Sci. Rep.* **2022**, *12*, 1366. [CrossRef]
14. Zhang, C.; Zhao, Y.; Han, P.; Bai, Q. Coal pillar failure analysis and instability evaluation methods: A short review and prospect. *Eng. Fail. Anal.* **2022**, *138*, 106344. [CrossRef]
15. Medhurst, T.P.; Brown, E.T. A study of the mechanical behaviour of coal for pillar design. *Int. J. Rock Mech. Min. Sci.* **1998**, *35*, 1087–1105. [CrossRef]
16. Adeleke, A.A. *Mineral Processing Technology: A Concise Introduction*; CRC Press: Boca Raton, FL, USA, 2023.
17. Wang, R.; Bai, J.B.; Yan, S.; Chang, Z.G.; Wang, X.Y. An innovative approach to theoretical analysis of partitioned width & stability of strip pillar in strip mining. *Int. J. Rock Mech. Min. Sci.* **2020**, *129*, 104301.
18. Gu, H.; Shang, X.; Zhao, H. Theory and technology for the prevention of mine water disasters. *Water* **2024**, *16*, 2952. [CrossRef]
19. Fan, J.; Li, Z.; Feng, G.; Zhang, H.; Qi, C.; Zhang, J. Failure analysis of coal pillars and overburden from underground water reservoir under the mining-water invasion coupling effect. *Eng. Fail. Anal.* **2023**, *151*, 107406. [CrossRef]
20. Han, P.; Zhang, C.; Wang, W. Failure analysis of coal pillars and gateroads in longwall faces under the mining-water invasion coupling effect. *Eng. Fail. Anal.* **2022**, *131*, 105912. [CrossRef]
21. Gu, H.L.; Lai, X.P.; Tao, M.; Cao, W.Z.; Yang, Z.K. The role of porosity in the dynamic disturbance resistance of water-saturated coal. *Int. J. Rock Mech. Min. Sci.* **2023**, *166*, 105388. [CrossRef]
22. Xue, Y.; Ranjith, P.G.; Dang, F.; Liu, J.; Wang, S.; Xia, T.; Gao, Y. Analysis of deformation, permeability and energy evolution characteristics of coal mass around borehole after excavation. *Nat. Resour. Res.* **2020**, *29*, 3159–3177. [CrossRef]
23. Yue, J.; Xu, J.; Sun, Y.; Shi, B.; Wang, Q.; Wang, L.; Wang, C. Dynamic response characteristics of water-gas-coal in freezing process of gas-bearing coal under water infiltration condition. *Phys. Fluids.* **2024**, *36*, 043107. [CrossRef]
24. Poulsen, B.A.; Shen, B.; Williams, D.J.; Huddlestone-Holmes, C.; Erarslan, N.; Qin, J. Strength reduction on saturation of coal and coal measures rocks with implications for coal pillar strength. *Int. J. Rock Mech. Min. Sci.* **2014**, *71*, 41–52. [CrossRef]
25. Yao, Q.; HAO, Q.; Chen, X.; Zhou, B.; Fang, J. Design on the width of coal pillar dam in coal mine groundwater reservoir. *J. China Coal Soc.* **2019**, *44*, 891–899.
26. Zhang, C.; Bai, Q.; Han, P. A review of water rock interaction in underground coal mining: Problems and analysis. *Bull. Eng. Geol. Environ.* **2023**, *82*, 157. [CrossRef]

27. Yao, Q.; Tang, C.; Xia, Z.; Liu, X.; Zhu, L.; Chong, Z.; Hui, X. Mechanisms of failure in coal samples from underground water reservoir. *Eng. Geol.* **2020**, *267*, 105494. [CrossRef]
28. Liu, S.; Fei, Y.; Wang, J.; Wu, D.; Guo, D.; Dong, Y.; Liu, X.; Shangguan, Y. A novel method to retain a waterproof coal pillar near water-conducting fault. *J. Geophys. Eng.* **2023**, *20*, 225–239. [CrossRef]
29. Hashikawa, H.; Mao, P.; Sasaoka, T.; Hamanaka, A.; Shimada, H.; Batsaikhan, U.; Oya, J. Numerical simulation on pillar design for longwall mining under weak immediate roof and floor strata in indonesia. *Sustainability* **2022**, *14*, 16508. [CrossRef]
30. Gu, W.; Xu, D.; Wang, Y.; Miao, K.; Yao, S.; Zhang, H.; Han, Z. Research on the stability and water Isolation of waterproof coal pillars between adjacent working faces under the Influence of water ponding goaf-A case study. *Appl. Sci.* **2024**, *14*, 884. [CrossRef]
31. Hu, M.; Zhao, W.; Lu, Z.; Ren, J.; Miao, Y. Research on the reasonable width of the waterproof coal pillar during the mining of a shallow coal seam located close to a reservoir. *Adv. Civ. Eng.* **2019**, *2019*, 3532784. [CrossRef]
32. Sun, H.; Pan, R.; Li, J.; Yang, B. Research on the mechanism of fault activation and water inrush across variable coal pillar widths. *Adv. Civ. Eng.* **2024**, *2024*, 8557425. [CrossRef]
33. Zhao, J.; Wang, X.; Bai, J.; Wang, G.; Chen, D.; Li, G. Study on instability mechanisms and control of coal pillar spalling and coal crumbs leakage during working face crossing faults. *Sci. Rep.* **2024**, *14*, 26667. [CrossRef]
34. Butchibabu, B.; Khan, P.K.; Jha, P.C. Geophysical investigations for stability and safety mitigation of regional crude-oil pipeline near abandoned coal mines. *J. Geophys. Eng.* **2021**, *18*, 145–162. [CrossRef]
35. Li, Y.; Wang, J.; Chen, Y.; Wang, Z. Overlying strata movement with ground penetrating radar detection in close-multiple coal seams mining. *Int. J. Distrib. Sens. Netw.* **2019**, *15*, 1550147719869852. [CrossRef]
36. Zhu, Z.D.; Xing, F.D.; Wang, S.J. Analysis on strength softening of argillite under groundwater by damage mechanics. *Chin. J. Rock Mech. Eng.* **2004**, *23*, 4739–4743.
37. Fan, G.Q.; Yan, R.B.; Lei, Z.Y. Design of "stress-seepage-damage" analysis width of section coal pillar. *J. Xi'an Univ. Sci. Technol.* **2017**, *37*, 164–170.
38. Deng, J.; Wang, J.R.; Ren, S.J.; He, Q.; Wang, C.P.; Qv, G.Y. Application and prospect of acoustic detection in the mining sector. *Coal Geol. Explor.* **2023**, *51*, 15.
39. Yang, Y.G.; Fan, N. Experimental study on surrounding rock loosing circle by single−hole acoustic wave testing method. *Coal Sci. Technol.* **2019**, *47*, 93–100.

MDPI AG
Grosspeteranlage 5
4052 Basel
Switzerland
Tel.: +41 61 683 77 34

Water Editorial Office
E-mail: water@mdpi.com
www.mdpi.com/journal/water

www.ingramcontent.com/pod-product-compliance
Lightning Source LLC
Chambersburg PA
CBHW061516210926
51850CB00008B/1941
* 9 7 8 3 7 2 5 8 6 5 7 4 1 *